# FinFETs and Transistors

# Series on Integrated Circuits and Systems

Series Editor:             Anantha Chandrakasan
                                Massachusetts Institute of Technology
                                Cambridge, Massachusetts

---

FinFETs and Other Multi-Gate Transistors
Jean-Pierre Colinge (Ed.)
ISBN 978-0-387-71751-7

Design for Manufacturability and Statistical Design: A Constructive Approach
Michael Orshansky, Sani R. Nassif, and Duane Boning
ISBN 978-0-387-30928-6

Low Power Methodology Manual: For System-on-Chip Design
Michael Keating, David Flynn, Rob Aitken, Alan Gibbons, and Kaijian Shi
ISBN 978-0-387-71818-7

Modern Circuit Placement: Best Practices and Results
Gi-Joon Nam and Jason Cong
ISBN 978-0-387-36837-5

CMOS Biotechnology
Hakho Lee, Donhee Ham and Robert M. Westervelt
ISBN 978-0-387-36836-8

SAT-Based Scalable Formal Verification Solutions
Malay Ganai and Aarti Gupta
ISBN 978-0-387-69166-4, 2007

Ultra-Low Voltage Nano-Scale Memories
Kiyoo Itoh, Masashi Horiguchi and Hitoshi Tanaka
ISBN 978-0-387-33398-4, 2007

Routing Congestion in VLSI Circuits: Estimation and Optimization
Prashant Saxena, Rupesh S. Shelar, Sachin Sapatnekar
ISBN 978-0-387-30037-5, 2007

Ultra-Low Power Wireless Technologies for Sensor Networks
Brian Otis and Jan Rabaey
ISBN 978-0-387-30930-9, 2007

Sub-Threshold Design for Ultra Low-Power Systems
Alice Wang, Benton H. Calhoun and Anantha Chandrakasan
ISBN 978-0-387-33515-5, 2006

High Performance Energy Efficient Microprocessor Design
Vojin Oklibdzija and Ram Krishnamurthy (Eds.)
ISBN 978-0-387-28594-8, 2006

Abstraction Refinement for Large Scale Model Checking
Chao Wang, Gary D. Hachtel, and Fabio Somenzi
ISBN 978-0-387-28594-2, 2006

A Practical Introduction to PSL
Cindy Eisner and Dana Fisman
ISBN 978-0-387-35313-5, 2006

Thermal and Power Management of Integrated Systems
Arman Vassighi and Manoj Sachdev
ISBN 978-0-387-25762-4, 2006

*Continued after index*

Jean-Pierre Colinge
Editor

# FinFETs and Other Multi-Gate Transistors

 Springer

Jean-Pierre Colinge
Tyndall National Institute
Cork, Ireland


*Series Editor:*
Anantha Chandrakasan
Department of Electrical Engineering and Computer Science
Massachusetts Institute of Technology
Cambridge, MA 02139
USA


ISBN 978-1-4419-4409-2     e-ISBN 978-0-387-71752-4

© 2008 Springer Science+Business Media, LLC
Softcover reprint of the hardcover 1st edition 2008
All rights reserved. This work may not be translated or copied in whole or in part without the written permission of the publisher (Springer Science+Business Media, LLC, 233 Spring Street, New York, NY 10013, USA), except for brief excerpts in connection with reviews or scholarly analysis. Use in connection with any form of information storage and retrieval, electronic adaptation, computer software, or by similar or dissimilar methodology now known or hereafter developed is forbidden. The use in this publication of trade names, trademarks, service marks and similar terms, even if they are not identified as such, is not to be taken as an expression of opinion as to whether or not they are subject to proprietary rights.

9 8 7 6 5 4 3 2 1

springer.com

# Preface

*The* adoption of Silicon-on-Insulator (SOI) substrates for the manufacturing of mainstream semiconductor products such as microprocessors has given SOI research an unprecedented impetus. In the past, novel transistor structures proposed by SOI scientists were often considered exotic and impractical, but the recent success of SOI in the field of microprocessor manufacturing has finally given this technology the credibility and acceptance it deserves.

The classical CMOS structure is reaching its scaling limits and "end-of-roadmap" alternative devices are being investigated. Amongst the different types of SOI devices proposed, one clearly stands out: the multigate field-effect transistor (multigate FET). This device has a general "wire-like" shape with a gate electrode that controls the flow of current between source and drain. Multigate FETs are commonly referred to as "multi(ple)-gate transistors", "wrapped-gate transistors", "double-gate transistors", "FinFETs", "tri(ple)-gate transistors", "Gate-all-Around transistors", etc. The International Technology Roadmap for Semiconductors (ITRS) recognizes the importance of these devices and calls them "Advanced non-classical CMOS devices".

There exists a number of textbooks on SOI technology. Some of these books tackle the subject of multigate FETs, but there is no book that contains a comprehensive description of the physics, technology and circuit applications of this new class of devices. This is why we decided to compile chapters dedicated to the different facets of multigate FET technology, written by world-leading experts in the field. This book contains seven chapters:

- *Chapter 1: **The SOI MOSFET: from Single Gate to Multigate***, by Jean-Pierre Colinge, is a general introduction that shows the evolution of the SOI MOS transistor and retraces the history of the multigate concept. The advantages of multigate FETs in terms of electrostatic integrity and short-channel control are described, and the challenges posed by the appearance of novel effects, some of quantum-mechanical origin, are outlined.

- *Chapter 2: **Multigate MOSFET Technology***, by Weize (Wade) Xiong, outlines the issues associated with multigate FET manufacturing. This chapter describes thin-fin formation techniques, advanced gate stack deposition and source/drain resistance reduction techniques. Issues related to fin crystal orientation and mobility enhancement via strain engineering are tackled as well.

- *Chapter 3: **BSIM CMG: A Compact Model for Multi-Gate Transistors***, by Mohan Vamsi Dunga, Chung-Hsun Lin, Ali M. Niknejad and Chenming Hu, describes the physics behind the BSIM-CMG (Berkeley Short-channel IGFET Model – Common Multi-Gate) compact models for multigate MOSFETs. A compact model serves as a link between process technology and circuit design. It is a concise mathematical description of the device physics in the transistor. Some simplifications in the physics, however, can be made to enable fast computer analysis of device/circuit behavior.

- *Chapter 4: **Physics of the Multigate MOS System***, by Bogdan Majkusiak, analyzes the electrostatics of the multigate MOS system. Using quantum-mechanical concepts, it describes electron energy quantization and the properties of a one-dimensional and two-dimensional electron gas. The effects of tunneling through thin gate dielectrics on the electron population of a device are studied as well.

- *Chapter 5:* ***Mobility in Multigate MOSFETs***, by Francisco Gámiz and Andrés Godoy, analyzes the behavior of electron mobility in different multigate structures comprising double-gate transistors, FinFETs, and silicon nanowires. Mobility in multiple gate devices is compared to that in single-gate devices and different approaches for improving the mobility in these devices, such as different crystallographic orientations and strained Si channels, are studied.

- *Chapter 6:* ***Radiation Effects in Advanced Single- and Multi-Gate SOI MOSFETs***, by Véronique Ferlet-Cavrois, Philippe Paillet and Olivier Faynot, describes the effects of ionizing radiations such as gamma rays and cosmic rays on SOI MOSFETs. These effects are extremely important in military, space and avionics applications. Multi-gate FETs show exceptional resistance to total-dose and single-event effects and could become the new standard in radiation-hardened electronics.

- *Chapter 7:* ***Multigate MOSFET Circuit Design***, by Gerhard Knoblinger, Michael Fulde and Christian Pacha, describes the interrelationship between the multi-gate FET device properties and elementary digital and analog circuits, such as CMOS logic gates, SRAM cells, reference circuits, operational amplifiers, and mixed-signal building blocks. This approach is motivated by the observation that a cost-efficient, heterogeneous SoC integration is a key factor in modern IC design.

*Jean-Pierre Colinge, August 2007*

# Table of Content

**Preface** ................................................................................... **v**

**Table of Content** ................................................................... **ix**

**Contributors** ......................................................................... **xv**

**1 The SOI MOSFET: from Single Gate to Multigate** ............ **1**
  1.1 MOSFET scaling and Moore's law ................................................. 1
  1.2 Short-Channel Effects .................................................................... 2
  1.3 Gate Geometry and Electrostatic Integrity ..................................... 4
  1.4 A Brief History of Multiple-Gate MOSFETs ................................. 8
    1.4.1 Single-gate SOI MOSFETs ..................................................... 9
    1.4.2 Double-gate SOI MOSFETs .................................................. 11
    1.4.3 Triple-gate SOI MOSFETs .................................................... 12
    1.4.4 Surrounding-gate (quadruple-gate) SOI MOSFETs ............... 13
    1.4.5 Other multigate MOSFET structures ..................................... 14
    1.4.6 Multigate MOSFET memory devices .................................... 16
  1.5 Multigate MOSFET Physics .......................................................... 17
    1.5.1 Classical physics .................................................................... 17
      1.5.1.1 Natural length and short-channel effects ....................... 17
      1.5.1.2 Current drive .................................................................. 23
      1.5.1.3 Corner effect .................................................................. 26
    1.5.2 Quantum effects ..................................................................... 28
      1.5.2.1 Volume inversion ........................................................... 28
      1.5.2.2 Mobility effects .............................................................. 31
      1.5.2.3 Threshold voltage .......................................................... 31
      1.5.2.4 Inter-subband scattering ................................................. 35
  References ........................................................................................... 37

## 2 Multigate MOSFET Technology ......................................................... 49
2.1 Introduction ............................................................................... 49
2.2 Active Area: Fins ....................................................................... 52
   2.2.1 Fin Width .......................................................................... 52
   2.2.2 Fin Height and Fin Pitch .................................................. 57
   2.2.3 Fin Surface Crystal Orientation ....................................... 61
   2.2.4 Fin Surface Preparation ................................................... 64
   2.2.5 Fins on Bulk Silicon ........................................................ 65
   2.2.6 Nano-wires and Self-Assembled Wires .......................... 66
2.3 Gate Stack .................................................................................. 70
   2.3.1 Gate Patterning ................................................................. 70
   2.3.2 Threshold Voltage and Gate Workfunction Requirements ..... 71
      2.3.2.1 Polysilicon Gate ...................................................... 74
      2.3.2.2 Metal Gate .............................................................. 75
      2.3.2.3 Tunable Workfunction Metal Gate ......................... 75
   2.3.3 Gate EWF and Gate Induced Drain Leakage (GIDL) ..... 79
   2.3.4 Independently Controlled Gates ...................................... 82
2.4 Source/Drain Resistance and Capacitance ................................. 85
   2.4.1 Doping the Thin Fins ....................................................... 85
   2.4.2 Junction Depth ................................................................. 87
   2.4.3 Parasitic Resistance/Capacitance and Raised Source and Drain Structure ............................................................ 87
2.5 Mobility and Strain Engineering ............................................... 91
   2.5.1 Introduction ...................................................................... 91
   2.5.2 Wafer Bending Experiment ............................................. 92
   2.5.3 Nitride Stress Liners ........................................................ 93
   2.5.4 Embedded SiGe and SiC Source and Drain .................... 94
   2.5.5 Local Strain from Gate Electrode .................................... 95
   2.5.6 Substrate Strain: Strained Silicon on Insulator ............... 97
2.6 Contacts to the Fins ................................................................... 98
   2.6.1 Dumbbell source and drain contact ................................. 99
   2.6.2 Saddle contact .................................................................. 99
   2.6.3 Contact to merged fins .................................................... 100
Acknowledgments ............................................................................ 100
References ........................................................................................ 101

## 3 BSIM-CMG: A Compact Model for Multi-Gate Transistors ......... 113
3.1 Introduction .................................................................................. 113
3.2 Framework for Multigate FET Modeling ..................................... 114
3.3 Multigate Models: BSIM-CMG and BSIM-IMG ........................ 115
   3.3.1 The BSIM-CMG Model ........................................................ 116
   3.3.2 The BSIM-IMG Model .......................................................... 116
3.4 BSIM-CMG ................................................................................. 117
   3.4.1 Core Model ............................................................................. 117
      3.4.1.1 Surface Potential Model ................................................. 117
      3.4.1.2 I-V Model ....................................................................... 123
      3.4.1.3 C-V Model ...................................................................... 126
   3.4.2 Modeling Physical Effects of Real Devices .......................... 130
      3.4.2.1 Quantum Mechanical Effects (QME) ............................. 131
      3.4.2.2 Short-channel Effects (SCE) ........................................... 137
   3.4.3 Experimental Verification ...................................................... 144
3.5 The BSIM-IMG Model ................................................................ 147
   3.5.1 Surface Potential of independent DG-FET ............................ 147
   3.5.2 BSIM-IMG features ............................................................... 150
3.6 Summary ...................................................................................... 151
References ......................................................................................... 151

## 4 Physics of the Multigate MOS System ............................................. 155
4.1 Device electrostatics .................................................................... 155
4.2 Double gate MOS system ............................................................ 163
   4.2.1 Modeling assumptions ........................................................... 163
   4.2.2 Gate voltage effect ................................................................. 167
   4.2.3 Semiconductor thickness effect ............................................. 169
   4.2.4 Asymmetry effects ................................................................. 174
   4.2.5 Oxide thickness effect ........................................................... 178
   4.2.6 Electron tunnel current .......................................................... 180
4.3 Two-dimensional confinement ..................................................... 184
References ......................................................................................... 185

**5 Mobility in Multigate MOSFETs ........................................................ 191**
  5.1 Introduction ................................................................................ 191
  5.2 Double-Gate MOSFETs and FinFETs......................................... 192
    5.2.1 Phonon-limited mobility...................................................... 197
    5.2.2 Confinement of acoustic phonons ....................................... 202
    5.2.3 Interface roughness scattering. ............................................ 205
    5.2.4 Coulomb scattering.............................................................. 209
    5.2.5 Temperature Dependence of Mobility.................................212
    5.2.6 Symmetrical and Asymmetrical Operation of DGSOI FETs 214
    5.2.7 Crystallographic orientation ................................................ 218
    5.2.8 High-k dielectrics. ............................................................... 224
    5.2.9 Strained DGSOI devices...................................................... 226
    5.2.10. Summary.......................................................................... 232
  5.3 Silicon multiple-gate nanowires ................................................. 233
    5.3.1 Introduction ......................................................................... 233
    5.3.2 Electrostatic description of Si nanowires. ........................... 235
    5.3.3 Electron transport in Si nanowires....................................... 239
    5.3.4 Surface roughness................................................................ 245
    5.3.5 Experimental results and conclusions.................................. 246
  References ......................................................................................... 247

**6 Radiation Effects in Advanced Single- and Multi-Gate SOI MOSFETs............................................................................................. 257**
  6.1. A brief history of radiation effects in SOI .................................. 257
  6.2. Total Ionizing Dose Effects ........................................................ 259
    6.2.1 A brief overview of Total Ionizing Dose effects................. 259
    6.2.2 Advanced Single-Gate FDSOI devices ............................... 260
      6.2.2.1 Description of Advanced FDSOI Devices................... 261
      6.2.2.2 Front-gate threshold voltage shift................................ 261
      6.2.2.3 Single-transistor latch .................................................. 264
    6.2.3 Advanced Multi-Gate devices ............................................. 266
      6.2.3.1 Devices and process description .................................. 267
      6.2.3.2 Front-gate threshold voltage shift................................ 269
  6.3 Single-Event Effects ................................................................... 271
    6.3.1 Background.......................................................................... 271
    6.3.2 Effect of ion track diameter in nanoscale devices ............... 274
    6.3.3 Transient measurements on single-gate and FinFET SOI transistors...................................................................................... 279
    6.3.4 Scaling effects ..................................................................... 285
  References ......................................................................................... 287

## 7 Multi-Gate MOSFET Circuit Design .................................................. 293
### 7.1 Introduction ......................................................................... 293
### 7.2 Digital Circuit Design ........................................................... 294
#### 7.2.1 Impact of device performance on digital circuit design ........ 294
#### 7.2.2 Large-scale digital circuits.................................................. 302
#### 7.2.3 Leakage-performance trade off and energy dissipation......... 306
#### 7.2.4 Multi-$V_T$ devices and mixed-$V_T$ circuits ............................... 311
#### 7.2.5 High-temperature circuit operation ..................................... 312
#### 7.2.6 SRAM design .................................................................... 313
### 7.3 Analog Circuit Design .......................................................... 315
#### 7.3.1 Device figures of merit and technology related design issues 315
##### 7.3.1.1 Transconductance ....................................................... 315
##### 7.3.1.2 Intrinsic transistor gain ............................................... 316
##### 7.3.1.3 Matching behavior...................................................... 317
##### 7.3.1.4 Flicker noise............................................................... 318
##### 7.3.1.5 Transit and maximum oscillation frequency ............... 319
##### 7.3.1.6 Self-heating................................................................ 320
##### 7.3.1.7 Charge trapping in high-k dielectrics.......................... 320
#### 7.3.2 Design of analog building blocks ........................................ 321
##### 7.3.2.1 $V_T$-based current reference circuit .............................. 321
##### 7.3.2.2 Bandgap voltage reference.......................................... 322
##### 7.3.2.3 Operational amplifier.................................................. 324
##### 7.3.2.4 Comparator ................................................................ 325
#### 7.3.3 Mixed-signal aspects .......................................................... 326
##### 7.3.3.1 Current steering DAC ................................................. 327
##### 7.3.3.2 Successive approximation ADC ................................. 327
#### 7.3.4 RF circuit design................................................................ 329
### 7.4 SoC Design and Technology Aspects.................................... 331
### Acknowledgments ...................................................................... 332
### References .................................................................................. 333

## Index .............................................................................................. 336

# Contributors

**Jean-Pierre Colinge**
University of California
Davis, California, USA
*and*
Tyndall National Institute
Cork, Ireland

**Mohan Vamsi Dunga**
University of California
Berkeley, California, USA

**Olivier Faynot**
Commissariat à l'Energie Atomique
LETI
Grenoble, France

**Véronique Ferlet-Cavrois**
Commissariat à l'Energie Atomique
DIF
Bruyères-le-Châtel, France

**Michael Fulde**
Infineon Technologies Austria
Villach, Austria
*and*
Technical University of Munich
Munich, Germany

**Francisco Gámiz**
Department of Electronics and
Computer Science
University of Granada
Granada, Spain

**Andrés Godoy**
Department of Electronics and
Computer Science
University of Granada
Granada, Spain

**Chenming Hu**
University of California
Berkeley, California, USA

**Gerhard Knoblinger**
Infineon Technologies Austria
Villach, Austria

**Chung-Hsun Lin**
University of California
Berkeley, California, USA

**Bogdan Majkusiak**
Institute of Microelectronics and
Optoelectroncs
Warsaw University of Technology
Warsaw, Poland

**Ali M. Niknejad**
University of California
Berkeley, California, USA

**Christian Pacha**
Infineon Technologies
Munich, Germany

**Philippe Paillet**
Commissariat à l'Energie Atomique
DIF
Bruyères-le-Châtel, France

**Weize (Wade) Xiong**
Silicon Technology Development
Texas Instruments, Inc.
Dallas, TX, USA

# 1 The SOI MOSFET: from Single Gate to Multigate

Jean-Pierre Colinge

## 1.1 MOSFET Scaling and Moore's Law

In 1965 Gordon Moore published his famous paper describing the evolution of the transistor density in integrated circuits. He predicted that the number of transistors per chip would quadruple every three years.[1] This prediction became known as Moore's law and has been remarkably followed by the semiconductor industry for the last forty years (Figure 1.1).

Since the early 1990's semiconductor companies and academia have teamed up to predict more precisely the future of the industry. This initiative gave birth to the International Technology Roadmap for Semiconductors (ITRS) organization.[2] Every year, the ITRS issues a report that serves as a benchmark for the semiconductor industry. These reports describe the type of technology, design tools, equipment and metrology tools that have to be developed in order to keep pace with the exponential progress of semiconductor devices predicted by Moore's law. Figure 1.1 shows the evolution of the number of transistors per chip predicted by the ITRS 2005 for DRAMs and high-performance microprocessors.

The semiconductor industry's workhorse technology is silicon CMOS, and the building block of CMOS is the MOS transistor, or MOSFET (MOS field-effect transistor). In order to keep up with the frantic pace imposed by Moore's law, the linear dimensions of transistors have reduced by half every three years. The sub-micron dimension barrier was overcome in the early 1980's, and in 2010 semiconductor manufacturers will produce transistors with a 20nm gate length on a regular basis. Since the first integrated circuit transistors were fabricated on "bulk" silicon wafers. At the end of the 1990's, however, it became apparent that significant performance improvement could be gained by switching to a new type of substrate, called SOI (Silicon-On-Insulator) in which transistors are made in a thin silicon layer sitting on top of a silicon dioxide layer. SOI technology brings about improvements in both circuit speed and power consumption. In the early 2000's major semiconductor companies, including IBM, AMD and Freescale, began manufacturing microprocessors using SOI substrates on an industrial scale. SOI devices offer the advantage of reduced parasitic capacitances and enhanced current drive.

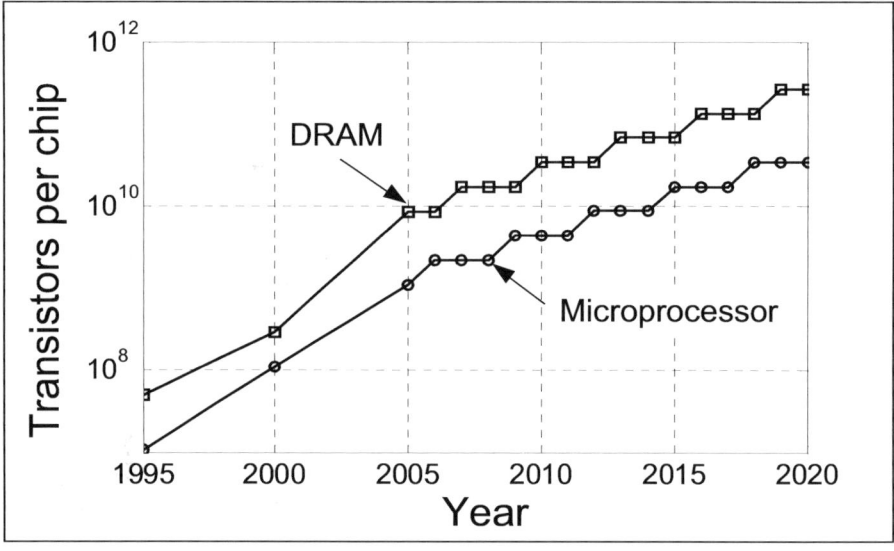

**Fig. 1.1.** Evolution of the number of transistors per chip (Moore's law) predicted by the ITRS 2005 for DRAMs and high-performance microprocessors.

## 1.2 Short-Channel Effects

As the dimensions of transistors are shrunk, the close proximity between the source and the drain reduces the ability of the gate electrode to control the potential distribution and the flow of current in the channel region, and undesirable effects, called the "short-channel effects" start plaguing MOSFETs. For all practical purposes, it seems impossible to scale the dimensions of classical "bulk" MOSFETs below 20nm. If that limitation cannot be overcome, Moore's law would reach an end around year 2012.

There exists a simple tool, called the Voltage-Doping Transformation model (VDT) [3], that can be used to translate the effects of shrinking device parameters such as gate length or drain voltage into electrical parameters. In the particular case of the Short-Channel Effect (SCE) and the Drain-Induced Barrier Lowering (DIBL), the following expressions can be derived from the VDT model: [4]

$$SCE = 0.64 \frac{\varepsilon_{Si}}{\varepsilon_{ox}} \left[ 1 + \frac{x_j^2}{L_{el}^2} \right] \frac{t_{ox}}{L_{el}} \frac{t_{dep}}{L_{el}} V_{bi} \equiv 0.64 \frac{\varepsilon_{Si}}{\varepsilon_{ox}} EI \, V_{bi} \quad (1.1)$$

and

$$DIBL = 0.80 \frac{\varepsilon_{Si}}{\varepsilon_{ox}} \left[ 1 + \frac{x_j^2}{L_{el}^2} \right] \frac{t_{ox}}{L_{el}} \frac{t_{dep}}{L_{el}} V_{DS} \equiv 0.80 \frac{\varepsilon_{Si}}{\varepsilon_{ox}} EI \, V_{DS} \quad (1.2)$$

where $L_{el}$ is the electrical (effective) channel length, $V_{bi}$ is the source or drain built-in potential, $t_{ox}$ is the gate oxide thickness, $x_j$ is the source and drain junction depth and $t_{dep}$ is the penetration depth of the gate field in the channel region, which is equal to the depth of the depletion region underneath the gate in a bulk MOSFET. The parameter $EI$ is called the "Electrostatic Integrity" factor. It depends on the device geometry and is a measure of the way the electric field lined from the drain influence the channel region, thereby causing SCE and DIBL effects. Based on the above expressions, the threshold voltage of a MOSFET with a given channel length $L_{el}$ can be calculated using the following relationship:

$$V_{TH} = V_{TH\infty} - SCE - DIBL \quad (1.3)$$

where $V_{TH\infty}$ is the threshold voltage of a long-channel device. The decrease of threshold voltage with decreased gate length is a well-known short-channel effect called the "threshold voltage roll-off".

As can be seen from these expressions, short-channel effects can be minimized by reducing the junction depth and the gate oxide thickness. They can also be minimized by reducing the depletion depth through an increase in doping concentration. For many years, designers have implicitly observed design rules that would ensure the fabrication of devices free of short-channel effects. For example, using $(x_j/L_{el})^2 = 1/3$, $t_{ox}/L_{el} = 1/30$ and $t_{dep}/L_{el} = 1/3$ we obtain a DIBL of 29 mV at $V_{DS}=1$V. In modern devices, however, practical limits on the scaling of junction depth and gate oxide thickness lead to a significant increase of short channel effects and excessively large values of DIBL can quickly be reached.

## 1.3 Gate Geometry and Electrostatic Integrity

Short-channel effects arise when control of the channel region by the gate is affected by electric field lines from source and drain. These field lines is illustrated graphically in Figure 1.2. In a bulk device (Fig. 1.2.A), the electric field lines propagate through the depletion regions associated with the junctions. Their influence on the channel can be reduced by increasing the doping concentration in the channel region. In very small devices, unfortunately, the doping concentration becomes too high ($10^{19}$ $cm^{-3}$) for proper device operation.

In a fully depleted SOI (FDSOI) device, most of the field lines propagate trough the buried oxide (BOX) before reaching the channel region (Fig. 1.2.B). Short channel effects in FDSOI devices may be better or worse than in bulk MOSFETs, depending on the silicon film thickness, buried oxide thickness, and doping concentrations. Short-channel effects can be reduced in FDSOI MOSFETs by using a thin buried oxide and an underlying ground plane. In that case, most of the electric field lines from the source and drain terminate on the buried ground plane instead of the channel region (Figure 1.2.C). This approach, however, has the inconvenience of increased junction capacitance and body effect. [5]

A much more efficient device configuration is obtained by using the double-gate transistor structure. This device structure was first proposed by Sekigawa and Hayashi in 1984 and was shown to reduce threshold voltage roll-off in short-cannel devices.[6-7] In a double-gate device, both gates are connected together. The electric field lines from source and drain underneath the device terminate on the bottom gate electrode and cannot, therefore, reach the channel region (Fig. 1.2.D). Only the field lines that propagate through the silicon film itself can encroach on the channel region and degrade short-channel characteristics. This encroachment can be reduced by reducing the silicon film thickness.

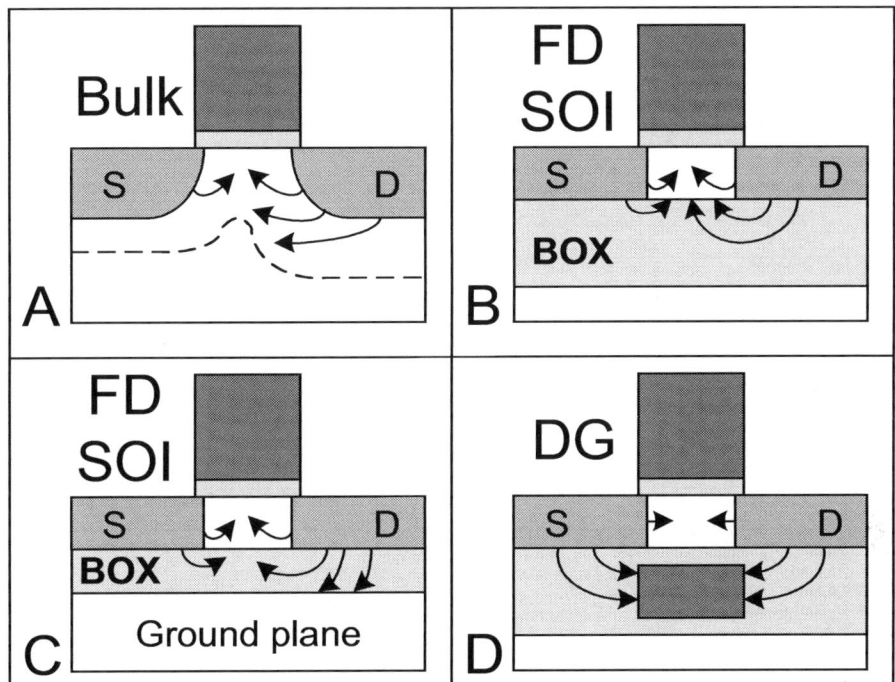

**Fig. 1.2.** Encroachment of electric field lines from source and drain on the channel region in different types of MOSFETs: A: Bulk MOSFET; B: Fully depleted SOI MOSFET, C: Fully depleted SOI MOSFET with thin buried oxide and ground plane; D: Double-gate MOSFET.

The Voltage-Doping Transformation model can readily be applied to fully depleted SOI and double-gate MOSFETs.[8] Using relationships (a) and (b), the Electrostatic Integrity factor of a *bulk device* can be written:

$$EI = \left[1 + \frac{x_j^2}{L_{el}^2}\right] \frac{t_{ox}}{L_{el}} \frac{t_{dep}}{L_{el}} \tag{1.4}$$

The equivalent factor can be obtained for a *fully depleted SOI device* (FDSOI) by noting that the junction depth is equal to the silicon film thickness, $t_{Si}$, and by noting that the gate field in the channel region penetrates the entirety of the depleted silicon film, $t_{Si}$, and extends to some depth in the buried oxide, $\lambda t_{BOX}$:

$$EI = \left[1 + \frac{t_{Si}^2}{L_{el}^2}\right] \frac{t_{ox}}{L_{el}} \frac{t_{Si} + \lambda t_{BOX}}{L_{el}} \tag{1.5}$$

In a *double-gate device*, the effective junction depth and the effective gate field penetration for each gate is equal to $t_{Si}/2$, which yields:

$$EI = \frac{1}{2}\left[1 + \frac{t_{Si}^2/4}{L_{el}^2}\right]\frac{t_{ox}}{L_{el}}\frac{t_{Si}/2}{L_{el}} \quad (1.6)$$

These expressions for the Electrostatic Integrity and the associated device cross sections are summarized in Figure 1.3. It is clear from relationships (1.1) and (1.2) that we want to minimize the *EI* factor in a device in order to keep short-channel effects under control. This can be achieved by reducing the thickness of the device, either by reducing the junction depth, $x_j$, and the depletion depth, $t_{dep}$, in a bulk device, by reducing the silicon film, $t_{si}$, and BOX thickness, $t_{BOX}$, in a fully depleted SOI device, or by reducing the silicon film thickness in a double-gate MOSFET. From an Electrostatic Integrity point of view, the double-gate device has the natural advantage of looking twice as thin the equivalent FDSOI transistor.

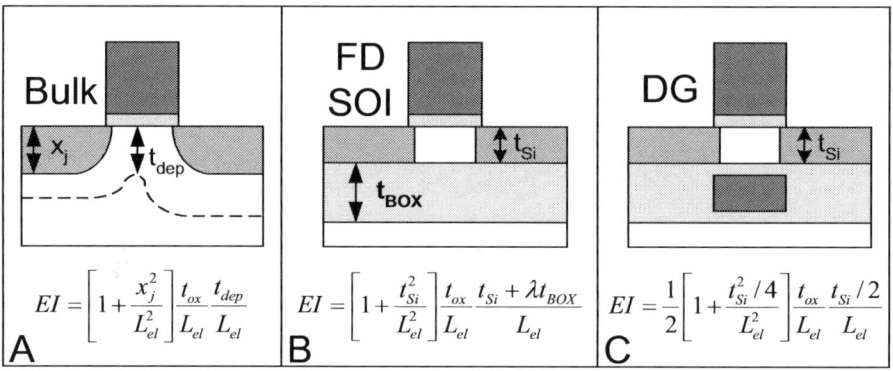

**Fig. 1.3.** Electrostatic Integrity in A: bulk, B: fully depleted SOI, and C: double-gate MOSFETs.

The VDT model has been implemented in a software package called MASTAR (Model for Analog and digital Simulation of mos TrAnsistoRs [9]). MASTAR has been extensively used in the ITRS 2005 Process Integration, Devices and Structures report to calculate the impact of transistor scaling on electrical characteristics.[10-11] Figure 1.4 shows typical values for DIBL in bulk, FDSOI and double-gate MOSFETs, as a function of gate length. Because thin-film SOI devices deliver better Electrostatic Integrity than bulk MOSFETs, SOI devices can be used at shorter channel lengths while keeping acceptable DIBL values (*e.g.* below

100 mV). The use of double-gate devices allows one to reduce gate length even further.[12-14]

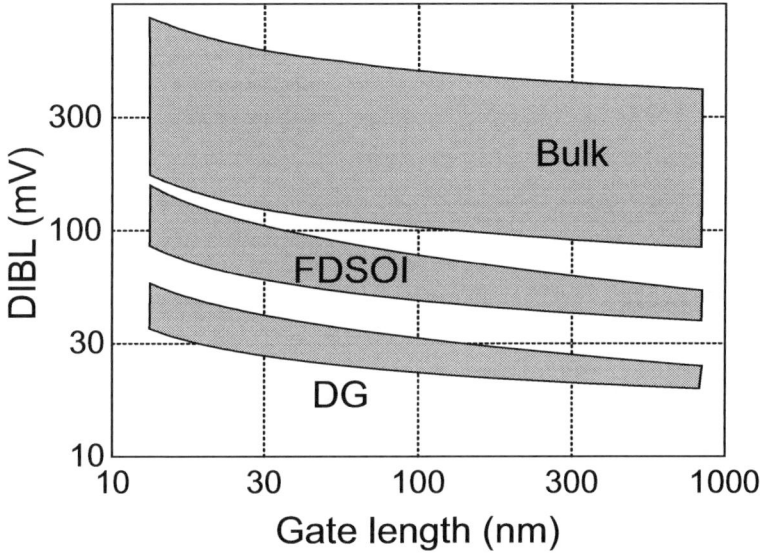

**Fig. 1.4.** Typical drain-induced barrier lowering in bulk, fully depleted SOI (FDSOI) and double-gate (DG) MOSFETs calculated by MASTAR.

Based on short-channel and DIBL considerations, the minimum gate length that can be used with the different technologies has been calculated with MASTAR and published in the 2005 ITRS. The result of these calculations is shown in Figure 1.5 for three different types of CMOS circuits: high-performance (HP), low operating power (LOP), and low standby power (LSTP) digital circuits. HP circuits are optimized for speed performance and feature the smallest available gate length at any moment in time, while the LSTP devices place the accent on low leakage currents, which necessitates the use of longer-channel devices.

An important conclusion can be derived from the data presented in Figure 1.5: bulk transistors run out of steam once they reach a gate length of 15-20 nm. FDSOI can be used until 10 nm, but smaller gate lengths can be only achieved by the double-gate structure.

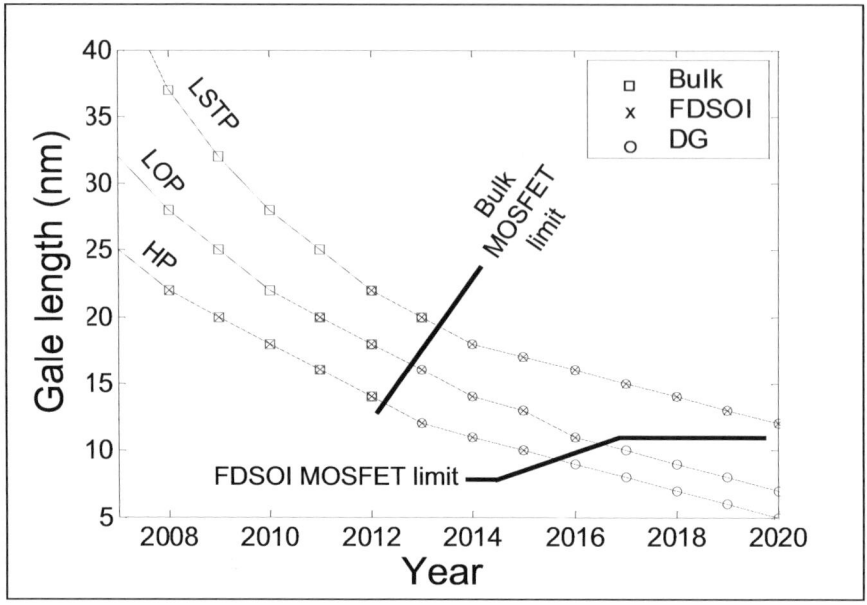

**Fig. 1.5.** Evolution of gate length predicted by the 2005 ITRS for high-performance (HP), low operating power (LOP), and low standby power (LSTP) digital circuits.

## 1.4 A Brief History of Multiple-Gate MOSFETs

In a continuous effort to increase current drive and better control short-channel effects, silicon-on-insulator MOS transistors have evolved from classical, planar, single-gate devices into three-dimensional devices with a multi-gate structure (double-, triple- or quadruple- gate devices). It is worth noting that, in most cases, the term "double gate" refers to a single gate electrode that is present on two opposite sides of the device. Similarly, the term "triple gate" is used for a single gate electrode that is folded over three sides of the transistor. One remarkable exception is the MIGFET (Multiple Independent Gate FET) where two separate gate electrodes can be biased with different potentials. It is also worth pointing out that one device may have several different names in the literature (Table 1.1).

**Table 1.1.** Device names found in the literature.

| Acronym | Also known as |
|---|---|
| MuGFET (Multiple-Gate FET) | Multi-gate FET, Multigate FET |
| MIGFET (Multiple Independent Gate FET) | Four-terminal (4T) FinFET |
| Triple-gate FET | Trigate FET |
| Quadruple-gate FET | Wrapped-Around Gate FET<br>Gate-All-Around FET<br>Surrounding-Gate FET |
| FinFET | DELTA (fully DEpleted Lean channel TrAnsistor) |
| FDSOI (Fully Depleted SOI) | Depleted Silicon Substrate |
| PDSOI (Partially depleted SOI) | Non-Fully Depleted SOI |
| Volume Inversion | Bulk Inversion |
| DTMOS (Dual Threshold Voltage MOS) | VTMOS (Varied Threshold MOS)<br>MTCMOS (Multiple Threshold CMOS)<br>VCBM (Voltage-Controlled Bipolar MOS)<br>Hybrid Bipolar-MOS Device |

### 1.4.1 Single-gate SOI MOSFETs

Figure 1.6 shows the "Family Tree" of SOI MOSFETs and shows the evolution from partially depleted, single-gate devices to multi-gate, fully depleted structures. Partially depleted silicon MOSFETs are the successors of earlier SOS (Silicon-On-Sapphire) devices. PDSOI MOSFETs were first used for niche applications such as radiation-hardened or high-temperature electronics. At the turn of the century PDSOI technology became mainstream as major semiconductor manufacturers started to use it to fabricate high-performance microprocessors. The low-voltage performance of PDSOI devices can be enhanced by creating a contact between the gate electrode and the floating body of the device. Such a contact improves the subthreshold slope, body factor and current drive, but limits the device operation to sub-1V supply voltages. [15-23] Fully depleted SOI devices have a better electrostatic coupling between the gate and the channel. This results in a better linearity, subthreshold slope, body coefficient and current drive. FDSOI technology is used in a number of applications ranging from low-voltage, low-power to RF integrated circuits. There exists a series of textbooks on SOI devices and technology, [24-26] as well as on SOI circuit design.[27-29]

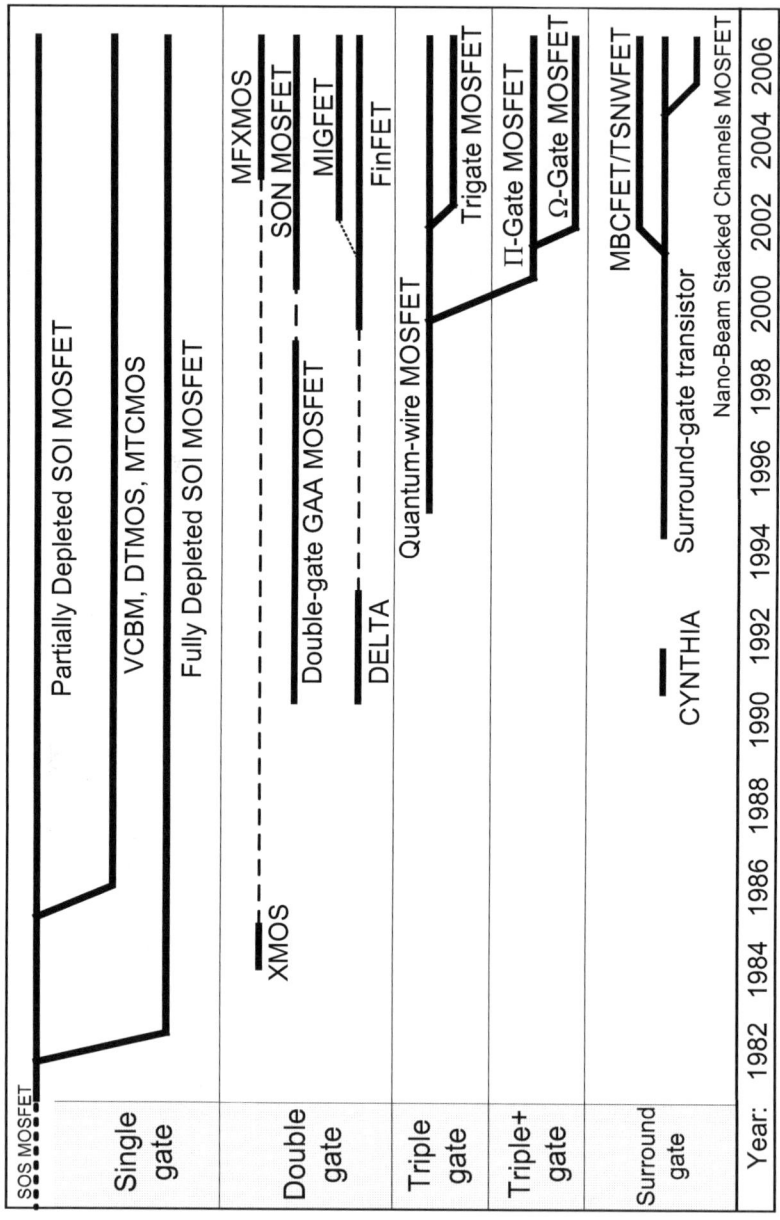

**Fig. 1.6.** "Family tree" of SOI and multigate MOSFETs.

## 1.4.2 Double-gate SOI MOSFETs

The first article on the double-gate MOS (DGMOS) transistor was published by T. Sekigawa and Y. Hayashi 1984.[30] That paper shows that one can obtain significant reduction of short-channel effects by sandwiching a fully depleted SOI device between two gate electrodes connected together. The device was called XMOS because its cross section looks like the Greek letter Ξ (*Xi*). Using this configuration, a better control of the channel depletion region is obtained than in a "regular" SOI MOSFET, and, in particular, the influence of the drain electric field on the channel is reduced, which reduces short-channel.[31] A more complete modeling that includes Monte-Carlo simulations, was published by Frank, Laux and Fischetti in 1992 in a paper that explores the ultimate scaling of the silicon MOSFET.[32] According to that article, the ultimate silicon device is a double-gate SOI MOSFET with a gate length of 30 nm, an oxide thickness of 3 nm, and a silicon film thickness of 5 to 20 nm. Such a (simulated) device shows no short-channel effects for gate lengths larger than 70 nm, and provides transconductance values up to 2300 mS/mm. The first fabricated double-gate SOI MOSFET was the "fully DEpleted Lean-channel TrAnsistor (DELTA, 1989)",[33] where the device is made in a tall and narrow silicon island called "finger", "leg" or "fin" (Figure 1.7). The FinFET structure is similar to DELTA, except for the presence of a dielectric layer called the "hard mask" on top of the silicon fin. [34-38] The hard mask is used to prevent the formation of parasitic inversion channels at the top corners of the device.

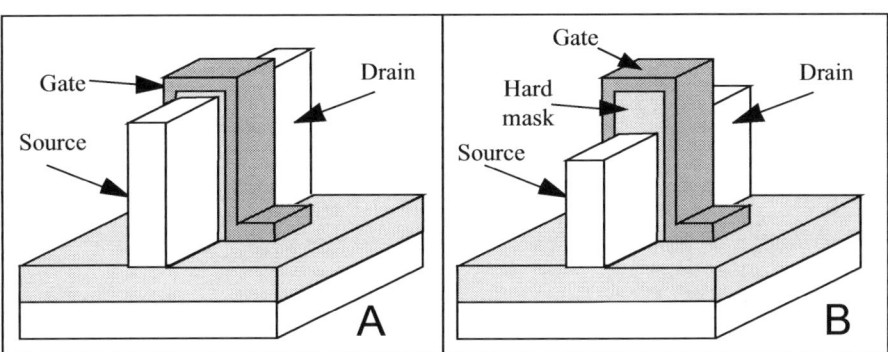

**Fig. 1.7.** Examples of double-gate MOS structure: A: DELTA MOSFET; B: FinFET.

Other implementations of vertical-channel, double-gate SOI MOSFETs include the "Gate-All-Around device" (GAA), which is a planar MOSFET with the gate electrode wrapped around the channel region (Figure

1.8.A)[39], the Silicon-On-Nothing (SON) MOSFET,[40-42] the Multi-Fin XMOS (MFXMOS) [43], the triangular-wire SOI MOSFET [44] and the Δ-channel SOI MOSFET.[45] It is worth noting that the original GAA device was a double-gate device, even though the gate was wrapped around all sides of the channel region, because the silicon island was much wider than thick. Nowadays, most people use the GAA acronym for quadruple-gate or surrounding-gate devices having a width-to-height ratio much closer to unity.[46-48]

The MIGFET (Multiple Independent Gate FET) is a double-gate device in which the two gate electrodes are not connected together and can, therefore, be biased with different potentials.[49-52,160] The main feature of the MIGFET is that the threshold voltage of one of the gates can be modulated by the bias applied to the other gate. This effect is similar to the body effect in FDSOI MOSFETs.[53] An application using MIGFET is signal modulation. A simple square law mixer can be formed using a single MIGFET by applying a small RF signal to one gate and a large low-frequency signal to the other gate. This single-device modulation is possible because the channel is fully depleted and the gates are perfectly symmetrical and aligned. This signal modulation circuit reduces transistor counts and rail-to-rail transistor stack, making it possible to design compact low-power mixers.[54]

## 1.4.3 Triple-gate SOI MOSFETs

The triple-gate MOSFET is a thin-film, narrow silicon island with a gate on three of its sides.[55] Implementations include the quantum-wire SOI MOSFET (Figure 1.8.B) [56-57] and the trigate MOSFET.[58-59]

The Electrostatic Integrity of triple-gate MOSFETs can be improved by extending the sidewall portions of the gate electrode to some depth in the buried oxide and underneath the channel region (Π-gate device [60-61] and Ω-gate device [62-64]). From an electrostatic point of view, the Π-gate and Ω-gate MOSFETs have an effective number of gates between three and four. The use of strained silicon, a metal gate and/or high-k dielectric as gate insulator can further enhance the current drive of the device.[65-68]

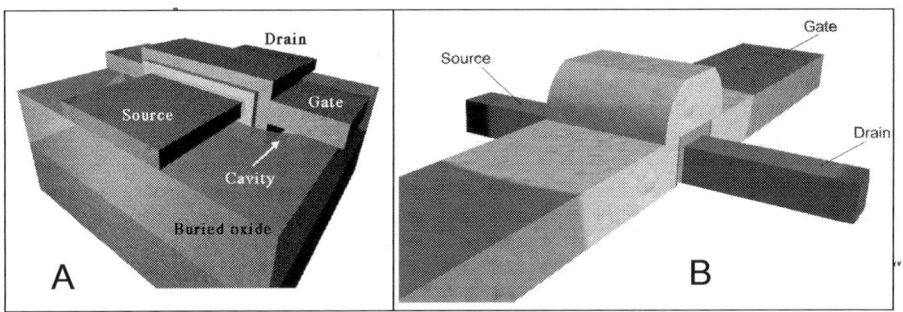

**Fig. 1.8.** A: Gate-All-Around (GAA) MOSFET; B: Triple-gate MOSFET.

### 1.4.4 Surrounding-gate (quadruple-gate) SOI MOSFETs

The structure that theoretically offers the best possible control of the channel region by the gate, and hence the best possible Electrostatic Integrity is the surrounding-gate MOSFET. The first surrounding-gate MOSFETs were fabricated by wrapping a gate electrode around a vertical silicon pillar. Such devices include the CYNTHIA device (circular-section device) [69-70] and the pillar surrounding-gate MOSFET (square-section device).[71] More recently, planar surrounding-gate devices with square or circular cross sections have reported.[72-73] Surrounding-gate SOI MOSFETs with a gate length as small as 5 nm and a diameter of 3 nm have shown to be fully functional.[74-75]

To increase the current drive per unit area, multiple surrounding-gate channels can be stacked on top of one another, while sharing common gate, source and drain. Such devices are called the Multi-Bridge Channel MOSFET (MBCFET), [76-77] the Twin-Silicon-Nanowire MOSFET (TSNWFET) [78], or the Nano-Beam Stacked Channels (GAA) MOSFET.[79] Analytical models for the electrical characteristics of cylindrical surrounding-gate MOSFETs can be found in the literature.[80-83]

Schematic cross sections corresponding to the different gate structures described in the previous sections are shown in Figure 1.9.

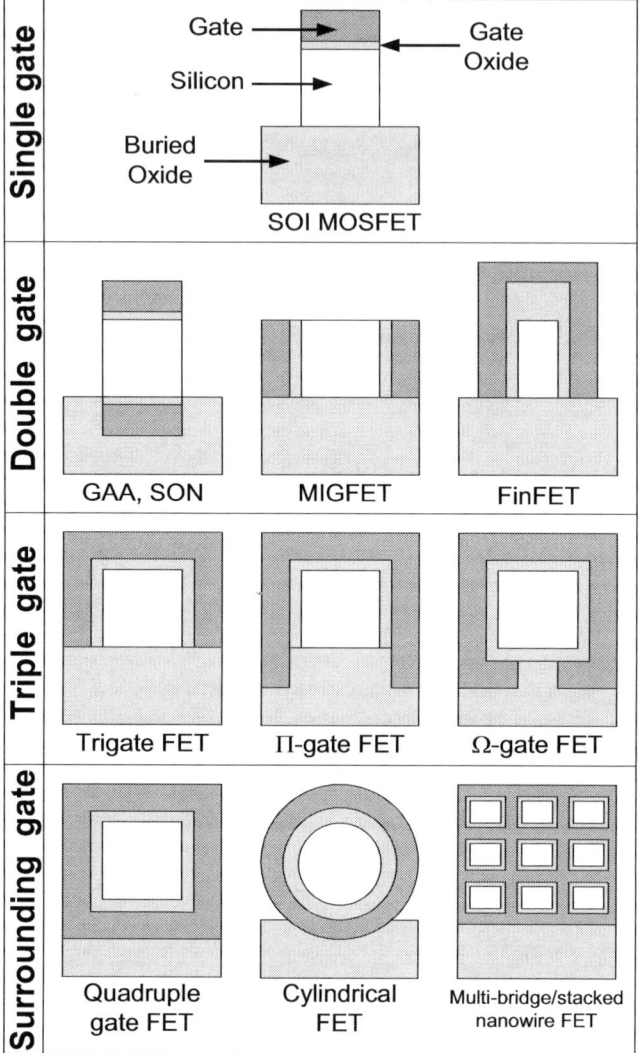

**Fig. 1.9.** Different gate structures.

### 1.4.5 Other multigate MOSFET structures

The Inverted T-channel FET (ITFET) combines a thin-film planar SOI device with a trigate transistor (Figure 1.10.A). [84-85] It comprises planar horizontal channels and vertical channels in a single device. The devices have multi-gate control around these channels. The Inverted T-gate structure has several advantages: the large base helps the fins from falling over during processing; it also allows for transistor action in the space between the fins, which is left unused in other MuGFET configurations. These additional channels increase the current drive. Numerical simulation of an N-channel ITFET reveals different turn-on mechanisms in different parts of the device. The corners of the device turn on first, immediately followed by the surface of the planar regions and the vertical channel. Since each ITFET has about seven corner elements they constitute a significant current to each ITFET device and in a well-designed device can yield substantially more current than a planar device of equivalent area.

The bulk FinFET is a FinFET made on bulk silicon instead of an SOI wafer. Fins are etched on a bulk silicon wafer and trimmed using an oxidation step. Field oxide is deposited to avoid inversion between the fins (Figure 1.10.B). Device with fin width down to 10 nm have shown to have good punchthrough immunity down to the sub-20nm gate length regime.[86-87]

The multi-channel Field Effect Transistor (McFET) is a modified bulk FinFET where a trench is etched in the center of the fin.[88] The trench is filled by the growth of a gate oxide and the deposition of gate material. This process produces a device having two very thin "twin" fins running from source to drain (Figure 1.10.C).

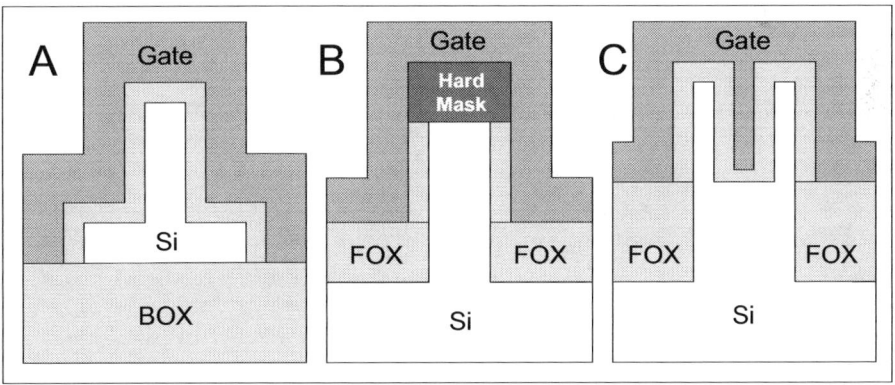

**Fig. 1.10.** Cross section of A: Inverted T channel FET; B: Bulk FinFET; C: Multi-channel Field Effect Transistor.

The 4-gate transistor ($G^4$ FET) has two (top and bottom) MOS gates and two (lateral) JFET gates. It is operated in accumulation mode and has the same structure as an inversion-mode partially depleted SOI MOSFET with two independent body contacts. These lateral contacts play the role of source and drain in the $G^4$ FET, while the junctions are used as lateral gates. The current in the body of the device is controlled by the front and back MOS gates, and by the two lateral junctions. The height and width of the conductive path is modulated by a mix of MOS and JFET effects. Each gate has the ability of switching the transistor on and off.[89-91] A model for the $G^4$ FET has been published In the literature.[92]

### 1.4.6 Multigate MOSFET memory devices

SONOS (Silicon-Oxide-Nitride-Oxide-Silicon) devices are non-volatile flash memory devices. They are essentially used in mobile applications. Flash devices have a small form factor, high storage density and low power consumption. For logic applications FinFET type devices are known to have good scalability down to 10-nm gate length. The FinFET device architecture combined with an ONO trapping layer as gate dielectric enables memory cells with feature sizes well below 50-nm. SONOS FinFET cells are programmed and erased using Fowler–Nordheim tunneling. SONOS FinFET memory devices show excellent functionality down to 20 nm channel length. [93-94] As an alternative to the ONO layer, nanocrystals embedded within the gate dielectric can be used to trap charges and achieve a similar memory effect. FinFET flash memory devices with a $V_{TH}$ window larger than 1 volt have been demonstrated using silicon nanocrystals embedded in the gate oxide.[95]

The SOI Zero-capacitor RAM (ZRAM) memory cell consists of a single PDSOI transistor in the floating body of which a charge can be stored. Such a capacitorless, one-transistor memory cell is, of course much smaller than a classical DRAM cell that requires the use of both a transistor and a storage capacitor. In the ZRAM memory cell a "1" binary state is stored by biasing the transistor in saturation and injecting holes in the floating body. Applying a negative bias to the bit line (device source) removes the charge from the floating body and, therefore, stores a binary "0" in the device. Since the threshold voltage of the FET varies with the charge stored in the body the logic state can easily be read by measuring the drain current when the device is turned on. Even though the storage of a charge requires a floating body, and, therefore, a partially depleted device, it is possible to realize ZRAM cells in fully depleted FinFETs of trigate devices. By applying a back-gate bias to the device, a floating body can be created at the bottom of the otherwise fully depleted fin and ZRAM operation can be activated.[96] Retention times of few tens of milliseconds are observed at room temperature on FinFETs with 300 nm channel length.

These values suit well typical embedded DRAM specification. FinFETs provide better retention characteristics than trigate devices which have a retention time of approximately 2 ms. Retention time decreases at shorter channel length but even 100 nm devices still show retention time of few milliseconds.[97]

When a positive bias is applied to the underlying silicon substrate (back gate) and a negative bias is applied to the gate of a MuGFET, it is possible to create an inversion layer at the bottom of the device and accumulation at the other interfaces. In that case, the so-called "MetaStable Dip" (MSD) effect can be observed, in which the transconductance as a function of the gate voltage increases, then drops, and then increases again. This effect is time-dependent and presents an hysteresis effect which depends whether the gate voltage is scanned in the forward or reverse mode.[98] The hysteretic nature of the phenomenon makes it useful for potential single-transistor memory applications.[99]

## 1.5 Multigate MOSFET Physics

### 1.5.1 Classical physics

#### 1.5.1.1 Natural length and short-channel effects

The Electrostatic Integrity (*EI*), which we know to be related to short channel effects, was compared between a bulk transistor, a FDSOI MOSFET, and a double-gate device in Figure 1.3. It was shown that the *EI* can be improved by reducing the gate oxide thickness and/or by decreasing the silicon film thickness (the junction depth in the case of a bulk device). We are now going to extend the analysis to all types of multigate MOSFETs.

The potential distribution in the channel of a fully depleted multigate MOSFET can be obtained by solving Poisson's equation using the depletion approximation:

$$\frac{d^2\Phi(x,y,z)}{dx^2} + \frac{d^2\Phi(x,y,z)}{dy^2} + \frac{d^2\Phi(x,y,z)}{dz^2} = \frac{qN_a}{\varepsilon_{si}} \quad (1.7)$$

Let us first understand the meaning of this equation. It can be rewritten in the form:

$$\frac{dE_x(x,y,z)}{dx} + \frac{dE_y(x,y,z)}{dy} + \frac{dE_z(x,y,z)}{dz} = C \qquad (1.8)$$

This relationship means that, at any point $(x,y,z)$ in the channel, the sum of the variations of the electric field components in the $x$, $y$ and $z$ directions is equal to a constant. Thus, if one of the components increases, the other ones (or, to be more precise, their sum) must decrease. The x-component of the electric field, $E_x$, represents the encroachment of the drain electric field on the channel region, and, therefore, short-channel effects. The influence of $E_x$ on a small element of the channel region located at coordinates $(x,y,z)$ (Figure 1.11) can be reduced by either increasing the channel length, $L$, or by increasing the control exerted by the top/bottom gates, $\frac{dE_y(x,y,z)}{dy}$, or the lateral gates, $\frac{dE_z(x,y,z)}{dy}$, on the channel. This can be achieved by reducing the silicon fin thickness, $t_{si}$, and/or the fin width, $W_{si}$. In addition, an increase of $\frac{dE_y(x,y,z)}{dy} + \frac{dE_z(x,y,z)}{dz}$, and, hence, in a better control of the channel by the gates and less short-channel effects, can also be obtained by increasing the number of gates: $\frac{dE_y(x,y,z)}{dy}$ can be increased by having two gates (top and bottom gates) instead of a single gate, and $\frac{dE_z(x,y,z)}{dy}$, is increased by the presence of lateral gates . Figure 1.11 illustrates how the electric field from the gates and from the drain compete for the control of the channel.

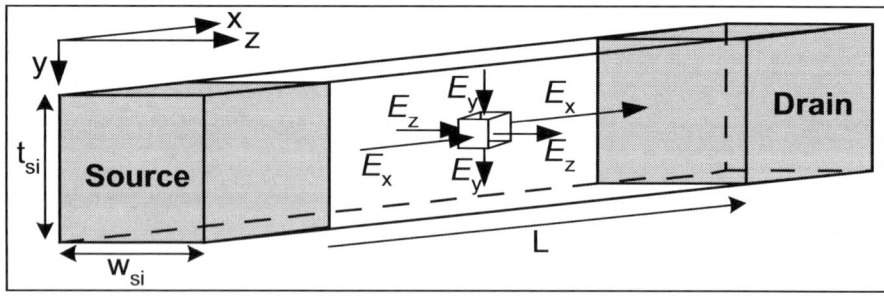

**Fig. 1.11.** Coordinate system and electric field components in a multiple-gate device.

In the case of a *wide* single- or double-gate device, we have $\frac{d\Phi}{dz} = 0$, and Poisson's equation becomes:

$$\frac{d^2\Phi(x,y)}{dx^2} + \frac{d^2\Phi(x,y)}{dy^2} = \frac{qN_a}{\varepsilon_{si}} \qquad (1.9)$$

Simplified one-dimensional analysis of a fully depleted device yields a parabolic potential distribution in the silicon film in the *y* (vertical) direction.[100] Assuming a similar distribution in the *y*-direction for a two-dimensional analysis Poisson's equation becomes:[101]

$$\Phi(x,y) = c_o(x) + c_1(x)y + c_2(x)y^2 \qquad (1.10)$$

**1.5.1.1.1 Single-gate SOI MOSFET**

In the case of a single-gate SOI device the boundary conditions to equation 1.10 are:[102]

1. $\Phi(x,0) = \Phi_f(x) = c_o(x)$ where $\Phi_f(x)$ is the front surface potential;

2. $\left.\dfrac{d\Phi(x,y)}{dy}\right|_{y=0} = \dfrac{\varepsilon_{ox}}{\varepsilon_{si}} \dfrac{\Phi_f(x) - \Phi_{gs}}{t_{ox}} = c_1(x)$

   where $\Phi_{gs} = V_{gs} - V_{FBF}$ is the front gate voltage, $V_{gs}$, minus the front-gate flat-band voltage, $V_{FBF}$.

3. If we assume that the buried oxide is very thick the potential difference across any finite distance in the BOX is negligible in the *y* direction we can write $\dfrac{d\Phi(x,y)}{dy} \cong 0$ in the BOX. Therefore, we have: $\left.\dfrac{d\Phi(x,y)}{dy}\right|_{y=t_{si}} = c_1(x) + 2t_{si}c_2(x) \cong 0$ and thus $c_2(x) \cong -\dfrac{c_1(x)}{2t_{si}}$.

Introducing these three boundary conditions in equation 1.10 we obtain:

$$\Phi(x,y) = \Phi_f(x) + \frac{\varepsilon_{ox}}{\varepsilon_{si}} \frac{\Phi_f(x) - \Phi_{gs}}{t_{ox}} y - \frac{1}{2t_{si}} \frac{\varepsilon_{ox}}{\varepsilon_{si}} \frac{\Phi_f(x) - \Phi_{gs}}{t_{ox}} y^2 \qquad (1.11)$$

Substituting equation 1.11 into equation 1.9, and setting *y=0*, at which depth $\Phi(x,y) = \Phi_f(x)$ we obtain:

$$\frac{d^2\Phi_f(x)}{dx^2} - \frac{\varepsilon_{ox}}{\varepsilon_{si}} \frac{\Phi_f(x) - \Phi_{gs}}{t_{si} t_{ox}} = \frac{qN_a}{\varepsilon_{si}} \quad (1.12)$$

Once $\Phi_f(x)$ is determined from equation 1.12, $\Phi(x,y)$ can be calculated using equation 1.11. Equation 1.12, however, can be used for another purpose. If we define:

$$\lambda_1 = \sqrt{\frac{\varepsilon_{si}}{\varepsilon_{ox}} t_{ox} t_{si}} \quad (1.13)$$

and

$$\varphi(x) = \Phi_f(x) - \Phi_{gs} + \frac{qN_a}{\varepsilon_{si}} \lambda_1^2 \quad (1.14)$$

then equation 1.12 can be re-written as follows:

$$\frac{d^2\varphi(x)}{dx^2} - \frac{\varphi(x)}{\lambda_1^2} = 0 \quad (1.15)$$

This equation has a solution in the form $\varphi(x) = \varphi_0 \exp\left(\pm \frac{x}{\lambda_1}\right)$ where $\lambda_1$ is a parameter that represents the spread of the electric potential in the $x$-direction. Note that $\varphi(x)$ differs from $\Phi_f(x)$ only by an $x$-independent term. Parameter $\lambda_1$ is called the "*natural length*" of the device. It depends on the gate oxide thickness and the silicon film thickness. The thinner the gate oxide and/or the silicon film, the smaller the natural length and, hence, the influence of the drain electric field on the channel region. Numerical simulations show that the effective gate length of a MOS device must be larger than 5 to 10 times the natural length to avoid short-channel effects.

### 1.5.1.1.2 Double-gate SOI MOSFET

In the case of a double-gate device the boundary conditions to equation 1.10 are:

1. $\Phi(x,0) = \Phi(x,t_{si}) = \Phi_f(x) = c_o(x)$ where $\Phi_f(x)$ is the front surface potential;

2. $\left.\dfrac{d\Phi(x,y)}{dy}\right|_{y=0} = \dfrac{\varepsilon_{ox}}{\varepsilon_{si}} \dfrac{\Phi_f(x)-\Phi_{gs}}{t_{ox}} = c_1(x)$ where $\Phi_{gs} = V_{gs} - V_{FBF}$ is the front gate voltage, $V_{gs}$, minus the front gate flat-band voltage, $V_{FBF}$.

3. $\left.\dfrac{d\Phi(x,y)}{dy}\right|_{y=t_{si}} = -\dfrac{\varepsilon_{ox}}{\varepsilon_{si}} \dfrac{\Phi_f(x)-\Phi_{gs}}{t_{ox}} = c_1(x) + 2 t_{si} c_2(x) = -c_1(x)$ and thus $c_2(x) = \dfrac{-c_1(x)}{t_{si}}$.

Substituting these boundary conditions into equation (1.15) yields:

$$\Phi(x,y) = \Phi_f(x) + \dfrac{\varepsilon_{ox}}{\varepsilon_{si}} \dfrac{\Phi_f(x)-\Phi_{gs}}{t_{ox}} y - \dfrac{1}{t_{si}} \dfrac{\varepsilon_{ox}}{\varepsilon_{si}} \dfrac{\Phi_f(x)-\Phi_{gs}}{t_{ox}} y^2 \quad (1.16)$$

The key difference between this expression and equation 1.10 is that the term $1/2t_{si}$ has now been replaced by $1/t_{si}$. In other words it looks as if the double-gate device was twice as thin as the single-gate transistor. The natural length of the double-gate device can be derived the same way it was done for the single-gate case, which yields:

$$\lambda_2 = \sqrt{\dfrac{\varepsilon_{si}}{2\varepsilon_{ox}} t_{ox} t_{si}} \quad (1.17)$$

### 1.5.1.1.3 Quadruple-gate SOI MOSFET

The original publication on the natural length concept analyzes single- and double-gate structures only. The concept can, however, be extended to quadruple-gate devices with a square cross section by noting that $\dfrac{d^2\Phi}{dy^2} = \dfrac{d^2\Phi}{dz^2}$ in the center of the device, where the influence of the electric field lines from the drain on the device body is the strongest. In that case the Poisson equation becomes:

$$\dfrac{d^2\Phi(x,y,z)}{dx^2} + 2\dfrac{d^2\Phi(x,y,z)}{dy^2} = \dfrac{qN_a}{\varepsilon_{si}} \quad (1.18)$$

and the natural length is equal to:

$$\lambda_4 = \sqrt{\dfrac{\varepsilon_{si}}{4\varepsilon_{ox}} t_{ox} t_{si}} \quad (1.19)$$

The natural length in a cylindrical surrounding-gate device can be found in Reference [103]. The natural length corresponding to different device geometries are summarized in Table 1.2.

**Table 1.2.** Natural length in devices with different geometries.

| Geometry | Natural length |
|---|---|
| Single gate | $\lambda_1 = \sqrt{\dfrac{\varepsilon_{si}}{\varepsilon_{ox}} t_{si} t_{ox}}$ |
| Double gate | $\lambda_2 = \sqrt{\dfrac{\varepsilon_{si}}{2\varepsilon_{ox}} t_{si} t_{ox}}$ |
| Quadruple gate (square channel cross section) | $\lambda_4 \cong \sqrt{\dfrac{\varepsilon_{si}}{4\varepsilon_{ox}} t_{si} t_{ox}}$ |
| Surrounding gate (circular channel cross section) | $\lambda_o = \sqrt{\dfrac{2\varepsilon_{si} t_{si}^2 \ln\left(1 + \dfrac{2t_{ox}}{t_{si}}\right) + \varepsilon_{ox} t_{si}^2}{16\varepsilon_{ox}}}$ |

The following observations can be made: the natural length (and hence short-channel effects) can be reduced by decreasing the gate oxide thickness, the silicon film thickness and by using a high-k gate dielectric instead of $SiO_2$. In addition, the natural length is reduced when the number of gates is increased. In very small devices, the reduction of oxide thickness below 1.5 nm causes gate tunneling current problems. Using multi-gate devices, it is possible to trade a thin gate oxide for thin silicon film/fin thinning since $\lambda$ is proportional to the product $t_{si} \times t_{ox}$.

The concept of an *"equivalent number of gates"* (*ENG*) can be introduced at this stage.[104] It is basically equal to the number of gates (a square cross section is assumed) but is also equal to the number that divides $\dfrac{\varepsilon_{si}}{\varepsilon_{ox}} t_{si} t_{ox}$ in the equations defining the natural length, $\lambda$. Thus we have *ENG*=1 for a single-gate FDSOI MOSFET, *ENG*=2 for a double-gate device and *ENG*=4 for a quadruple-gate MOSFET. *ENG*=3 for a triple-gate device and, by some strange twist of fate, *ENG* is close to π in a Π-gate device. In the Ω-gate device the value of *ENG* ranges between 3 and 4 depending on the extension of the gate under the fin.

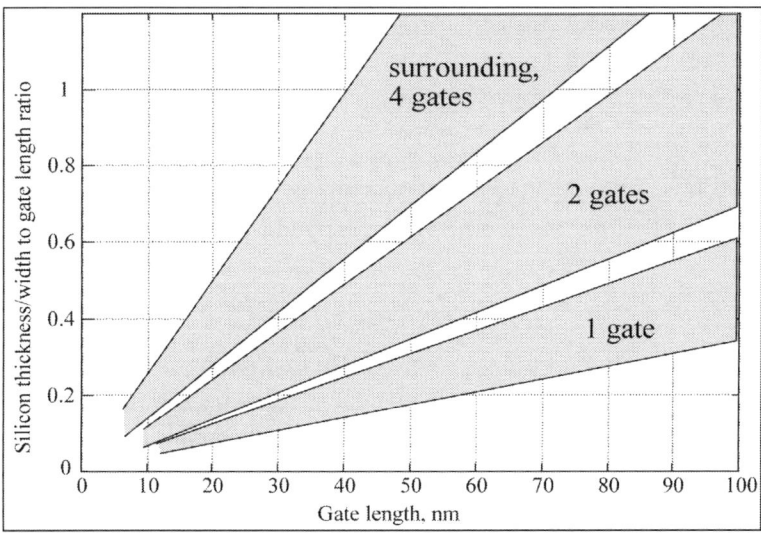

**Fig. 1.12.** Maximum allowed silicon film thickness and device width *vs.* gate length to avoid short-channel effects.

The natural length can be used to estimate the maximum silicon film thickness and device width that can be used in order to avoid short-channel effects. Figure 1.12 shows the maximum allowed silicon film thickness (and device width in a triple-gate device with a square cross section) to avoid short-channel effects. The calculation assumes a gate oxide thickness of 1.5 nm. It reveals that for a gate length of 50 nm, for instance, the thickness of the silicon film in a single-gate, fully depleted device needs to be 3 to 5 times smaller than the gate length. If a double-gate structure is used, the requirements on silicon film thickness are more relaxed and the film needs to be thinned to only half the gate length.[105] Further relaxation is obtained using a surrounding-gate structure, where the silicon film thickness/width/diameter can be as large as the gate length. The film thickness requirements for triple-gate, Π-gate and Ω-gate devices are located between those for double-gate and surrounding-gate devices.

### 1.5.1.2 Current drive

In a multigate FET the current drive is essentially equal to the sum of the currents flowing along all the interfaces covered by the gate electrode. It is, therefore, equal to the current in a single-gate device multiplied by the equivalent number of gates (a square cross section is assumed) if carriers have the same mobility at each interfaces. For instance, the current drive of a double-gate device is double that of a single-gate transistor of equivalent gate length and width. In triple-gate and vertical double-gate structures all individual fins have the same thickness and width. As a result the current drive is fixed to a single, discrete value, for a given gate length. To drive larger currents multi-fin devices are used. The current drive of a multi-fin MOSFET is equal to the current of an individual fin multiplied by the number of fins (also sometimes referred to as "fingers" or "legs").

Let us now compare the current drive of a single-gate, planar MOSFET with that of a multi-fin multigate FET having the same gate area, $W \times L$ (Figure 1.13). Let us assume the planar FET is made on (100) silicon and that the surface mobility is $\mu_{top}$. Let us also assume the multigate FET is made on (100) silicon and that the top surface mobility is $\mu_{top}$. The sidewall interface mobility may be different from the top mobility, depending on the sidewall crystal orientation, usually (100) or (110), and is noted $\mu_{side}$.[106]

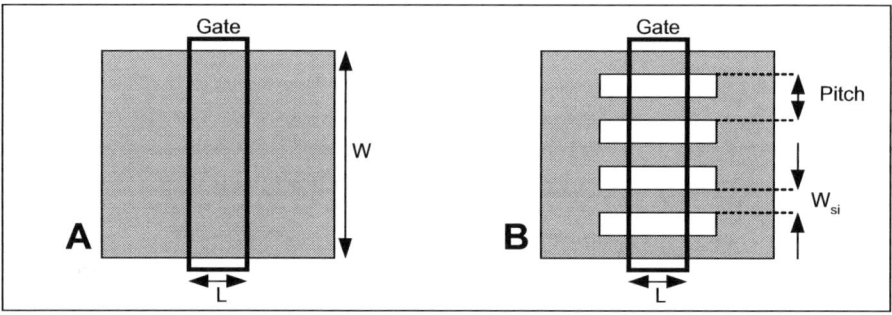

**Fig. 1.13.** A: Single-gate, planar MOSFET layout; B: Multi-fin multigate FET layout.

Considering a pitch $P$ for the fins, the current in the multigate device is given by:

$$I_D = I_{Do} \frac{\theta \mu_{top} W_{si} + 2\mu_{side} t_{si}}{\mu_{top} P} \quad (1.20)$$

where $I_{Do}$ is the current in the single-gate, planar device, $W_{si}$ is the width of each individual fin, $t_{si}$ is the silicon film thickness (Figure 1.14); $\theta=1$ in a triple-gate device where conduction occurs long three interfaces, and $\theta=0$ in a FinFET where channels are formed at the sidewall interfaces only.[107]

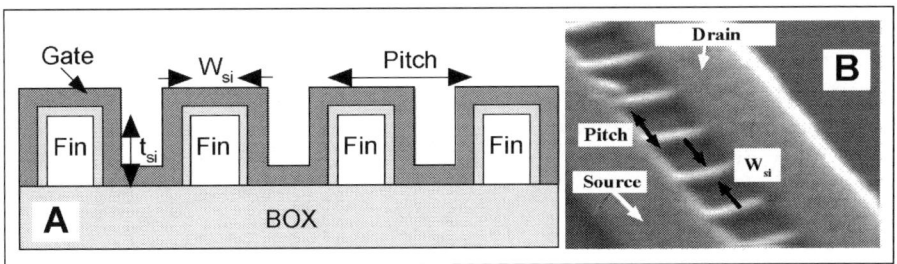

**Fig. 1.14.** A: Cross section of a multi-fin multigate MOSFET; B: SEM picture of the fins.

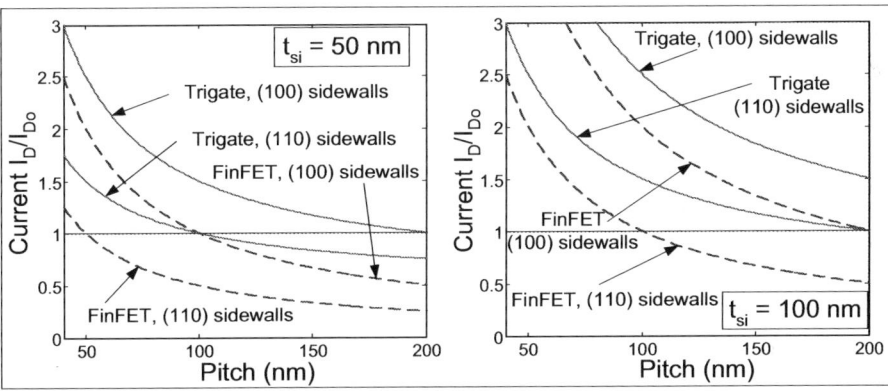

**Fig. 1.15.** Normalized current drive of a 50nm-thick FinFET and triple-gate MOSFET vs. pitch. $W_{si}$=pitch/2; The (100)-interface electron mobility is 300 cm$^2$/Vs and the (110)-interface mobility is 150 cm$^2$/Vs. Left: $t_{si}$=50nm; Right: $t_{si}$=100nm.

Figure 1.15 shows the current drive in a multi-fin, trigate MOSFET and a multi-fin FinFET as a function of the fin pitch, $P$, assuming $W_{si}$=$P/2$. The fin height is either 50nm or 100 nm. The drain current is normalized to that of a single-gate planar device occupying the same silicon real estate. This figure shows that a multigate device can deliver significantly more current that single-gate planar MOSFET, provided a small enough fin pitch can be achieved. It also shows the impact of device orientation ((100) or (110) sidewalls) on current drive. The current drive can be increased by increasing the fin height, $t_{si}$, but the use of tall fins often raises difficulties during device processing.

It is worth noting that gate capacitance increases with the Effective Number of Gates (ENG). As a result, the gate delay $C_g V_{DD}/I_{ON}$ does not improve when the ENG is increased. On the contrary, the delay increases with the ENG and is, therefore, larger in GAA than in trigate devices and larger in double-gate FETs than in single-gate devices.[108]

### 1.5.1.3 Corner effect

Devices with a triple, quadruple, Π or Ω gate structure present a non-planar silicon/gate oxide interface with corners. It has been known for long that premature inversion can form in at the corners of SOI structures because of charge-sharing effects between two adjacent gates. In particular, one can observe the presence of two different threshold voltages (one in the corners and at the top or sidewall Si-SiO$_2$ interfaces), as well as a kink in the subthreshold $I_D(V_G)$ characteristics. [109] The presence of corners can degrade the subthreshold characteristics of a device. Avoiding that problem is the reason why there is a hard mask at the top of FinFETs. To complicate he matter, the radius of curvature of the corners has a significant impact on the device electrical characteristics and can decide whether or not a different threshold voltage will be measured at the corners and at the planar interfaces of the device.

In classical single-gate SOI MOSFETs, corner effects are purely parasitical. They are not part of intrinsic device operation and they can usually be eliminated by increasing the doping concentration in the corners. In a multigate device, on the other hand, the corners are part of the *intrinsic* transistor structure. Therefore, it is worth understanding the relationship and interaction between currents in the corner currents and currents in the planar surfaces of the device.[110] To illustrate the corner effect, let us use the Ω-gate device shown in Figure 1.16. The thickness and width of the device are $t_{si}$ and $W_{si}$, and the radius of curvature of the top and the bottom corners is noted $r_{top}$ and $r_{bot}$, respectively. The gate oxide thickness is 2 nm, and $t_{si}=W_{si}=30$ nm. Because the gate material is N$^+$ polysilicon, high doping concentrations have to be used to achieve useful threshold voltage values (N-channel device).

Figure 1.17 presents the simulated $dg_m/dV_G$ characteristics of the device at $V_{DS}=0.1$V for different doping concentrations and a top and bottom corner radius of curvature of either 1 or 5 nm. The $dg_m/dV_G$ characteristics have been used by several authors to identify the different threshold voltage(s) single- and double-gate SOI devices.[111-114] The humps of the $dg_m/dV_G$ curve correspond to the formation of channels in the device (*i.e.* they correspond to threshold voltages).

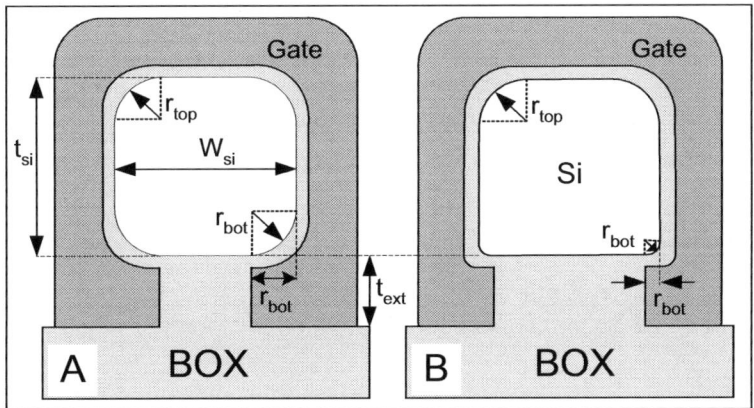

**Fig. 1.16.** $\Omega$-gate device cross section. A: $r_{top} = r_{bot}$ ; B: $r_{top} \neq r_{bot}$.

When the corner radius of curvature is equal to 1 nm the devices with the lowest doping concentrations exhibit a single hump, indicating that both corners and edges build up channels at the same time. The devices with the more heavily doped channels have two humps. The first of these two humps corresponds to inversion in the top corners, and the second one to top and sidewall channel formation.

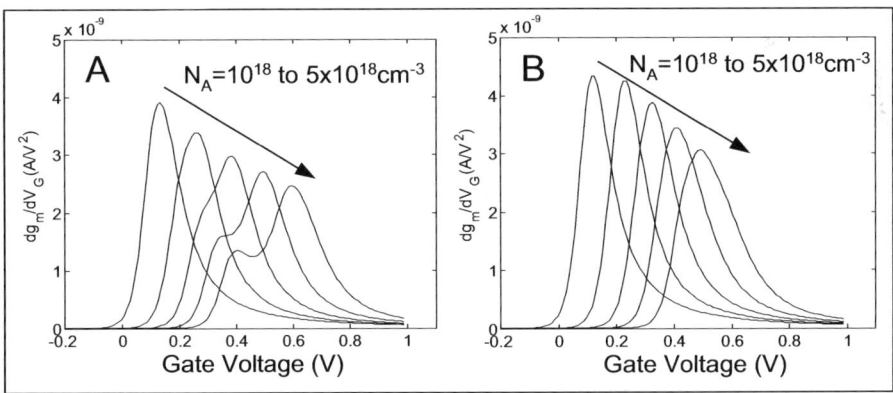

**Fig. 1.17.** $dg_m/dV_G$ in the $\Omega$-gate MOSFET of Figure 1.16. Gate is $N^+$ polysilicon. A: $r_{top} = r_{bot} = 1$ nm; B: $r_{top} = r_{bot} = 5$ nm.

When the corner radius of curvature is 5 nm, a single peak is observed for all doping concentrations, which indicates that premature corner inversion has been eliminated. In that case all devices reach a subthreshold swing of 60 mV/decade over a significant range of their subthreshold current. The corner effect can thus be eliminated by using either a low doping concentration in the channel, or corners with a large enough radius of curvature.[115-117] The general manufacturing trend being to use undoped channels in conjuction with a midgap metal gate, the corner effect is not expected to pose any problem to the use of MuGFET technology.[118-120]

### 1.5.2 Quantum effects

The thickness and/or width of multi-gate FETs is reaching values that are less than 10 noanometers. Unde these conditions the electrons in the "channel" (if we take the example of an n-channel device) form either a two-Dimensional Electron Gas (2DEG) if we consider a double-gate device or a one-Dimensional Electron Gas (1DEG) if we consider a triple or quadruple-gate MOSFET. Consider a double-gate device made in a thin silicon film. Adopting the coordinate system used in Figure 1.11, the film thickness being $t_{si}$, electrons are free to move in the $x$ and $z$ direction, but they are confined in the $y$ direction. In a thin and narrow triple- or quadruple-gate device electrons are free to move in the $x$ direction only, and are confined in the $y$ and $z$ directions. This results in the formation of energy subbands and in electron distributions in the silicon film that can be significantly different from what is predicted by classical theory. In particular, inversion layers need not to be localized at the surface of the silicon film, but can instead be found in the "depth" of the film, giving rise to volume inversion. The confinement of electrons is also at the origin of new and previously unexpected mobility and threshold voltage behavior.

### 1.5.2.1 Volume inversion

Volume inversion was discovered in 1987 by Balestra *et al.* [121] and was first observed in double-gate GAA MOSFETs in 1990.[122] Volume inversion is a phenomenon that appears in very thin (or narrow) film multigate SOI MOSFETs due to the fact that inversion carriers are not confined near the Si/SiO$_2$ interface, as predicted by classical device physics, but rather at the center of the film. To correctly predict volume inversion one needs to solve both the Schrödinger and the Poisson equation in a self-consistent manner.

Volume inversion can be modeled using a Poisson-Schrödinger solver. It was first measured on double-gate devices in 1994 [123-124] and has been confirmed by many research groups since. [125-130] Volume inversion can be observed in triple-gate SOI MOSFETs as well.[131-133] When a multigate MOSFET operates in the volume inversion regime, the electrons form a low-dimensional electron gas (a two-Dimensional Electron Gas (2DEG) for a double-gate device and a one-Dimensional Electron Gas (1DEG) for a trigate, Π-gate, Ω-gate or surrounding-gate FET). As a result, energy subbands are formed. The $j$-th electron wave function and the corresponding energy level $E_j$ can be found solving the Schrödinger equation using the effective mass approximation:

$$\left(-\frac{\hbar^2}{2m^*}\nabla^2 - q\Phi\right)\Psi_j = E_j\Psi_j \tag{1.21}$$

and the Poisson equation:

$$\nabla^2\Phi = -\frac{q}{\varepsilon_{si}}[p - n + N_D - N_A] \tag{1.22}$$

in a self-consistent manner. The electron concentration being given by:

$$n = \sum_j \left[(\Psi_j \times \Psi_j^*) \times \int_{E_j}^{\infty} \rho_j(E) f_{FD}(E) dE\right] \tag{1.23}$$

where $\rho_j(E)$ is the density of states as a function of the energy and $f_{FD}(E)$ is the Fermi-Dirac distribution function. In a 2DEG the density of states is a constant (independent of energy), and in a 1DEG it is a function of $(E - E_j)^{-\frac{1}{2}}$.[134-136] Please note that Equation 1.21 is anisotropic and is equal to

$$\left(-\frac{\hbar^2}{2}\left(\frac{\partial}{\partial x}\left(\frac{1}{m_x^*}\frac{\partial}{\partial x}\right) + \frac{\partial}{\partial y}\left(\frac{1}{m_y^*}\frac{\partial}{\partial y}\right) + \frac{\partial}{\partial z}\left(\frac{1}{m_z^*}\frac{\partial}{\partial z}\right)\right) - q\Phi\right)\Psi_j = E_j\Psi_j \tag{1.24}$$

where the effective masses $m_x^*$, $m_y^*$ and $m_z^*$ correspond to different valleys and depend on crystal orientation.[137] An example of electron concentration in double-gate FETs for different silicon film thickness, $t_{si}$, and for $V_G > V_{TH}$ is presented in Figure 1.18.[138] The electron concentration in a FinFET, a trigate FET and a gate-all-around device is shown in Figure 1.19.[139] In each case one can observe the high electron concentration in the center of the silicon film or fin, corresponding to volume inversion. A direct consequence of volume inversion is the increase of inversion carrier mobility in thin-film devices.

30    Jean-Pierre Colinge

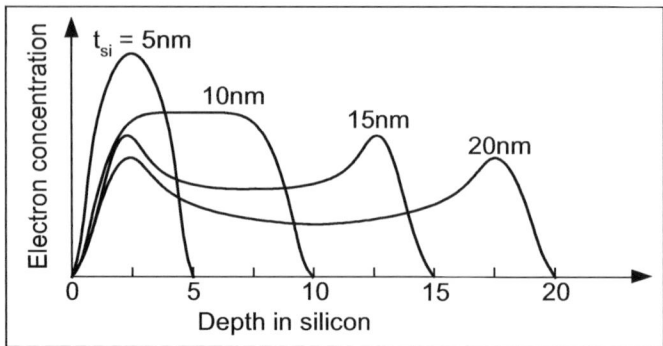

**Fig. 1.18.** Electron concentration profile in double-gate MOSFETs with different silicon film thickness ($t_{si}$) values.

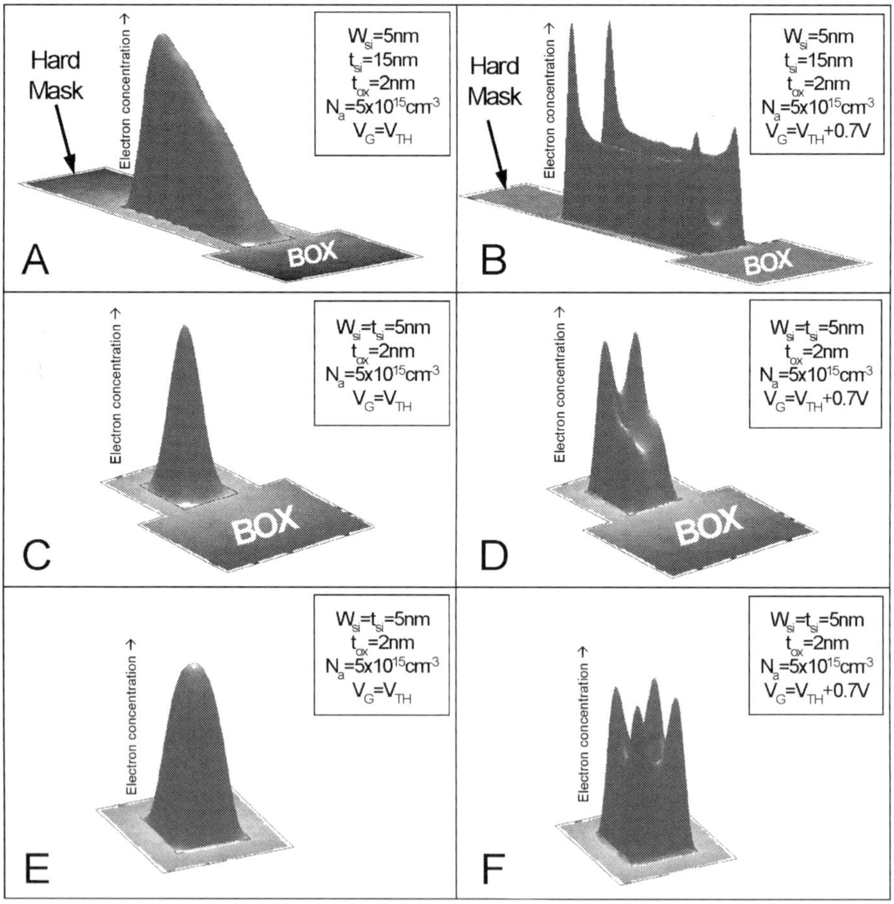

**Fig. 1.19.** Electron concentration profile in a FinFET (A,B), a trigate FET (C,D) and a gate-all-around device (E,F) at threshold (A,C,E) and above threshold (B,D,F). The vertical scale (electron concentration) is different for each plot.

### 1.5.2.2 Mobility effects

Volume-inversion carriers experience less interface scattering than carriers in a surface inversion layer. As a result an increase of the mobility and transconductance is observed in double-gate devices. Furthermore, the phonon scattering rate is lower in double-gate devices than in single-gate transistors. The dependence of mobility on film thickness in double-gate MOSFETs is illustrated in figure 1.20. In thick films there is no interaction between the front and the back channel and there is no volume inversion. The mobility is identical to that in a bulk MOSFET. If the film gets thinner volume inversion appears and the mobility is increased because of reduced Si-SiO$_2$ interface scattering.

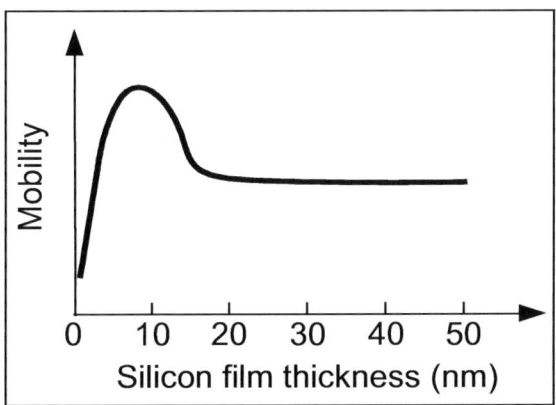

**Fig. 1.20.** Mobility *vs.* silicon film thickness in a double-gate transistor.

In thicker films the inversion carriers are concentrated near the interfaces, but in thinner films most of the carriers are concentrated near the center of the silicon film, further away from the interface scattering centers (Figure 1.18), which increases their mobility. In very thin silicon films, however, the inversion carriers in the volume inversion layer do experience surface scattering because of their physical proximity to the interfaces, and mobility drops with any decrease in film thickness.[140-142]

### 1.5.2.3 Threshold voltage

Classical theory predicts that the threshold voltage decreases in a fully depleted SOI MOSFET when silicon film thickness is decreased, assuming the doping concentration $N_a$ is held constant.[143] This is due to the reduction of depletion charge $qN_a t_{si}$ in when the film thickness is deceased.

When the film thickness is below 10 nm, however, the depletion charge is very small and can usually be neglected. On the other hand, two non-classical contributions to threshold voltage have to be taken into account. The first contribution comes from the fact that the concentration of inversion carriers needs to be bigger that what classical theory predicts in order to reach threshold. Thus the potential $\Phi$ in the thin silicon film is larger than the classical $2\Phi_F$. The second contribution arises from the splitting of the conduction band splits into subbands: the mimimum energy of the subbands (and thus the minimum energy in the conduction band) increases when the film thickness is decreased, which increases the gate voltage needed to reach any particular inversion carrier concentration. This also causes the threshold voltage to increase. This quantum phenomenon was first reported in 1993 by Omura et al. [144-145] and has since been confirmed and measured by several research groups.[146-149]

In the particular case of a double-gate device, and neglecting any potential variation in the silicon film ($\Phi=0$), Equation 1.21 can be used to find the minimum energy of the first conduction subband. The solution to Equation 1.21 is:

$$E_n = \frac{\pi^2 \hbar^2 n^2}{2 m^* t_{si}^2} \quad (n=1,2,3,...) \tag{1.25}$$

The energy of the lowest subband is found by letting $n=1$:

$$E = E_{co} + \frac{\pi^2 \hbar^2}{2 m^* t_{si}^2} \tag{1.26}$$

where $E_{co}$ is the classical, "three-dimensional" minimum energy of the conduction band. The threshold voltage of a single-gate device can be derived from the basic capacitance relationship:

$$C_{inv} = C_{ox} + C_{depl} \tag{1.27}$$

In a lightly doped (ideally undoped), fully depleted device, the depletion capacitance can be neglected, such that:

$$C_{inv} = C_{ox} \tag{1.28}$$

Under the same conditions, and in subthreshold operation, the inversion charge and capacitance in a double-gate device are given by:

$$Q_{inv} = -q\, n_i\, t_{si}\, e^{\left(\frac{q\Phi}{kT}\right)} \quad \text{and} \quad C_{inv} = -\frac{1}{2}\frac{dQ_{inv}}{d\Phi} = -\frac{1}{2}\frac{q}{kT}Q_{inv} \qquad (1.29)$$

where $\Phi$ is the potential in the channel and the coefficient $1/2$ is due to the double gate. Using relationships (1.28) and (1.29) we can write:

$$Q_{inv} = -2\, C_{ox}\, \frac{kT}{q} \qquad (1.30)$$

Combining (1.29) and (1.30), potential $\Phi$ can be found. Adding the workfunction difference between the gate and the silicon film, $\Phi_{MS}$, and the increase of bandgap (1.25) to the channel potential, we find the threshold voltage: [150]

$$V_{TH} = \Phi_{MS} + \frac{kT}{q}\ln\left(\frac{2\,C_{ox}\,kT}{q^2\,n_i\,t_{si}}\right) + \frac{\pi^2\,\hbar^2}{2\,q\,m^*\,t_{si}^2} \qquad (1.31)$$

The first term of (1.31) is the workfunction difference between the gate and the silicon film. The second term of the equation represents the potential $\Phi$ in the channel. It is inversely proportional to the silicon film thickness $t_{si}$. In very thin films, $\Phi$ can be significantly larger than $2\Phi_F$, and as a result, the inversion carrier concentration at threshold can be much larger in a thin-film device than in a thicker one.[151]

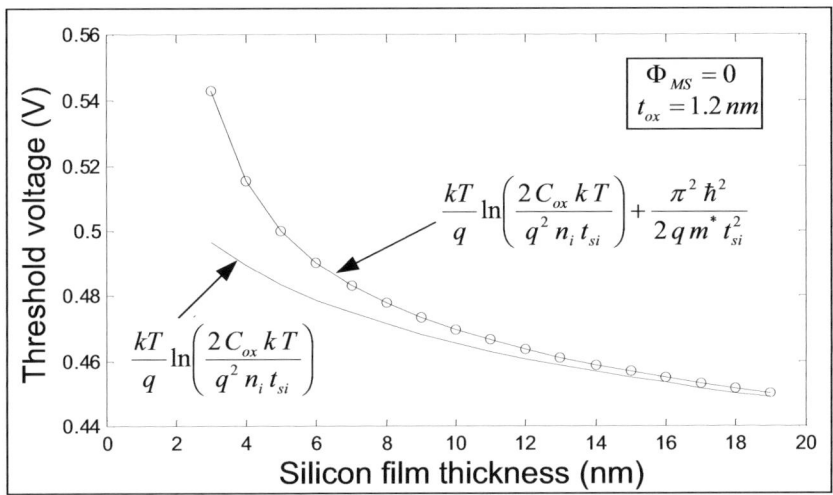

**Fig. 1.21.** Threshold voltage dependence on silicon film thickness in a long-channel, lightly-doped/undoped double-gate transistor. The lower curve represents the "classical part of Equation (1.31) and the upper curve includes quantum-mechanical considerations.

This increase of threshold voltage is correctly predicted by classical simulators since it does not involve the Schrödinger equation. The third term of (1.31) is related to the variation of the minimum energy in the conduction band with silicon film thickness, which can only be predicted through quantum-mechanical calculations (Figure 1.21).

A similar increase of threshold voltage is observed in trigate, Π-gate, Ω-gate and GAA transistors when the section of the silicon fin is reduced. Both an increase of threshold voltage and inversion carrier concentration at threshold are observed when the fin section is reduced. The effect is predicted by classical simulation using Poisson's equation, but the increase of $V_{TH}$ is more dramatic when Schrödinger's equation is also used. Figure 1.22 shows the drain current vs. gate voltage in trigate transistors with various fin width and height values.[152] Clearly, the threshold voltage increases as the cross sectional area of the device is reduced.

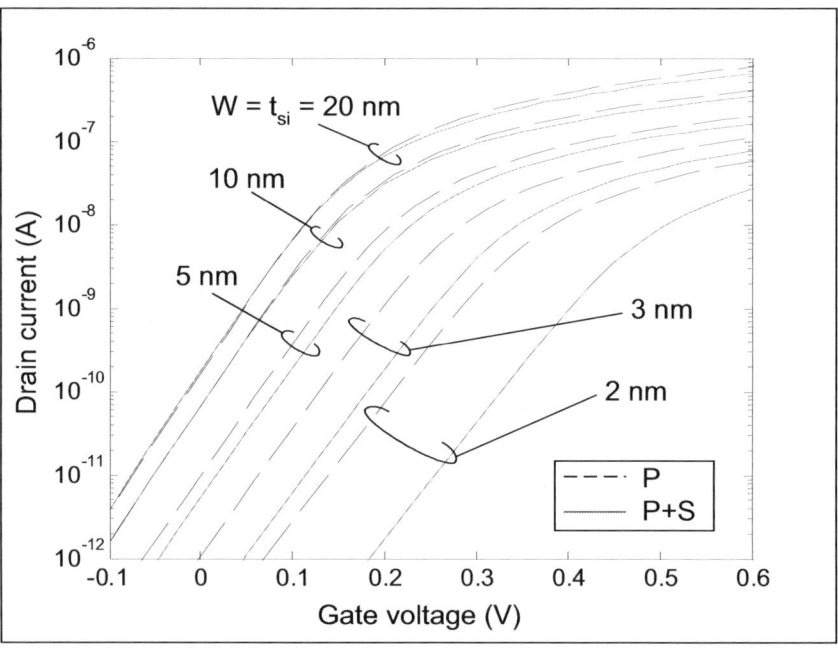

**Fig. 1.22.** Drain current vs. gate voltage in trigate MOSFETs with different cross sections. Devices are simulated using either Poisson's equation only (P) or a Poisson+Schrödinger solver (P+S). $V_{DS}$ = 50 mV, $t_{ox}$ = 2 nm, $N_a$ = 5x10$^{17}$ cm$^{-3}$.

### 1.5.2.4 Inter-subband scattering

The formation of subbands in thin or narrow multigate FETs is at the origin of some interesting electrical properties that are not observed in classical devices. The Density of States (DoS) in the conduction band of trigate MOSFETs is shown in Figure 1.23.A for a fin width of 5nm and a fin height of 5 nm. In that case the DoS is clearly one-dimensional as is constituted by a succession of peaks, each followed by a distribution in the shape of $1/\sqrt{E}$.[153-155] Each peak corresponds to an individual subband. Figure 1.23.B shows the DoS in the same structure, but for a fin height of 100 nm. Since $t_{si} \gg W_{si}$ the device now exhibits the staircase-like DoS distribution that is characteristic of a two-dimensional electron gas.[156]

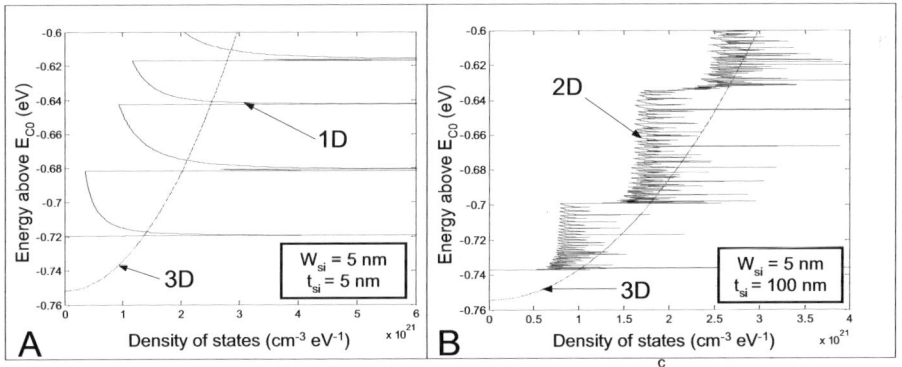

**Fig. 1.23.** Density of states in the conduction band of a trigate MOSFET in strong inversion. A: 1D DoS distribution calculated for $t_{si}=W_{si}=5$nm, and B: 2D DoS distribution calculated for $t_{si}=100$nm and $W_{si}=5$nm. The 3D DoS is shown for comparison.

Inter-subband scattering occurs between electrons that belong to different energy subbands. These scattering events reduce the electron mobility. By definition, there is no intersubband scattering if only one subband is occupied, which occurs right above threshold.

As the gate voltage and the electron concentration are increased, however, a larger number of subbands become populated, and scattering occurs between electrons belonging to different subbands. If the temperature is not too high (compared to $\Delta E/k$, where $\Delta E$ is the energy separation between two subbands, and $k$ is Boltzmann's constant) and if the drain voltage not much larger than $\Delta E/q$, inter-subband scattering phenomena can be directly observed in the form of oscillations of drain current amplitude when the gate voltage is increased. This effect can be seen in Figure 1.24, in which each "dip" of the curve corresponds to a reduction of mobility caused by scattering due to the population of each new subband. This effect is more pronounced at low temperature, but can be observed at room temperature provided the cross-section of the devices is small enough. Table 1.3 shows the maximum temperature and maximum drain voltage at which current oscillations due to inter-subband scattering have been experimentally observed in MuGFETs.

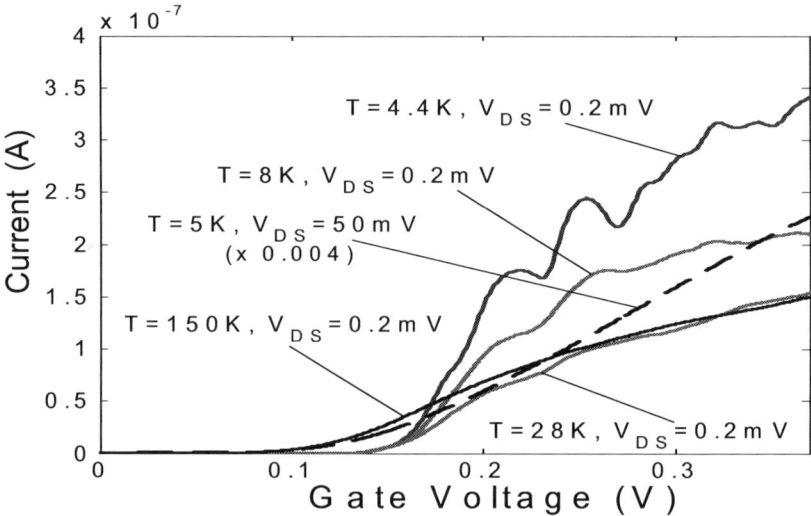

**Fig. 1.24.** Drain current *vs.* gate voltage for different temperature and drain voltage values. The amplitude of the curve for $V_{DS}$=50mV is multiplied by a factor 200µV/50mV=0.004 to fit in the same graph as the curves measured at $V_{DS}$=200µV.

**Table 1.3.** Subband energy separation, $\Delta E$, maximum temperature and drain voltage at which current oscillations due to intersubband scattering have been observed.

| $W_{si} \times t_{si}$ | $\Delta E$ | Temperature | Drain voltage | Reference |
|---|---|---|---|---|
| 45nm × 82nm | 0.15 meV | 28 K | 0.2 mV | 157 |
| 11nm × 58nm | 1-2 meV | 300K | 1 mV | 158 |
| 6nm × 6nm | 35 meV | 5 K | 100 mV | 159 |
| 6nm × 6nm | 35 meV | 200K | 50 mV | 159 |

# References

1. G. Moore: Cramming more components onto integrated circuits. Electronics, **38**, 114 (1965)
2. http://www.itrs.net/
3. T. Skotnicki, G. Merckel, T. Pedron: The voltage-doping transformation: a new approach to the modeling of MOSFET short-channel effects. IEEE Electron Device Letters **9**, 109 (1988)
4. T. Skotnicki : Heading for decananometer CMOS - is navigation among icebergs still a viable strategy? *Proceedings of the 30th European Solid-State Device Research Conference.* Frontier Group, Gif-sur-Yvette, France, 19 (2000)
5. W.Xiong, K. Ramkumar, S.J. Jamg, J.T. Park, J.P. Colinge: Self-aligned ground-plane FDSOI MOSFET. *Proceedings of the IEEE International SOI Conference*, 23 (2002)
6. T. Sekigawa and Y. Hayashi: Calculated threshold-voltage characteristics of an XMOS transistor having an additional bottom gate. Solid-State Electronics **27**, 827 (1984)
7. T. Sekigawa, Y. Hayashi, K. Ishii: Feasibility of very-short-channel MOS transistors with double-gate structure. Electronics and Communications in Japan, Part 2. **76-10**, 39 (1993)
8. T. Skotnicki: Ultimate scaling of SOI MOSFETs. *MIGAS Short Course*, Villard de Lans, France (2004)
9. T. Skotnicki, C. Denat, P. Senn, G. Merckel, B. Hennion: A new analog/digital CAD model for sub-halfmicron MOSFETs. *Technical Digest of the International Electron Devices Meeting (IEDM)*, 165 (1994)
10. T. Skotnicki, F. Boeuf, R. Cerutti, S. Monfray, C. Fenouillet-Beranger, M. Muller, A. Pouydebasque: New materials and device architectures for the end-of-roadmap CMOS nodes. Materials Science & Engineering B (Solid-State Materials for Advanced Technology) **124-125**, 3 (2005)
11. C. Fenouillet-Beranger, T. Skotnicki, S. Monfray, N. Carriere, F. Boeuf: Requirements for ultra-thin-film devices and new materials for the CMOS roadmap. Solid-State Electronics **48-6**, 961(2004)
12. T. Skotnicki, J.A. Hutchby, Tsu-Jae King, H.-S.P. Wong, F. Boeuf: The end of CMOS scaling: toward the introduction of new materials and structural changes to improve MOSFET performance. IEEE Circuits and Devices Magazine **21-1**, 16 (2005)
13. C. Fenouillet-Beranger, T. Skotnicki, S. Monfray, N. Carriere, F. Boeuf: Requirements for ultra-thin-film devices and new materials for the CMOS roadmap. Solid-State Electronics **48-6**, 961 (2004)
14. T. Skotnicki and F. Boeuf: CMOS Technology Roadmap – Approaching Uphill Specials, *Proceedings of the $9^{th}$ Intl. Symp. On Silicon Materials Science and Technology*, ECS Volume **2002-2**, 720, 2002
15. J.P. Colinge: An SOI voltage-controlled bipolar-MOS device. IEEE Trans. Electron Devices. **34**, 845, (1987)

16  S.A. Parke, C. Hu, P.K. Ko: Bipolar-FET hybrid-mode operation of quarter-micrometer SOI MOSFETs [MESFETs read MOSFETs. IEEE Electron Device Lett. **14**, 234 (1993)
17  M. Matloubian: Analysis of hybrid-mode operation of SOI MOSFETs. *IEEE International SOI Conference Proceedings*, 106, (1993)
18  T. Douseki, S. Shigematsu, J. Yamada, M. Harada, H. Inokawa, T. Tsuchiya: A 0.5-V MTCMOS/SIMOX logic gate. IEEE Journal of Solid-State Circuits **32**, 1604 (1997)
19  F. Assaderaghi, D. Sinitsky, S. A. Parke, J. Bokor, P.K. Ko, C. Hu: A dynamic threshold voltage MOSFET (DTMOS) for ultra-low voltage operation. *Tech. Digest of IEDM*, 809 (1994)
20  Z. Xia, Y. Ge, Y. Zhao: A study of varied threshold voltage MOSFET (VTMOS) performance and principle. *Proceedings 22nd International Conference on Microelectronics (MIEL)*, **1**, 159 (2000)
21  V. Ferlet-Cavrois, A. Bracale, N. Fel, O. Musseau, C. Raynaud, O. Faynot, J.L. Pelloie: High frequency characterization of SOI dynamic threshold voltage MOS (DTMOS) transistors. *Proceedings IEEE Intl. SOI Conference*, 24 (1999)
22  A. Yagishita, T. Saito, S. Inumiya, K. Matsuo, Y. Tsunashima, K. Suguro: Dynamic threshold voltage damascene metal gate MOSFET (DT-DMG-MOS) technology for very low voltage operation of under 0.7 V. IEEE Transactions on Electron Devices **49-3**, 422 (2002)
23  J.P. Colinge, J.T. Park: Application of the EKV model to the DTMOS SOI transistor. Journal of Semiconductor Technology and Science, **3-4**, 223 (2003)
24  J.P. Colinge: *Silicon-on-Insulator Technology: Materials to VLSI ($3^{rd}$ Ed.)* (Springer, New York, 2004)
25  S. Cristoloveanu, S. Li: *Electrical Characterization of Silicon-on-Insulator Materials and Devices.* (Springer, New York, 1995)
26  J.B. Kuo, Ker-Wei Su: *CMOS VLSI Engineering.* (Springer, New York, 1998)
27  T. Sakurai, A. Matsuzawa, T. Douseki: *Fully-Depleted SOI CMOS Circuits and Technology for Ultralow-Power Applications.* (Springer, New York, 2006)
28  A. Marshall, S. Natarajan: *SOI Design: Analog, Memory and Digital Techniques.* (Springer, New York, 2002)
29  K. Bernstein, N. Rohrer: *SOI Circuit Design Concepts.* (Springer, New York, 2000)
30  T. Sekigawa and Y. Hayashi: Calculated threshold-voltage characteristics of an XMOS transistor having an additional bottom gate. Solid-State Electronics **27**, 827 (1984)
31  B. Agrawal, V.K. De, J.M. Pimbley, J.D. Meindl: Short channel models and scaling limits of SOI and bulk MOSFETs. IEEE Journal of Solid-State Circuits **29-2**, 122 (1994)

32  D.J. Frank, S.E. Laux, M.V. Fischetti: Monte Carlo simulation of a 30 nm dual-gate MOSFET: how short can Si go? *Technical Digest of IEDM*, 553 (1992)
33  D. Hisamoto, T. Kaga, Y. Kawamoto, E. Takeda: A fully depleted lean-channel transistor (DELTA)-a novel vertical ultra thin SOI MOSFET. *Technical Digest of IEDM*, 833 (1989)
34  Xuejue Huang, Wen-Chin Lee, C. Kuo, D. Hisamoto, Leland Chang, J. Kedzierski, E. Anderson, H. Takeuchi, Yang-Kyu Choi, K. Asano, V. Subramanian, Tsu-Jae King, J. Bokor, Chenming Hu: Sub 50-nm FinFET: PMOS. *Technical Digest of IEDM*, 67 (1999)
35  D. Hisamoto, Wen-Chin Lee, J. Kedzierski, H. Takeuchi, K. Asano, C. Kuo, E. Anderson, Tsu-Jae King, J. Bokor, Chenming Hu: FinFET-a self-aligned double-gate MOSFET scalable to 20 nm. IEEE Transactions on Electron Devices **47-12**, 2320 (2000)
36  Bin Yu, Leland Chang, S. Ahmed, Haihong Wang, S. Bell, Chih-Yuh Yang, C. Tabery, Chau Ho, Qi Xiang, Tsu-Jae King, J. Bokor, Chenming Hu, Ming-Ren Lin, D. Kyser: FinFET scaling to 10 nm gate length. *Technical Digest of IEDM*, 251 (2002)
37  L. Chang, Yang-kyu Choi, D. Ha, P. Ranade, Shiying Xiong, J. Bokor, Chenming Hu, T.J. King: Extremely scaled silicon nano-CMOS devices. Proceedings of the IEEE **91-11**, 1860 (2003)
38  Yang-Kyu Choi: FinFET for Terabit era. Journal of Semiconductor Technology and Science **4-1**, 1 (2004)
39  J.P. Colinge, M.H. Gao, A. Romano, H. Maes, C. Claeys: Silicon-on-insulator gate-all-around device. *Technical Digest of IEDM*, 595 (1990)
40  M. Jurczak, T. Skotnicki, M. Paoli, B. Tormen, J. Martins, J.L. Regolini, D. Dutartre, P. Ribot, D. Lenoble, R. Pantel, S. Monfray: Silicon-on-Nothing (SON)-an innovative process for advanced CMOS. IEEE Transactions on Electron Devices **47-11**, 2179 (2000)
41  S. Harrison, P. Coronel, F. Leverd, R. Cerutti, R. Palla, D. Delille, S. Borel, S. Jullian, R. Pantel, S. Descombes, D. Dutartre, Y. Morand, M.P. Samson, D. Lenoble, A. Talbot, A. Villaret, S. Monfray, P. Mazoyer, J. Bustos, H. Brut, A. Cros, D. Munteanu, J.L. Autran, T. Skotnicki: Highly performant double gate MOSFET realized with SON process. *Technical Digest of IEDM*, 18.6.1 (2003)
42  J. Pretet, S. Monfray, S. Cristoloveanu, T. Skotnicki: Silicon-on-nothing MOSFETs: performance, short-channel effects, and backgate coupling. IEEE Transactions on Electron Devices **51-2**, 240 (2002)
43  Y. Liu, K. Ishii, T. Tsutsumi, M. Masahara, E. Suzuki: Ideal rectangular cross-section Si-Fin channel double-gate MOSFETs fabricated using orientation-dependent wet etching. IEEE Electron Device Letters **24-7**, 484 (2003)
44  T. Hiramoto: Nano-scale silicon MOSFET: towards non-traditional and quantum devices. *IEEE International SOI Conference Proceedings*, 8 (2001)
45  Z. Jiao and A.T. Salama: A Fully Depleted Delta Channel SOI NMOSFET. *Electrochem. Society Proceedings* **2001-3**, 403 (2001)

46  N. Singh, A. Agarwal, L.K. Bera, T.Y. Liow, R. Yang, S.C. Rustagi, C.H. Tung, R. Kumar, G.Q. Lo, N. Balasubramanian, D.L. Kwong: High-performance fully depleted silicon nanowire (diameter<5 nm) gate-all-around CMOS devices. IEEE Electron Device Letters **27-5**, 383 (2006)
47  K.E. Moselund, P. Dainesi, M. Declercq, M. Bopp, P. Coronel, T. Skotnicki, A.M. Ionescu: Compact gate-all-around silicon light modulator for ultra high speed operation. Sensors and Actuators A (Physical), **130-131**, 220 (2006)
48  K. Castellani-Coulie, D. Munteanu, J.L. Autran, V. Ferlet-Cavrois, P. Paillet, J. Baggio: Investigation of 30 nm gate-all-around MOSFET sensitivity to heavy ions: a 3-D simulation study. IEEE Transactions on Nuclear Science **53-4**, 1950 (2006)
49  L. Mathew, Yang Du, A.V.-Y. Thean, M. Sadd, A. Vandooren, C. Parker, T. Stephens, R. Mora, Raghav Rai, M. Zavala, D. Sing, S. Kalpai, J. Hughes, R. Shimer, S. Jallepalli, G. Workman, B.E. White, B.Y. Nguyen, A. Mogab: Multi gated device architectures advances, advantages and challenges. *International Conference on Integrated Circuit Design and Technology*, 97 (2004)
50  Weimin Zhang, J.G. Fossum, L. Mathew, Yang Du: Physical insights regarding design and performance of independent-gate FinFETs. IEEE Transactions on Electron Devices **52-10**, 2198 (2005)
51  S. Eminente, Kyoung-Il Na, S. Cristoloveanu, L. Mathew, A. Vandooren: Lateral and vertical coupling effects in MIGFETs. *Proceedings of the IEEE International SOI Conference*, 94 (2005)
52  K. Endo, Y. Liu, M. Masahara, T. Matsukawa, S. O'uchi, E. Suzuki: Fabrication and power-management demonstration of four-terminal FinFEts. ECS Transactions **6-4**, 71 (2007)
53  J.P. Colinge: Fully-depleted SOI CMOS for analog applications. IEEE Transactions on Electron Devices **45-5**, 1010 (1998)
54  L. Mathew, Y. Du, A.V.Y. Thean, M. Sadd, A. Vandooren, C. Parker, T. Stephens, R. Mora, R. Rai, M. Zavala, D. Sing, S. Kalpat, J. Hughes, R. Shimer, S. Jallepalli, G. Workman, W. Zhang, J.G. Fossum, B.E. White, B.Y. Nguyen, J. Mogab: CMOS Vertical Multiple Independent Gate Field Effect Transistor (MIGFET). *Proceedings IEEE International SOI Conference*, 187 (2004)
55  M.C. Lemme, T. Mollenhauer, W. Henschel, T. Wahlbrink, M. Baus, O. Winkler, R. Granzner, F. Schwierz, B. Spangenberg and H. Kurz: Subthreshold behavior of triple-gate MOSFETs on SOI Material. Solid State Electronics 48-4, 529 (2004)
56  X. Baie, J.P. Colinge, V. Bayot, E. Grivei: Quantum-wire effects in thin and narrow SOI MOSFETs. *Proceedings IEEE International SOI Conference*, 66 (1995)
57  J.P. Colinge, X. Baie, V. Bayot, E. Grivei: A silicon-on-insulator quantum wire. Solid-State Electronics **39**, 49 (1996)
58  R. Chau, B. Doyle, J. Kavalieros, D. Barlage, A. Murthy, M. Doczy, R. Arghavani, S. Datta: Advanced depleted-substrate transistors: single-gate,

double-gate and tri-gate. *Extended Abstracts of the International Conference on Solid State Devices and Materials (SSDM)*, 68 (2002)

59 B.S. Doyle, S. Datta, M. Doczy, B. Jin, J. Kavalieros, T. Linton, A. Murthy, R. Rios, R. Chau: High performance fully-depleted tri-gate CMOS transistors. IEEE Electron Device Letters **24-4**, 263 (2003)

60 J.T. Park, J.P. Colinge, C. H. Diaz: Pi-gate SOI MOSFET. IEEE Electron Device Letters **22**, 405 (2001)

61 J.T. Park and J.P. Colinge: Multiple-gate SOI MOSFETs: device design guidelines. IEEE Transactions on Electron Devices **49-12**, 2222 (2002)

62 F.L. Yang, H.Y. Chen, F.C. Cheng, C.C. Huang, C.Y. Chang, H.K. Chiu, C.C. Lee, C.C. Chen H.T. Huang, C.J. Chen, H.J. Tao, Y.C. Yeo, M.S. Liang, C. Hu: 25 nm CMOS Omega FETs. *Technical Digest of IEDM*, 255 (2002)

63 Fu-Liang Yang, Di-Hong Lee, Hou-Yu Chen, Chang-Yun Chang, Sheng-Da Liu, Cheng-Chuan Huang, Tang-Xuan Chung, Hung-Wei Chen, Chien-Chao Huang, Yi-Hsuan Liu, Chung-Cheng Wu, Chi-Chun Chen, Shih-Chang Chen, Ying-Tsung Chen, Ying-Ho Chen, Chih-Jian Chen, Bor-Wen Chan, Peng-Fu Hsu, Jyu-Horng Shieh, Han-Jan Tao, Yee-Chia Yeo, Yiming Li, Jam-Wem Lee, Pu Chen, Mong-Song Liang, Chenming Hu: 5nm-gate nanowire FinFET. *Symposium on VLSI Technology,* 196 (2004)

64 R. Ritzenthaler, C. Dupré, X. Mescot, O. Faynot, T. Ernst, J.C. Barbé, C. Jahan, L. Brévard, F. Andrieu, S. Deleonibus, S. Cristoloveanu: Mobility behavior in narrow $\Omega$-gate FET devices. *Proceedings IEEE International SOI Conference*, 77 (2006)

65 Z. Krivokapic, C. Tabery, W. Maszara, Q. Xiang, M.R. Lin: High-performance 45-nm CMOS technology with 20-nm multi-gate devices. *Extended Abstracts of the International Conference on Solid State Devices and Materials (SSDM)*, 760 (2003)

66 F. Andrieu, C. Dupré, F. Rochette, O. Faynot, L. Tosti, C. Buj, E. Rouchouze, M. Cassé, B. Ghyselen, I. Cayrefourcq, L. Brévard, F. Allain, J.C. Barbé, J. Cluzel, A. Vandooren, S. Denorme, T. Ernst, C. Fenouillet-Béranger, C. Jahan, D. Lafond, H. Dansas, B. Previtali, J.P. Colonna, H. Grampeix, P. Gaud, C. Mazuré, S. Deleonibus: 25nm Short and Narrow Strained FDSOI with TiN/HfO$_2$ Gate Stack. *Symposium on VLSI Technology*, paper 16.4 (2006)

67 J. Kavalieros, B. Doyle, S. Datta, G. Dewey, M. Doczy, B. Jin, D. Lionberger, M. Metz, W. Rachmady, M. Radosavljevic, U. Shah, N. Zelick and R. Chau: Tri-Gate Transistor Architecture with High-k Gate Dielectrics, Metal Gates and Strain Engineering. *Symposium on VLSI Technology*, paper 7.1 (2006)

68 T. Irisawa, T. Numata, T. Tezuka, N. Sugiyama, S. Takagi: Electron transport properties of ultrathin-body and tri-gate SOI nMOSFETs with biaxial and uniaxial strain. *Technical Digest of IEDM*, paper 17.2 (2006)

69 S. Miyano, M. Hirose, F. Masuoka: Numerical analysis of a cylindrical thin-pillar transistor (CYNTHIA). IEEE Transactions on Electron Devices **39-8**, 1876 (1992)

70 T. Ohba, H. Nakamura, H. Sakuraba, F. Masuoka: A novel tri-control gate surrounding gate transistor (TCG-SGT) nonvolatile memory cell for flash memory. Solid-State Electronics **50-6**, 924 (2005)

71 A. Nitayama, H. Takato, N. Okabe, K. Sunouchi, K. Hieda, F. Horiguchi, F. Masuoka: Multi-pillar surrounding gate transistor (M-SGT) for compact and high-speed circuits. IEEE Transactions on Electron Devices **38-3**, 579 (1991)

72 V. Passi, B. Olbrechts, J.P. Raskin: Fabrication of a Quadruple Gate MOSFET in Silicon-on-Insulator technology. *Abstracts of the NATO Advanced Research Workshop on Nanoscaled Semiconductor-on-Insulator Structures and Devices*, 11 (2006)

73 N. Singh, A. Agarwal, L.K. Bera, T.Y. Liow, R. Yang, S.C. Rustagi, C.H. Tung, R. Kumar, G.Q.Lo, N. Balasubramanian, D.L. Kwong: High-performance fully depleted silicon nanowire (diameter<5 nm) gate-all-around CMOS devices", IEEE Electron Device Letters **27-5**, 383 (2006)

74 Hyunjin Lee, Lee-Eun Yu, Seong-Wan Ryu, Jin-Woo Han, Kanghoon Jeon, Dong-Yoon Jang, Kuk-Hwan Kim, Jiye Lee, Ju-Hyun Kim, Sang Cheol Jeon, Gi Seong Lee, Jae Sub Oh, Yun Chang Park, Woo Ho Bae, Hee Mok Lee, Jun Mo Yang, Jung Jae Yoo, Sang Ik Kim, Yang-Kyu Choi: Sub-5nm all-around gate FinFET for ultimate scaling. *Symposium on VLSI Technology*, paper 7.5 (2006)

75 N. Singh, F.Y. Lim, W.W. Fang, S.C. Rustagi, L.K. Bera, A. Agarwal, C.H. Tung, K.M. Hoe, S.R. Omampuliyur, D. Tripathi, A.O. Adeyeye, G.Q. Lo, N. Balasubramanian, D.L. Kwong: Ultra-narrow silicon nanowire gate-all-around CMOS device: impact of diameter, channel-orientation and low temperature on device performance. *Technical Digest of IEDM*, paper 20.4 (2006)

76 Sung-Young Lee, Sung-Min Kim, Eun-Jung Yoon, Chang Woo Oh, Ilsub Chung, Donggun Park, Kinam Kim: Three-dimensional MBCFET as an ultimate transistor. IEEE Electron Device Letters **25-4**, 217 (2004)

77 Eun-Jung Yoon, Sung-Young Lee, Sung-Min Kim, Min-Sang Kim, Sung Hwan Kim, Li Ming, Sungdae Suk, Kyounghawn Yeo, Chang Woo Oh, Jung-dong Choe, Donguk Choi, Dong-Won Kim, Donggun Park, Kinam Kim, Byung-Il Ryu: Sub 30 nm multi-bridge-channel MOSFET (MBCFET) with metal gate electrode for ultra high performance application. *Technical Digest of IEDM*, 627 (2005)

78 Donggun Park: 3 dimensional GAA transitors: twin silicon nanowire MOSFET and multi-bridge-channel MOSFET. *Proceedings IEEE International SOI Conference*, 131 (2006)

79 T. Ernst, C. Dupre, C. Isheden, E. Bernard, R. Ritzenthaler, V. Maffini-Alvaro, J. Cluzel, A. Toffoli, C. Vizioz, S. Borel, F. Andrieu, F. De Crecy, V. Delaye, D. Lafond, G. Rabille, J.-M. Hartmann, M. Rivoire, B. Guillaumot, A. Suhm, P. Rivallin, O. Faynot, G. Ghibaudo, S. Deleonibus: Novel 3D Integration Process for Highly Scalable Nano-Beam Stacked-channels GAA (NBG) CMOSFETs with $HfO_2$/TiN Gate Stack. *Technical Digest of IEDM*, paper 38.4 (2006)

80 B. Iñiguez, H. Abd-Elhamid, D. Jimenez, J. Roig, J. Pallares, L.F. Marsal: Two-dimensional analytical threshold voltage roll-off and subthreshold swing models for undoped cylindrical gate all around MOSFET. Solid-State Electronics **50-5**, 805 (2006)

81 C.P. Auth, J.D. Plummer: Scaling theory for cylindrical, fully-depleted, surrounding-gate MOSFETs. IEEE Electron Device Letters **18-2**, 74 (1997)

82 C.P. Auth, J.D. Plummer: A simple model for threshold voltage of surrounding-gate MOSFETs. IEEE Transactions on Electron Devices **45-11**, 2381 (1998)

83 S.L. Jang, S.S. Liu: An analytical surrounding gate MOSFET model. Solid-State Electronics **42-5**, 721 (1998)

84 W. Zhang, J. G. Fossum, L. Mathew: A Hybrid FinFET/SOI MOSFET. *Proceedings IEEE International SOI Conference*, 151 (2005)

85 L. Mathew, M. Sadd, S. Kalpat, M. Zavala, T. Stephens, R. Mora, S. Bagchi, C. Parker, J. Vasek, D. Sing, R. Shimer, L. Prabhu, G.O. Workman, G. Ablen, Z. Shi, J. Saenz, B. Min, D. Burnett, B.-Y. Nguyen, J. Mogab, M.M. Chowdhury, W. Zhang, J.G. Fossum: Inverted T channel FET (ITFET) – Fabrication and Characteristics of Vertical-Horizontal, Thin-Body, Multi-Gate, Multi-Orientation Devices, ITFET SRAM Bit-cell operation. A Novel Technology for 45nm and Beyond CMOS. *Technical Digest of IEDM*, 713 (2005)

86 K. Okano, T. Izumida, H. Kawasaki, A. Kaneko, A. Yagishita, T. Kanemura, M. Kondo, S. Ito, N. Aoki, K. Miyano, T. Ono, K. Yahashi, K. Iwade, T. Kubota, T. Matsushita, I. Mizushima, S. Inaba, K. Ishimaru, K. Suguro, K. Eguchi, Y. Tsunashima, H. Ishiuchi: Process Integration Technology and Device Characteristics of CMOS FinFET on Bulk Silicon Substrate with Sub-10nm Fin Width and 20nm Gate Length. *Technical Digest of IEDM*, 725 (2005)

87 Kyoung-Rok Han, Byung-Gil Choi, Jong-Ho Lee: Design considerations of body-tied FinFETs ($\Omega$MOSFETs) implemented on bulk Si wafers, Journal of Semiconductor Technology and Science **4-1**, 12 (2004)

88 Sung Min Kim, Eun Jung Yoon, Hye Jin Jo, Ming Li, Chang Woo Oh, Sung Young Lee, Kyoung Hwan Yeo, Min Sang Kim, Sung Hwan Kim, Dong Uk Choe, Jeong Dong Choe, Sung Dae Suk, Dong-Won Kim, Donggun Park, Kinam Kim, Byung-Il Ryu: A Novel Multi-channel Field Effect Transistor (McFET) on Bulk Si for High Performance Sub-80nm Application. *Technical Digest of IEDM*, 639 (2004)

89 G.K. Celler, S. Cristoloveanu: Frontiers of silicon-on-insulator. Journal of Applied Physics **93-9**, 4955 (2003)

90 B.J. Blalock, S. Cristoloveanu, B.M. Dufrene, F. Allibert, M.M. Mojarradi: The G4-FET: a universal and programmable logic gate. Journal of High Speed Electronics **12-2**, 511 (2002)

91 B. Dufrene, K. Akarvardar, S. Cristoloveanu, B.J. Blalock, P. Gentil, E. Kolawa, M.M. Mojarradi: Investigation of the four-gate action in G4 –FETs. IEEE Transactions on Electron Devices **51-11**, 1931 (2004)

92  B. Dufrene, B. Blalock, S. Cristoloveanu, M. Mojarradi, E.A. Kolawa: Saturation Current Model for the N-channel G$^4$FET. *Electrochemical Society Proceedings* **2003-05**, 367 (2003)
93  F. Hofmann, M. Specht, U. Dorda, R. Kömmling, L. Dreeskornfeld, J. Kretz, M. Städele, W. Rösner and L. Risch: NVM based on FinFET device structures. Solid-State Electronics **49-11**, 1799 (2005)
94  Chang Woo Oh, Sung Dae Suk, Yong Kyu Lee, Suk Kang Sung, Jung-Dong Choe, Sung-Young Lee, Dong Uk Choi, Kyoung Hwan Yeo, Min Sang Kim, Sung-Min Kim, Ming Li, Sung Hwan Kim, Eun-Jung Yoon, Dong-Won Kim, Donggun Park, Kinam Kim, Byung-Il Ryu: Damascene Gate FinFET SONOS Memory Implemented on Bulk Silicon Wafer. *Technical Digest of IEDM*, 893 (2004)
95  Sang Soo Kim, Won-Ju Cho, Chang-Geun Ahn, Kiju Im, Jong-Heon Yang, In-Bok Baek, Seongjae Lee, Koeng Su Lim: Fabrication of fin field-effect transistor silicon nanocrystal floating gate memory using photochemical vapor deposition. Applied Physics Letters **88**, 223502 (2006)
96  W. Xiong, C.R. Cleavelin, R. Wise, S. Yu, M. Pas, R.J. Zaman, M. Gostkowski, K. Matthews, C. Maleville, P. Patruno, T.J. King, J.P. Colinge: Full/partial depletion effects in FinFETs. Electronics Letters **41-8**, 504 (2005)
97  M. Nagoga, S. Okhonin, C. Bassin, P. Fazan, W. Xiong, C.R. Cleavelin, T. Schulz, K. Schruefer, M. Gostkowski, P. Patruno, C. Maleville: Retention characteristics of zero-capacitor RAM (Z-RAM) cell based on FinFET and tri-gate devices. *Proceedings IEEE International SOI Conference*, 203 (2005)
98  M. Bawedin,S. Cristoloveanu, J.G. Yun, D. Flandre. A new memory effect (MSD) in fully depleted SOI MOSFETs. Solid-Sate Electronics **49-9**, 1547 (2005)
99  J.G. Yun, M. Bawedin, S. Cristoloveanu, D. Flandre, H.-D. Lee. The meta-stable dip (MSD) effect in SOI FinFETs. Microelectronic Engineering **84-4**, 590 (2007)
100 H.K. Lim and J.G. Fossum: Threshold voltage of thin-film silicon-on-insulator (SOI) MOSFETs IEEE. Trans. on Electron Devices **30-10**, 1244 (1983)
101 K.K. Young: Analysis of conduction in fully depleted SOI MOSFETs. IEEE Transactions on Electron Devices **36-3**, 504 (1989)
102 R.H. Yan, A. Ourmazd, and K.F. Lee: Scaling the Si MOSFET: from bulk to SOI to bulk. IEEE Transactions on Electron Devices **39-7**, 1704 (1992)
103 C.P. Auth, J.D. Plummer: Scaling theory for cylindrical, fully-depleted, surrounding-gate MOSFET's. IEEE Electron Device Letters **18-2**, 74 (1997)
104 Chi-Woo Lee, Se-Re-Na Yun, Chong-Gun Yu, Jong-Tae Park, J.P. Colinge: Device design guidelines for nano-scale MuGFETs. Solid-State Electronics **51-3**, 505 (2007)
105 Q. Chen, E.M. Harrell II, J.D. Meindl: A physical short-channel threshold voltage model for undoped symmetric double-gate MOSFETs. IEEE Transactions on Electron Devices **50-7**, 1631 (2003)

106 E. Landgraf, W. Rösner, M. Städele, L. Dreeskornfeld, J. Hartwich, F. Hofmann, J. Kretz, T. Lutz, R.J. Luyken, T. Schulz, M. Specht, L. Risch: Influence of crystal orientation and body doping on trigate transistor performance. Solid-State Electronics **50-1**, 38 (2006)

107 J.P. Colinge: Novel Gate Concepts for MOS Devices. *Proceedings of ESSDERC*, 45 (2004)

108 J. Saint-Martin, A. Bournel, P. Dollfus: Comparison of multiple-gate MOSFET architectures using Monte Carlo simulation. Solid-State Electronics **50**, 94 (2006)

109 D.J. Foster: Subthreshold currents in CMOS transistors made on oxygen-implanted silicon. Electronics Letters, **19-17**, 684 (1983)

110 A. Burenkov, J, Lorenz: Corner effect in double and triple gate FinFETs. *Proceedings of ESSDERC*, 135 (2003)

111 H.S. Wong, M.H. White, T.J. Krutsck, R.V. Booth: Modeling of transconductance degradation and threshold voltage in thin oxide MOSFETs. Solid-State Electronics, **30-9**, 953 (1987)

112 P. Francis, A. Terao, D. Flandre, F. Van de Wiele: Modeling of ultrathin double-gate nMOS-SOI transistors. Solid-State Electronics **41-5**, 715 (1994)

113 A. Terao, D. Flandre, E. Lora-Tamayo, F. Van de Wiele, IEEE Electron Device Letters, **12-12**, 682 (1991)

114 E. Rauly, B. Iñiguez, D. Flandre, C. Raynaud: Investigation of single and double gate SOI MOSFETs in Accumulation Mode for enhanced performances and reduced technological drawbacks. *Proceedings of ESSDERC*, 540 (2000)

115 W. Xiong, J.W. Park, J.P. Colinge: Corner effect in multiple-gate SOI MOSFETs. *Proceedings IEEE International SOI Conference*, 111 (2003)

116 J.P. Colinge, J.W. Park, W. Xiong: Threshold voltage and subthreshold slope of multiple-gate SOI MOSFETs. IEEE Electron Device Letters **28-8**, 515 (2003)

117 Chi-Woo Lee, Se-Re-Na Yun, Chong-Gun Yu, Jong-Tae Park, J.P. Colinge: Device design guidelines for nano-scale MuGFETs. Solid-State Electronics **51-3**, 505 (2007)

118 Jung A Choi, Kwon Lee, You Seung Jin, Yong Jun Lee, Soo Yong Lee, Geon Ung Lee, Seung Hwan Lee, Min Chul Sun, Dong Chan Kim, Young Mi Lee, Su Gon Bae, Jeong Hwan Yang, Shigenobu Maeda, Nae In Lee, Ho Kyu Kang, Kwang Pyuk Suh: Large Scale Integration and Reliability Consideration of Triple Gate Transistors. *Technical Digest of IEDM*, paper 674 (2004)

119 M. Stadele, R.J. Luyken, M. Roosz, M. Specht, W. Rosner, L. Dreeskornfeld, J. Hartwich, F. Hofmann, J. Kretz, E. Landgraf, L. Risch: A comprehensive study of corner effects in tri-gate transistors. *Proceedings of ESSDERC*, 165 (2004)

120 J.G. Fossum, J.W. Yang, V.P. Trivedi: Suppression of corner effects in triple-gate MOSFETs. IEEE Electron Device Letters **24-12**, 745 (2003)

121 F. Balestra, S. Cristoloveanu, M. Benachir, J. Brini, T. Elewa: Double-gate silicon-on-insulator transistor with volume inversion: A new device with

greatly enhanced performance. IEEE Electron Device Letters **8-9**, 410 (1987)

122 J.P. Colinge, M.H. Gao, A. Romano, H. Maes, C. Claeys: Silicon-on-insulator 'gate-all-around' MOS device. *Technical Digest of IEDM*, 595, 1990

123 T. Ouisse: Self-consistent quantum-mechanical calculations in ultrathin silicon-on-insulator structures. Journal of Applied Physics, Part 1 **76-10**, 5989 (1994)

124 J.P. Colinge, X. Baie, V. Bayot: Evidence of two-dimensional carrier confinement in thin n-channel SOI gate-all-around (GAA) devices. IEEE Electron Device Letters **15-6**, 193 (1994)

125 A. Rahman, M.S. Lundstrom: A compact scattering model for the nanoscale double-gate MOSFET. IEEE Transactions on Electron Devices **49-3**, 481 (2002)

126 X. Baie, J.P. Colinge: Two-dimensional confinement effects in gate-all-around (GAA) MOSFETS. Solid-State Electronics **42-4**, 499 (1998)

127 G. Baccarani, S. Reggiani: A compact double-gate MOSFET model comprising quantum-mechanical and nonstatic effects. IEEE Transactions on Electron Devices **46-8**, 1656 (1999)

128 Lixin Ge, J.G. Fossum: Analytical modeling of quantization and volume inversion in thin Si-film DG MOSFETs. IEEE Transactions on Electron Devices **49-2**, 287 (2002)

129 S. Cristoloveanu, D.E. Ioannou: Adjustable confinement of the electron gas in dual-gate silicon-on-insulator MOSFETs. Superlattices & Microstructures **8-1**, 131 (1990)

130 T. Ouisse, D.K. Maude, S. Horiguchi, Y. Ono, Y. Takahashi, K. Murase, S. Cristoloveanu: Subband structure and anomalous valley splitting in ultra-thin silicon-on-insulator MOSFET's. Physica B **249-251**, 731 (1998)

131 X. Baie, J.P. Colinge, V. Bayot, E. Grivei: Quantum-wire effects in thin and narrow SOI MOSFETs. *Proceedings IEEE International SOI Conference*, 66, 1995

132 J.P. Colinge, X. Baie, V. Bayot, E. Grivei: A silicon-on-insulator quantum wire. Solid-State Electronics **39-1**, 49 (1996)

133 F. Gamiz, A. Godoy, J. Roldán, C. Sampedro, L. Donetti: Charge transport in nanoscaled SOI devices. *Abstracts of the NATO Advanced Research Workshop on Nanoscaled Semiconductor-on-Insulator Structures and Devices*, 73 (2006)

134 T. Ando, A.B. Fowler, F. Stern: Electronic properties of two-dimensional systems. *Review of Modern Physics* **54**, 437 (1982)

135 P.N. Butcher, *Physics of low-dimensional semiconductor structures*, Ed. by P.N. Butcher, N.H. March and M.P. Tosi, Plenum Press, 95 (1993)

136 F. Stern, *Physics of low-dimensional semiconductor structures*, Ed. by P.N. Butcher, N.H. March and M.P. Tosi, Plenum Press, 177, (1993)

137 Xue Shao, Zhiping Yu: Nanoscale FinFET simulation: A quasi-3D quantum mechanical model using NEGF. Solid-State Electronics **49**, 1435 (2005)

138 B. Majkusiak, T. Janik, J. Walczak: Semiconductor thickness effects in the double-gate SOI MOSFET. IEEE Transactions on Electron Devices **45-5**, 1127 (1998)
139 J.P. Colinge, J. C. Alderman, W. Xiong, C. R. Cleavelin: Quantum-Mechanical Effects in Trigate SOI MOSFETs. IEEE Transactions on Electron Devices **53-5**, 1131 (2006)
140 F. Gámiz, J.B. Roldán, J.A. López-Villanueva, P. Cartujo-Cassinello, J.E. Carceller, P. Cartujo: Monte Carlo simulation of electron transport in silicon-on-insulator devices. *Electrochemical Society Proceedings* **2001-3**, 157 (2001)
141 L. Ge, J.G. Fossum, F. Gámiz: Mobility enhancement via volume inversion in double-gate MOSFETs. *Proceedings of the IEEE International SOI Conference*, 153 (2003)
142 G. Tsutsui, M. Saitoh, T. Saraya, T. Nagumo, T. Hiramoto: Mobility enhancement due to volume inversion in [110]-oriented ultra-thin body double-gate nMOSFETs with body thickness less than 5 nm. *Technical Digest of IEDM*, 729 (2005)
143 H.K. Lim and J.G. Fossum: Threshold voltage of thin-film silicon-on-insulator (SOI) MOSFETs IEEE. Transactions on Electron Devices **30-10**, 1244 (1983)
144 Y. Omura, S. Horiguchi, M. Tabe, K. Kishi: Quantum-mechanical effects on the threshold voltage of ultrathin SOI nMOSFETs. IEEE Electron Device Letters **14-12**, 569 (1993)
145 B. Majkusiak, T. Janik, J. Walczak: Semiconductor thickness effects in the double-gate SOI MOSFET. IEEE Transactions on Electron Devices, **45-5**, 1127 (1998)
146 K. Uchida, J. Koga, R. Ohba, T. Numata, S.I. Takagi: Experimental evidences of quantum-mechanical effects on low-field mobility, gate-channel capacitance, and threshold voltage of ultrathin body SOI MOSFETs. *Technical Digest of IEDM*, 29.4.1 (2001)
147 J. Lolivier, S. Deleonibus, F. Balestra: Threshold voltage quantum simulations for ultra-thin silicon-on-insulator transistors. Electrochemical Society Proceedings **2003-05**, 379 (2003)
148 T. Ernst, D. Munteanu, S. Cristoloveanu, T. Ouisse, N. Hefyene, S. Horiguchi, Y. Ono, Y. Takahashi, K. Murase: Ultimately thin SOI MOSFETs: special characteristics and mechanisms. *Proceedings ot the IEEE International SOI Conference*, 92 (1999)
149 T. Ernst, S. Cristoloveanu, G. Ghibaudo, T. Ouisse, S. Horiguchi, Y. Ono, Y. Takahashi, K. Murase: Ultimately thin double-gate SOI MOSFETs. IEEE Transactions on Electron Devices **50-3**, 830 (2003)
150 T. Poiroux, M. Vinet, O. Faynot, J. Widiez, J. Lolivier, T. Ernst, B. Previtali, S. Deleonibus: Multiple gate devices: advantages and challenges. Microelectronic Engineering **80**, 378 (2005)
151 W. Xiong, C. R. Cleavelin, T. Schulz, K. Schrüfer, P. Patruno, J.P. Colinge: MuGFET CMOS process with midgap gate material. *Abstracts of NATO*

*International Advanced Research Workshop "Nanoscaled Semiconductor-on-Insulator Structures and Devices"*, 96 (2006)
152 J.P. Colinge, J.C. Alderman, W. Xiong, C.R. Cleavelin: Quantum-Mechanical Effects in Trigate SOI MOSFETs. IEEE Transactions on Electron Devices **53-5**, 1131 (2006)
153 A. Marchi, E. Gnani, S. Reggiani, M. Rudan, G. Baccarani: Investigating the performance limits of silicon-nanowire and carbon-nanotube FETs. Solid-State Electronics **50**, 78 (2006)
154 E. Gnani, A. Marchi, S. Reggiani, M. Rudan, G. Baccarani: Quantum-mechanical analysis of the electrostatics in silicon-nanowire and carbon-nanotube FETs. Solid-State Electronics **50**, 709 (2006)
155 J.P. Colinge: Nanowire Quantum Effects in Tri-Gate SOI MOSFETs. *Abstracts of the NATO Advanced Research Workshop on Nanoscaled Semiconductor-on-Insulator Structures and Devices, 85 (2006)*
156 T. Ando, A.B. Fowler, F. Stern: Electronic properties of two-dimensional systems. Reviews of Modern Physics **54-2**, 437 (1982)
157 J.P. Colinge, A.J. Quinn, L. Floyd, G. Redmond, J.C. Alderman, W. Xiong, C.R. Cleavelin, T. Schulz, K. Schruefer, G. Knoblinger, P. Patruno: Low-Temperature Electron Mobility in Trigate SOI MOSFETs. IEEE Electron Device Letters **27-2**, 120 (2006)
158 J.P. Colinge, W. Xiong, C.R. Cleavelin, T. Schulz, K. Schrüfer, K. Matthews, P. Patruno: Room-Temperature Low-Dimensional Effects in Pi-Gate SOI MOSFETs. IEEE Electron Device Letters **27-9**, 775 (2006)
159 N. Singh, F.Y. Lim, W.W. Fang, S.C. Rustagi, L.K. Bera, A. Agarwal, C.H. Tung, K.M. Hoe, S.R. Omampuliyur, D. Tripathi, A.O. Adeyeye, G.Q. Lo, N. Balasubramanian, D.L. Kwong: Ultra-narrow silicon nanowire gate-all-around CMOS device: impact of diameter, channel-orientation and low temperature on device performance. *Technical Digest of IEDM*, paper 20.4 (2006)
160 M. Masahara, R. Surdeanu, L. Witters, G. Doornbos, V.H. Nguyen, G. Van den bosch, C. Vrancken, K. Devriendt, F. Neuilly, E. Kunnen, E. Suzuki, M. Jurczak, S. Biesmans: Independent double-gate FinFETs with asymmetric gate stacks. Microelectronic Engineering, **84-9/10**, 2097 (2007)

# 2 Multigate MOSFET Technology

Weize (Wade) Xiong

## 2.1 Introduction

Manufacturing a self-aligned double-gate MOSFET has been the holy grail of device engineers and researchers ever since it was proposed by Sekigawa and Hayashi in 1984.[1] The first modern self-aligned vertical multi-gate MOSFET was called DELTA (fully **DE**pleted **L**ean channel **T**r**A**nsistor).[2] This device was proposed by D. Hisamoto *et al.* in 1989. Figure 2.1 shows a cross section of the DELTA MOSFET.

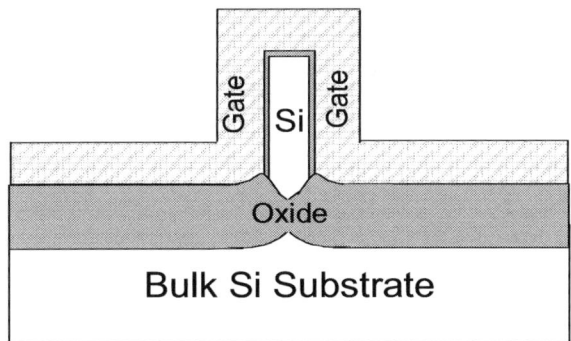

**Fig. 2.1.** DELTA MOSFET [1-2]. The oxide under the Si fin was formed through LOCOS oxidation while the Si fin was protected by the nitride hard mask and a nitride spacer prior to the oxidation process.

The two key features of the DELTA MOSFET are: 1) front and back MOSFET gates are inherently self-aligned and 2) the channels are on the sidewall of the silicon body/fin. The compatibility of the DELTA FET with existing planar CMOS processing is the main reason why it was adopted as the mainstream multi-gate structure. FinFETs, Tri-gate and Π/Ω gate devices have all evolved from of this original concept [3-8].

The critical fabrication steps in the front-end processing of a multi-gate MOSFET include, sequentially: 1) fin formation, 2) gate stack formation, 3) source and drain extension implant, 4) spacer formation, 5) epitaxial raised source/drain formation, and 6) deep source/drain implantation and activation anneal.

The fabrication flow of a Tri-gate MOSFET on an SOI substrate is shown in Figure 2.2. The SOI silicon top layer ($T_{si}$) thickness defines the fin height ($Fin_{HEIGHT}$). The fin pattern and the critical dimension of fin width ($Fin_{WIDTH}$) can be defined by optical lithography or by spacer image transfer (*SIT*) [9-10], followed by plasma etching.

After fin etch, the fin sidewall surfaces are rough. Therefore, oxidation and $H_2$ annealing are often used to smooth the sidewalls.[11-12] Next, the gate dielectric is grown and metal gate is deposited. It is suitable to tune the threshold voltage ($V_{th}$) of the MOSFET by using a gate material that has the appropriate effective workfunction rather than by doping the channel. Sections 2.3, 2.5 and 2.6 will show that it is highly desirable to have intrinsic or lightly doped channels.

Since the gate stack is over the fin topography, a planarization step is desirable to flatten the gate surface, which reduces the burden on photolithography and gate etch. Significant overetch of the gate material is required to clear the bottom of the fins. As a result, the gate etch must have a high selectivity to the gate dielectric on top of the fin, if one wants to avoid damage to the fin during gate etch. Source and drain (S/D) extensions are formed after gate patterning using low-energy and large-tilt angled implants.[13-14] Next, S/D offset spacers are formed along the sidewalls of the gate and fin. The sidewall spacers on the fins are subsequently removed to expose the fin to grow raised source and drain using selective epitaxy.[13-15] The raised source and drain structure helps to reduce the parasitic resistance associated with thin fins.[13]

The rest of this chapter is divided into sections according to multi-gate device fabrication sequences. Section 2.2 covers fin formation, Section 2.3 describes gate stack structures, Section 2.4 describes source and drain formation and Section 2.5 highlights mobility and strain engineering.

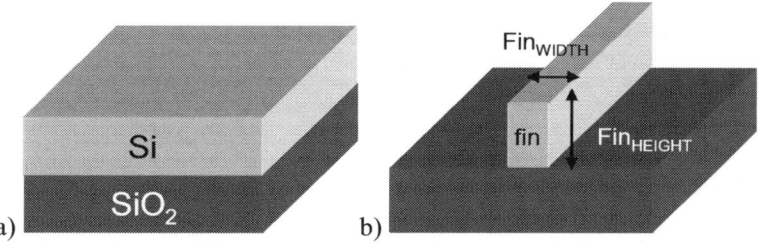

Starting SOI Material; fin patterning and etch. Define $Fin_{WIDTH}$; the fin height is the thickness of the silicon top SOI layer.

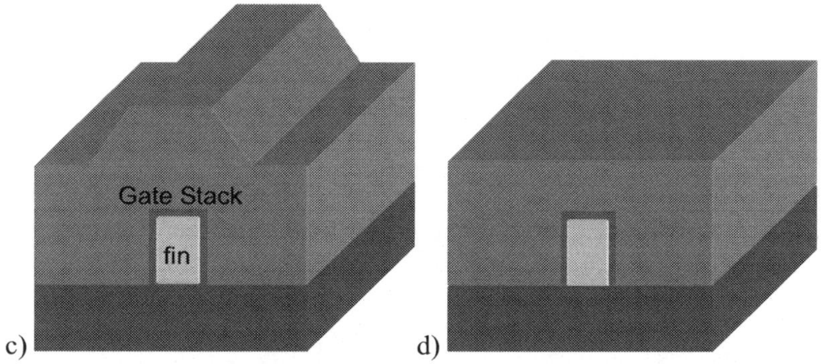

Gate Stack deposition and planarization

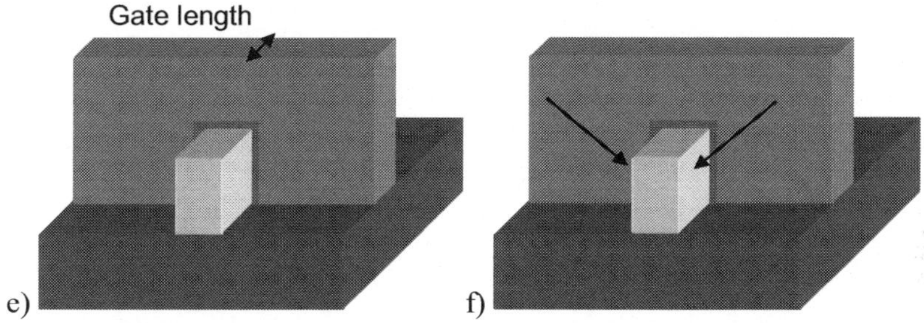

Gate etch and source and drain extension implant

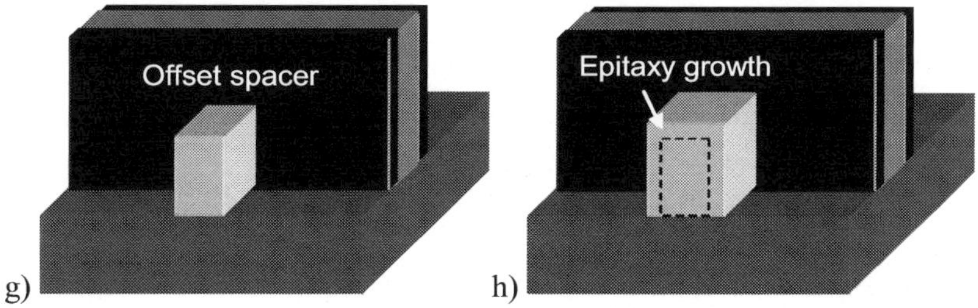

Spacer formation and selective epitaxal growth. Removing the spacer on the sidewall of the fin while keeping the gate spacer is a critical step.

**Fig. 2.2.** General Fabrication Sequence of a Tri-gate MOSFET.

## 2.2 Active Area: Fins

### 2.2.1 Fin Width

As the MOSFET gate length, $L_g$, continues to shrink, year after year, the drain starts to compete with the gate electrode for control of the channel potential. The resulting "short-channel effects" (SCE) are: 1) higher sub-threshold leakage, 2) a threshold voltage ($V_{th}$) roll-off, and 3) a form of punch-through between the drain and source when $V_{DS}$ is equal to the supply voltage.[16]

To reduce the drain influence in the channel, scaled planar bulk MOSFETs and PDSOI MOSFETs rely on: 1) gate oxide thickness reduction and 2) higher channel doping. The use of a thinner gate oxide increases the gate-to-channel capacitance. A higher channel doping concentration reduces charge sharing between the gate and drain in the channel, and it creates a large potential barrier between source and drain.

However, in modern MOSFETs, direct tunneling current through the gate dielectric prevents further scaling of the gate oxide thickness. The use of high channel doping concentrations reduces carrier mobility and increases Gate Induced Drain Leakage (GIDL).[17-18]

Another way to control SCEs is to use two or more gate electrodes and a thin, fully depleted semiconductor body. The idea is to create a large potential curvature in the source and drain direction by introducing a large electrical field gradient in the direction vertical to the gate. The key parameter in this approach is the thickness of the thin, fully depleted semiconductor body. In the case of the multi-gate technologies, the fin width is the thickness of the thin, fully depleted semiconductor body.

Chapter 1 introduced a powerful concept called natural length, $\lambda$, with:

$$\lambda = \sqrt{\frac{\varepsilon_{si}}{2\varepsilon_{ox}} t_{si} t_{ox}} \quad \text{for a double-gate MOSFET} \tag{2.1}$$

$\lambda$ is a measure of the short-channel effect. It represents the distance of penetration of the drain electric field in the channel. A small value $\lambda$ is desirable to minimize the SCE.

$\lambda$ is proportional to the square root of the product of the device body thickness $t_{si}$, (or $Fin_{WIDTH}$ in multi-gate technology) and gate oxide thickness $t_{ox}$. Thus, one can trade off $t_{ox}$ scaling with $Fin_{WIDTH}$ reduction. This approach is not possible for bulk or PDSOI MOSFETs.[17]

In a double-gate MOSFET, the center of the fin is the furthest away from the gate electrodes. Therefore, gate control is the weakest at this point. Should punchthrough occur, it would be in the center of the fin.[20] This implies using a slightly different boundary condition than Equation 2.1, which assumes the punch through is along the channel surfaces.

In this section, we will derive a new value for $\lambda$ based on the assumption that punchthrough occurs in the center of the channel. The new $\lambda$ has the same physical meaning as Equation 2.1, but it can better explain some of the experimental results shows later in this section. We will use the same coordinate system as in Figure 1.11. Starting with Poisson's equation, we can write:

$$\frac{d^2\Phi(x,y,z)}{dx^2} + \frac{d^2\Phi(x,y,z)}{dy^2} + \frac{d^2\Phi(x,y,x)}{dz^2} = \frac{qN_{sub}}{\varepsilon_{si}} \quad (2.2)$$

For single-gate or double-gate devices $\frac{d\phi}{dz} = 0$, such that Equation 2.2 is reduced to:

$$\frac{d^2\Phi(x,y)}{dx^2} + \frac{d^2\Phi(x,y)}{dy^2} = \frac{qN_{sub}}{\varepsilon_{si}} \quad (2.3)$$

Assuming a parabolic potential distribution in the y-direction [21] we can write:

$$\Phi(x,y) \approx c_0(x) + c_1(x)y + c_2(x)y^2 \quad (2.4)$$

We now apply the boundary condition $\left.\frac{d\Phi(x,y)}{dy}\right|_{y=0} = 0$ at $y = \frac{t_{si}}{2}$,

which yields:

$$\Phi(x,y) = \Phi_f(x) + \frac{\varepsilon_{ox}}{\varepsilon_{si}}\frac{\Phi_s(x) - \Phi_{gs}}{t_{ox}}y - \frac{1}{t_{si}}\frac{\varepsilon_{ox}}{\varepsilon_{si}}\frac{\Phi_s(x) - \Phi_{gs}}{t_{ox}}y^2 \quad (2.5)$$

where $\Phi_s(x)$ is the front- and back-surface potential and $\Phi_{gs} = V_{gs} - V_{FBF}$ is the front gate voltage $V_{gs}$ minus the flat-band voltage, $V_{FB}$. Since the potential at the center of the fin, $\Phi_c(x)$, is the most relevant to SCE, we extract a relationship between $\Phi_c(x)$ and $\Phi_s(x)$ from 2.5 by substituting $y = \dfrac{t_{si}}{2}$ as follows:

$$\Phi_s(x) = \dfrac{1}{1 + \dfrac{\varepsilon_{ox} t_{si}}{4\varepsilon_{si} t_{ox}}} \left( \Phi_c(x) + \dfrac{\varepsilon_{ox} t_{si}}{4\varepsilon_{si} t_{ox}} \Phi_{gs} \right) \qquad (2.6)$$

Expressing $\Phi(x, y)$ as a function of $\Phi_c(x)$ we obtain:

$$\Phi(x, y) = \left( 1 + \dfrac{\varepsilon_{ox}}{\varepsilon_{si}} \dfrac{y}{t_{ox}} - \dfrac{\varepsilon_{ox}}{\varepsilon_{si}} \dfrac{y^2}{t_{ox} t_{si}} \right) \dfrac{\Phi_c(x) + \dfrac{\varepsilon_{ox} t_{si}}{4\varepsilon_{si} t_{ox}} \Phi_{gs}}{1 + \dfrac{\varepsilon_{ox} t_{si}}{4\varepsilon_{si} t_{ox}}} \qquad (2.7)$$

$$- \left( \dfrac{\varepsilon_{ox}}{\varepsilon_{si}} \dfrac{y}{t_{ox}} \Phi_{gs} - \dfrac{\varepsilon_{ox}}{\varepsilon_{si}} \dfrac{y^2}{t_{ox} t_{si}} \Phi_{gs} \right)$$

Substituting 2.7 back in 2.3 we get:

$$\dfrac{d^2 \Phi_c(x)}{dx^2} + \dfrac{\Phi_{gs} - \Phi_c(x)}{\lambda^2} = \dfrac{qN_A}{\varepsilon_{si}} \qquad (2.8)$$

where natural length $\lambda$ is now written:

$$\lambda = \sqrt{\dfrac{\varepsilon_{si}}{2\varepsilon_{ox}} \left( 1 + \dfrac{\varepsilon_{ox} t_{si}}{4\varepsilon_{si} t_{ox}} \right) t_{si} t_{ox}} \qquad (2.9)$$

Numerical simulations in [17] show that for reasonable short-channel effect control, the gate length needs to be 5-10 times greater than $\lambda$.

We have established a theoretical relationship between $L_g$ and $\lambda$. For a device designer, it would be more relevant to establish a guideline for the $L_g$ to $t_{si}$ ratio to control short-channel effects. Equation 2.9 does not separate the effects of $t_{si}$ and $t_{ox}$. To walk around the problem, let us assume a good SCE control can be achieved with $L_g = n\lambda$, where $n$ is some arbitrary number. We now have:

$$\frac{L_g}{n} = \lambda = \sqrt{\frac{\varepsilon_{si}}{2\varepsilon_{ox}}\left(1 + \frac{\varepsilon_{ox}}{4\varepsilon_{si}}\frac{t_{si}}{t_{ox}}\right)t_{si}t_{ox}} \qquad (2.10)$$

In the limiting case when $L_g = 0$, we can write:

$$\frac{\varepsilon_{si}}{2\varepsilon_{ox}}\left(1 + \frac{\varepsilon_{ox}}{4\varepsilon_{si}}\frac{t_{si}}{t_{ox}}\right)t_{si}t_{ox} = 0 \qquad (2.11A)$$

which implies:

$$t_{si} = -4\frac{\varepsilon_{si}}{\varepsilon_{ox}}t_{ox} \qquad (2.11B)$$

The right-hand term of Equation 2.11B represents the intercept point of the $L_g$ vs. $Fin_{WIDTH}$ plot for $L_g = 0$. Thus, one can define an effective fin width, $FinW_{eff}$, as:

$$FinW_{eff} = Fin_{WIDTH} + 4\frac{\varepsilon_{si}}{\varepsilon_{ox}}t_{ox} \qquad (2.12)$$

Equation 2.12 shows that the impact of gate oxide can be considered equivalent to a constant offset of physical fin width. The validity of Equation 2.12 was demonstrated in reference [22]. Fig 2.3 shows contours of subthreshold swing on a $L_g$ vs. $Fin_{WIDTH}$ plot with 48 devices having 6 different $Fin_{WIDTH}$ values and 8 different gate lengths. Each line represents a fixed short channel effect control. Since all devices have the same gate oxide (1.6nm oxynitride), they all intercept $L_g = 0$ at the same point. This point is approximately equal to $-4\frac{\varepsilon_{si}}{\varepsilon_{ox}}t_{ox}$.

Figure 2.4 shows the experimental double-gate FinFET results of $L_g/FinW_{eff}$ vs. subthreshold swing (or Subthreshold Slope, SS).[22] If we assume that a SS value of 90mV/dec is acceptable, then $L_g/FinW_{eff}$ is ~ 1.5. A triple-gate MOSFET with enhanced gate electrostatic control of the channel provides a lower $L_g/Fin_{WIDTH}$ ratio than a double-gate FinFET for a same SCE control.[23] $L_g/Fin_{WIDTH}$ as low as 1.0 was reported for $\Omega$-Gate devices in [8]. *It is worthwhile noting that the minimum feature size in multi-gate technology is the fin width ($Fin_{WIDTH}$), and not the gate length.*

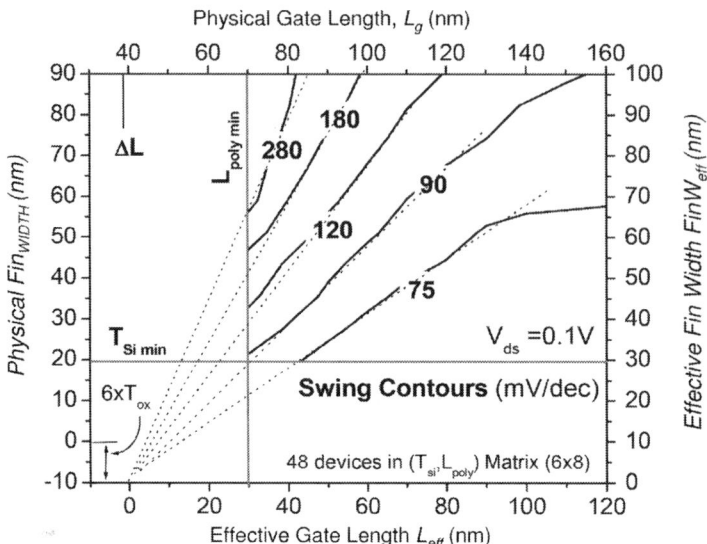

**Fig. 2.3.** Subthreshold swing contours for NMOS FinFET devices, $V_d$=0.1V. $Fin_{WIDTH}$ varies from 20nm to 82.5nm in steps of 12.5nm, physical gate length; $L_g$ varies from 70nm to 157.5nm in steps of 12.5nm. Contours are extrapolated with dotted lines to a common origin.[22] Copyright© 2001 IEEE.

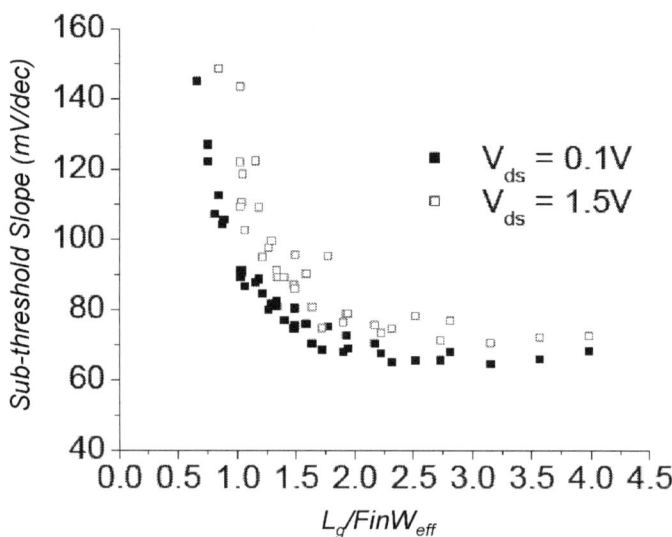

**Fig. 2.4.** Measured subthreshold swing vs. $L_g/FinW_{eff}$. $Fin_{WIDTH}$ varies from 20nm to 90nm; $L_g$ varies from 30nm to 120nm. The subthreshold slope is shown for both for $V_d$=0.1V and 1.5V.[22] Copyright© 2001 IEEE.

## 2.2.2 Fin Height and Fin Pitch

Having understood the requirements on $Fin_{WIDTH}$ for SCE control, let us look at how fin pitch and fin height affect the drive current. For a double-gate FinFET, the channels are on the sidewalls of the fins. The effective channel width is thus equal to: [24]

$$W_{eff} = 2n\, Fin_{HEIGHT} \qquad (2.13)$$

Where $n$ is an integer number that is equal to the number of fins in a device. $W_{eff}$ of multi-gate devices can only be increased or decreased by a discrete amount equal to $2 \times Fin_{HEIGHT}$.[25]

The fin pitch, $P_{FIN}$, is the spacing between the fins plus the fin width. $P_{FIN}$ is limited by the lithography pattern capability. Exactly one fin can be placed in one $P_{FIN}$. Therefore, the effective channel width per pitch is $2 \times Fin_{HEIGHT}$. Figure 2.5 shows the layout comparison between a planar device and a FinFET. For the planar MOSFET, the effective device channel width, $W_{eff}$, is equal to the footprint on the substrate, $W_{foot}$. For FinFET, $W_{eff}$ is related to $W_{foot}$ through:

$$W_{eff} = 2 Fin_{HEIGHT} \frac{W_{foot}}{P_{FIN}} \qquad (2.14)$$

If we want the FinFET layout to be "competitive", we need $W_{eff} \geq W_{foot}$, or:

$$2 Fin_{HEIGHT} \geq P_{FIN} \qquad (2.15)$$

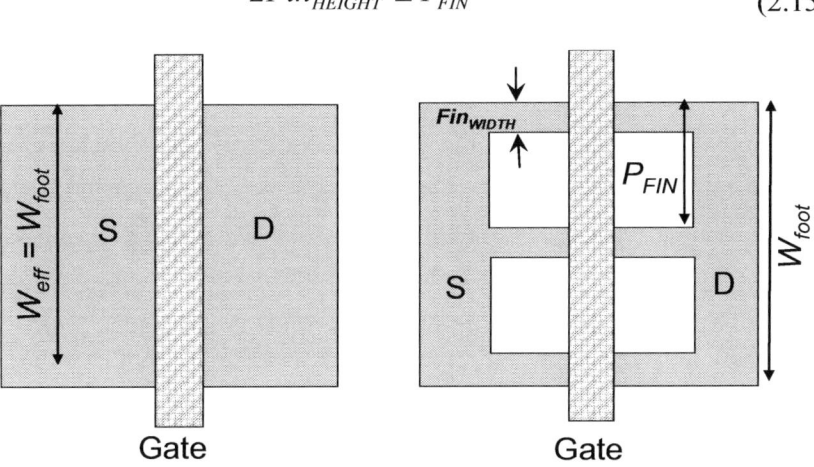

**Fig. 2.5.** Layout comparison between a planar MOSFET and a typical FinFET.

The analysis of the tri-gate channel width is further complicated by the "volume inversion" or "bulk inversion" effect reported in [26]. For thin fins, the inversion charge is not limited to the device surfaces. A large portion of the inversion charge is in the bulk of the fins, even at high gate voltages. Thus, without activating the top gate, multi-gate devices already have a significant charge at the top surface of the fins. Figure 2.6 shows the two-dimensional numerical simulation of carrier density on the top surface of 1) a tri-gate, 2) a double-gate and 3) a FinFET with a hard mask on top of the fin.

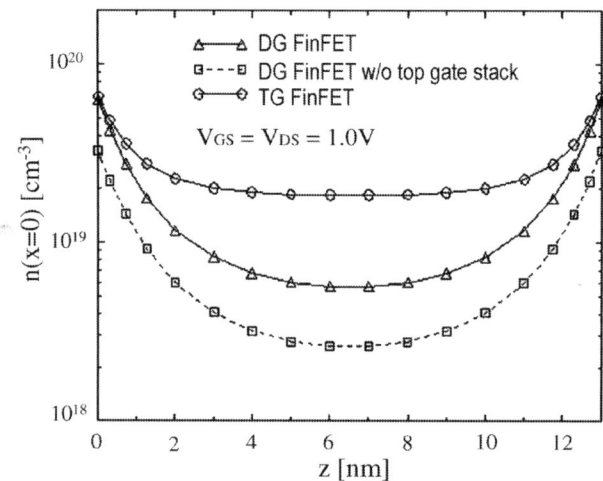

**Fig. 2.6.** Davinci prediction of on-state electron density along the top fin surface at the center of the channel ($x = 1/2\ L_{eff}$). DG stands for double gate. TG stands for tri-gate. Due to bulk inversion, the electron density on the top fin surface for DG FinFET is comparable to that in the TG device [26]. Copyright© 2005 IEEE.

Due to this "bulk inversion", the effective width, $\Delta W_{eff}$, of the top gate in a tri-gate MOSFET is only a fraction of the physical top gate width. Numerical simulation in [26] shows $\Delta W_{eff}$ is a function of fin height and fin width.

For tri-gate devices, Equation 2.12 becomes: $2 Fin_{HEIGHT} + \Delta W_{eff} \geq P_{FIN}$

Fins can be defined by conventional photolithography and reactive ion etch. Fins with width less than 20nm have been reported with 193nm dry photolithography processing.[27-28] Figure 2.7 shows the TEM cross section of some of these fins.[27-31] Continuing advances in lithography pitch and line dimension, including emerging 193nm photolithography and Extreme UV (EUV) lithography pave the way for further $Fin_{WIDTH}$ and $P_{FIN}$ reduction.

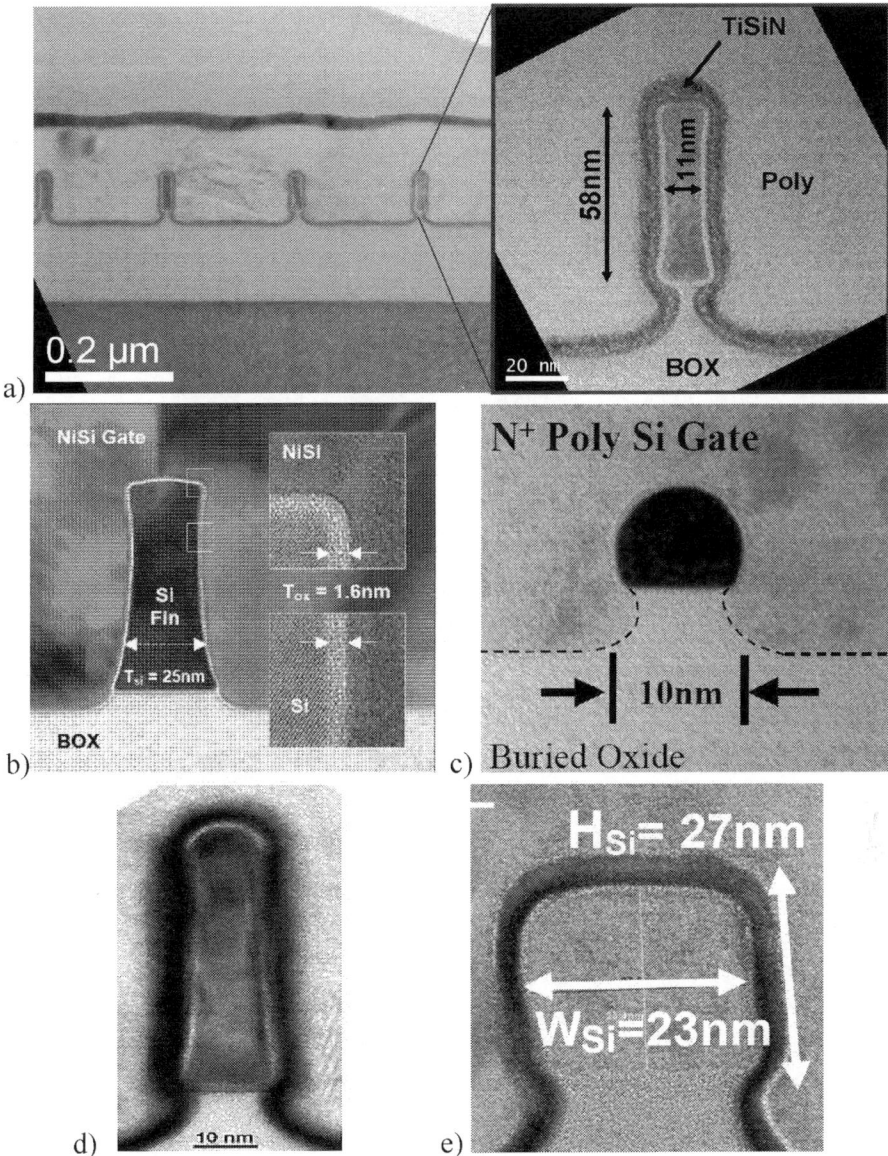

**Fig. 2.7.** Cross-sectional TEM of fins under gates. a) [27] Copyright© 2006 IEEE, b) [28] Copyright© 2002 IEEE, c) [29] Copyright© 2004 IEEE, d) [30] Copyright© 2005 IEEE, e) [31] Copyright© 2006 IEEE.

To increase the current drive per unit $W_{foot}$, we can increase $Fin_{HEIGHT}$, or reduce $P_{FIN}$. Both options have practical limitations. Increasing $Fin_{HEIGHT}$ undermines the fin stability and increases the difficulty of gate patterning.

$P_{FIN}$ is limited by the optical lithography capability. In order to improve $P_{FIN}$ beyond optical lithography capability, spacer-defined fin patterning has been proposed.[32] Figure 2.8 shows the general concept. First a sacrificial pattern is defined with conventional optical lithography. Then spacers are formed on the sidewalls of the sacrificial pattern through thin film deposition and anisotropic reactive ion etch (RIE). The sacrificial pattern is subsequently selectively removed leaving the remaining spacers to serve as hard mask for fin etch.

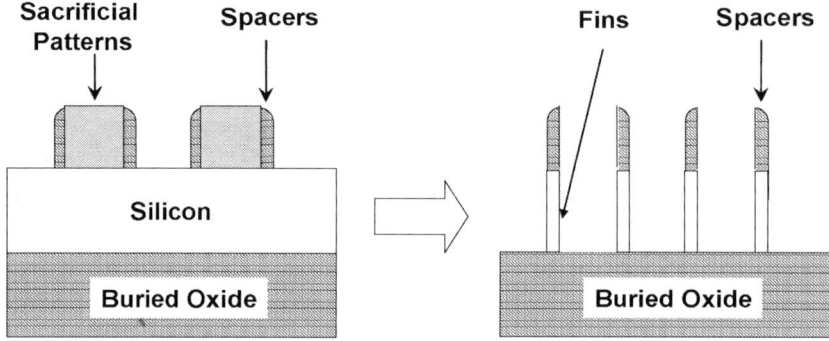

**Fig. 2.8.** Illustration of a spacer-defined fin formation process.

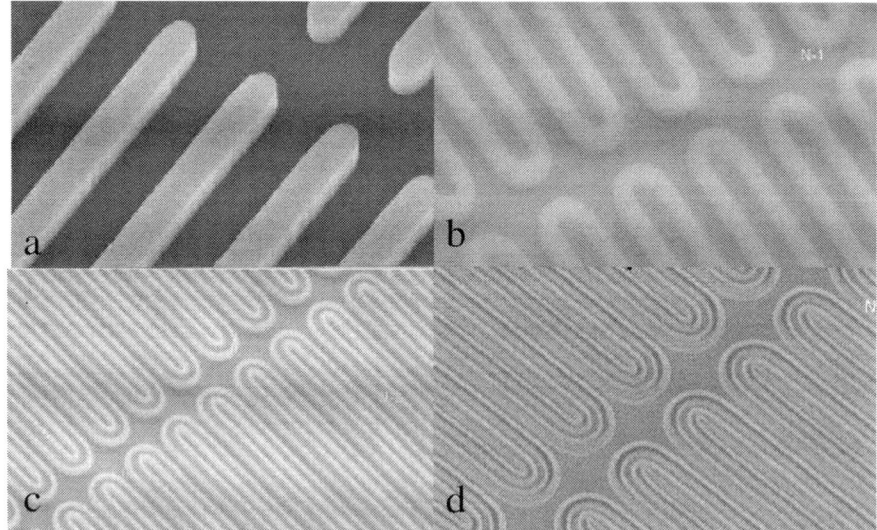

**Fig. 2.9.** SEM pictures of three progressive spacer-defined fin formation processes. a) initial optical lithography of sacrificial pattern. b) after first spacer imaging; c) after second spacer imaging; and d) after third spacer imaging. The final pattern pitch is 6 times smaller than the optical lithography pitch. REPRINTED WITH PERMISSION FROM [33]. COPYRIGHT 2003 AMERICAN INSTITUTE OF PHYSICS.

There are two advantages to spacer-defined fin formation. First, it doubles the number of fins for a given lithography pitch, *i.e.*: it reduces the fin pitch to half of the lithography pitch. In fact, when the spacer process is used twice, the pitch can be reduced by a factor of four. Figure 2.9 shows the SEM images of spacer imaging technique progressively being used three times.[33] Second, the fin width is defined by the thickness of deposited film, which has the potential to 1) be narrower than a lithography-defined fin, 2) have better CD control and 3) have reduced line-edge roughness.[32] The disadvantage of spacer-defined fin is that a clean-up mask is required to remove the fins where designers do not want them.[25] CMOS FinFET devices have been successfully fabricated using this imaging technique. Successful conversion of a planar microprocessor design into a spacer-defined FinFET microprocessor was also demonstrated in [25].

## 2.2.3 Fin Surface Crystal Orientation

Electrons and holes mobility depends on the orientation of the silicon surface as well as on the current flow direction.[34-36] Figure 2.10 shows the measured surface mobility of electrons and holes as a function of the inversion layer charge density.[37] For electrons, the highest mobility surface is (100) and the lowest mobility surface is (110). For holes, the (110) surface has the highest mobility and the (100) surface has the lowest mobility. Electrons and holes at a (111) surface have mobilities laying somewhere between those at (100) and (110) surfaces. The origins of the above differences are due to the anisotropy of effective mass and surface scattering.[35,36]

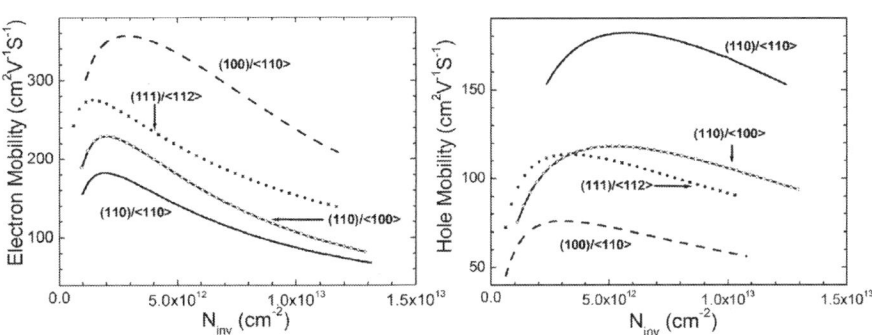

**Fig. 2.10.** Effective surface mobility of electrons and holes as a function of inversion layer charge density, for different surface orientations.[37] Copyright© 2003 IEEE.

Multi-gate MOSFETs have been fabricated on all 3 major Si surfaces [27,32,19]. The directions of the current flow for the respective surfaces are <100> for (100), <110> for (110) and <112> for (111). (110) Fin sidewall surfaces are made by patterning fins at 0 or 90 degrees with respect to the <110> notch on silicon wafers with a (001) surface orientation. This layout style is the same as for conventional planar devices. (100) fin surfaces are made by rotating the fin layout by a 45 degree angle with respect to the <110> notch direction. Figure 2.11 shows the layout directions of (100) fin surfaces and (110) fin surfaces.[38] Alternately, the Si top layer of the SOI wafers can be rotated by a 45 degree angle with respect to the substrate during bonding, to have the (100) surface fin layout at 0 and 90 degrees and the (110) surface at a 45 degree rotation.[39]

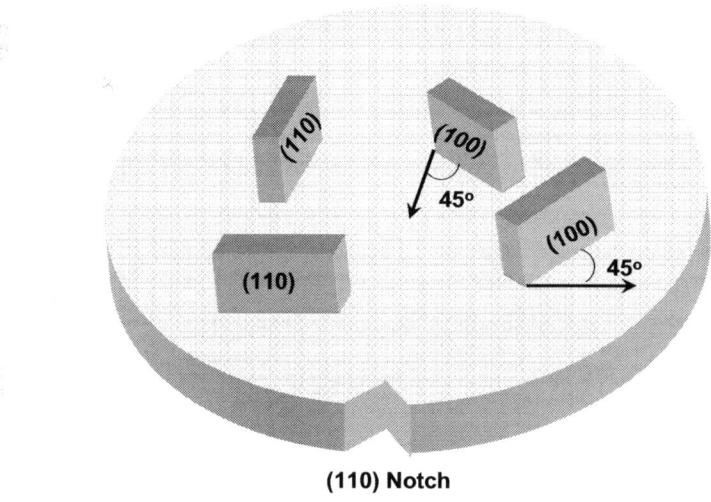

**(110) Notch**

**Fig. 2.11.** Fins with (110)/<110> and (100)/<100> surfaces/directions can be readily made using standard Si substrate by rotating the fin layout by 45 degrees.

Having fins with both (110) and (100) surface orientation on the same chip allows one to optimize electron (NMOS) and hole (PMOS) mobility simultaneously. The price to pay for such devices is an increased layout area [41] and a higher photolithography process complexity. Optimizing optical lithography lines at 0, 90 and 45 degree angles is challenging. Figure 2.12 shows the comparison of a 6-transistor SRAM cell layout containing both (110) and (100) fins directions (mixed layout) with a standard layout with only one fin orientation. The mixed layout results in a 16.6% larger cell area.[41]

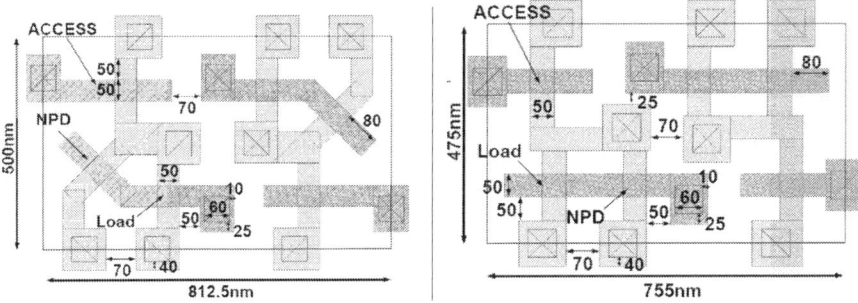

**Fig. 2.12.** 6-transistor SRAM cell layout comparison: standard layout (right) and layout with NMOS pull-down transistors rotated by 45 degrees (left).[41] Copyright© 2005 IEEE.

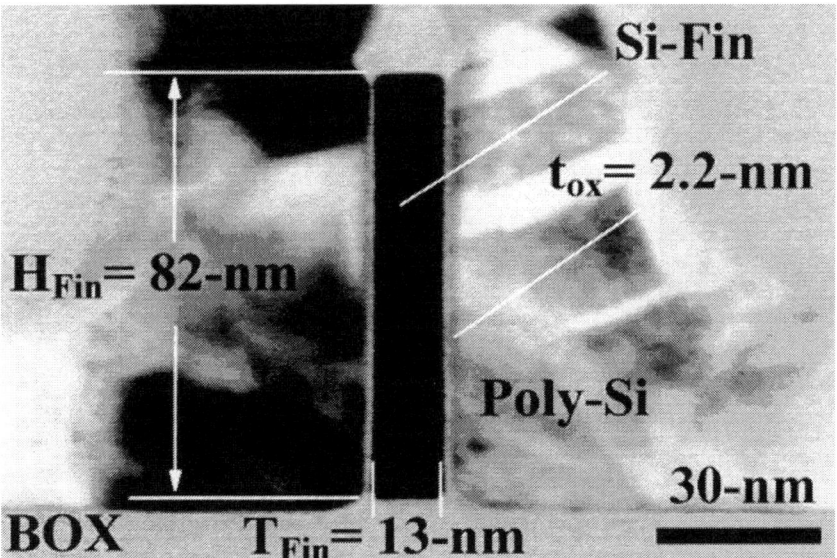

**Fig. 2.13.** Cross-sectional TEM of a (111)-surface sidewall fin. The smooth rectangular shaped fins were formed using surface-orientation dependent wet etch chemistry.[19] Copyright© 2003 IEEE.

The formation of (111) surface fins on (110) surface SOI substrates has been demonstrated.[19,42] In this experiment, fin patterns were aligned parallel to the <112> direction. The advantage of the (111) surface fins is that very smooth and straight fins can be formed by a wet etch process. The (111) surface silicon has an extremely low etch rate in tetramethylammonium hydroxide (TMAH) solution, compared with other crystallographic planes. The etch rate ration at 50°C for (110) planes versus (111) planes is 20:1. Thus, after initial plasma fin etch, an additional wet etch in TMAH can ensure very straight and smooth (111)-surface fins. Figure 2.13 shows the TEM of a smooth and vertical (111)-sidewall surface fin. One can appreciate the nearly perfect rectangular shape of the fin. The smoothness of the (111) surface fin using surface orientation dependent wet etch chemistry is very attractive. However, the drawback of this technique is that (111) surface carriers mobility is rather low. Furthermore, the mobility enhancement with strain engineering is not currently as well characterized on (111) as on the other two surfaces. Strain engineering will be further discussed in Section 2.5.

### 2.2.4 Fin Surface Preparation

The fin surfaces exhibit significant roughness right after plasma etch. In addition, the fin corners are sharp. Gate oxide grown on these surfaces is prone to defects and localized thinning. Without proper treatments, silicon crystal defects such as twin formation in the fin corners could happen.[40] Sharp corners also result in a kink in the subthreshold region if the fin doping level is high.[43] Such kinks are caused by parasitic corner devices, which have a lower threshold voltage than the sidewalls due to the higher electrical field in the corners. The fin corner problem is similar to the parasitic STI edge device effect that can be found in planar CMOS technologies.

Carrier mobility along the fin sidewalls without treatment is rather poor.[44] The simplest way to repair surface roughness and damage from the plasma etch is through oxidation. The requirement for fin sidewall smoothing oxidation is similar to STI liner oxidation. It is desirable to have conformal oxidation with a growth rate that is insensitive to the fin surface orientation. In-Situ-Steam-Generated (ISSG) rapid thermal oxidation has such desired properties.[45-46]. ISSG typically uses pressure above 5 Torr. Hydrogen and oxygen are introduced simultaneously into the chamber. The reactants remain as free species ($O_2$ and $H_2$) until they reach the wafer surface, where heat induces a local reaction. $O_2$ and $H_2$ react to form steam and oxygen radicals. The unstable oxygen radicals promote further oxidation. As a result, the oxidation is conformal and has no surface orientation preference. In addition, ISSG has the advantage of being insensitive to the doping level in the silicon and it provides better corner rounding than conventional oxidation.

The sacrificial smoothing oxide is typically removed using a hydro-fluoride (HF)-based isotropic wet etch step.[40] Some of the buried oxide in the SOI substrate is etched at the same time as the sacrificial oxide. As a result, buried oxide recesses below the bottom fin surface and creates the Π/Ω-shaped gate (See Fig. 2.8a). Π/Ω-Gate MOSFETs take advantage of such buried oxide recess to improve the gate electro-static control of the channel.[7-8] However, such a recess can undermine the fin stability if the fin width is scaled below 10nm.

$H_2$ annealing is another way to smooth the fin surfaces and round the corners. $H_2$ annealing has been shown to: 1) improve electron mobility, 2) reduce gate leakage and 3) lower 1/f noise.[11,12,47] The $H_2$ annealing process conditions have a great impact on the final shape of the fins. If the temperature is too high and/or the pressure is too low, the Si fins can agglomerate and break. The narrower the fins, the more sensitive they are to the $H_2$ annealing conditions.

## 2.2.5 Fins on Bulk Silicon

The fins we described so far are fabricated on SOI wafers. MuGFETs can also be made on bulk silicon wafers.[48-53] In some references they are called body-tied FinFETs or bulk FinFETs. The main advantages of using bulk silicon over SOI substrates are 1) lower wafer cost and 2) better substrate heat transfer rate. However, bulk multi-gate MOSFET require additional isolation steps, which increase the cycle time and cost. Furthermore, the fin height of bulk-based multi-gate devices are entirely determined by the fin etch step. This puts more pressure on etch variation control, since any fin height variation translates into transistor width variation.

The control of the electrical potential by the gate in bulk fins is as good as in their SOI counterparts. But underneath the fin, in bulk devices, the source and drain have to be separated by heavy channel stop implants to prevent sub-surface punchthrough. As a result, multi-gate devices made on bulk silicon suffer from higher source and drain diode leakage and higher junction capacitance than devices made in SOI.

The bulk silicon under the fin can be accessed by a body contact. But the body factor of a bulk FinFET is very low, since the electrostatic potential inside the fins is dominated by the gate, not by the body. Therefore, body bias is not effective in changing the threshold voltage of a bulk multi-gate MOSFET. Figure 2.14 illustrates the general fabrication process of a bulk MuGFET.

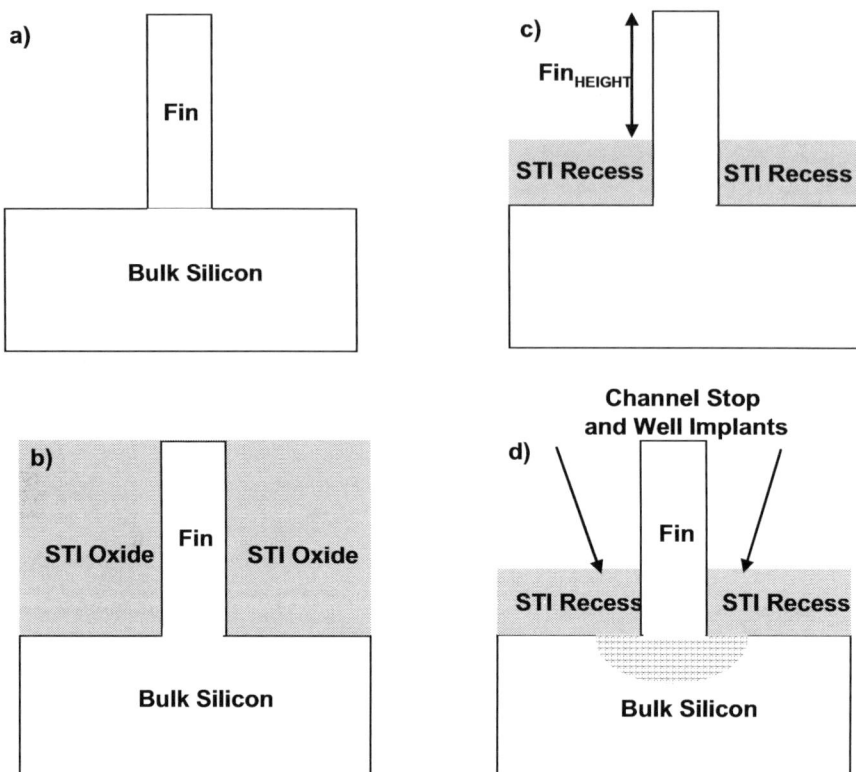

**Fig. 2.14.** General process flow for making a FinFET on a bulk silicon bulk wafer. a) Fin etched in bulk silicon, the initial $Fin_{HEIGHT}$ is greater than final device $Fin_{HEIGHT}$ to allow some STI (Shallow trench Isolation) oxide. b) STI oxide deposition and planarization. c) STI oxide recess to expose silicon fin with desired FinHEIGHT. d) Channel stop and Well implants. Channel stop is needed to prevent the source and drain punch through below the fin in the bulk silicon.

### 2.2.6 Nano-wires and Self-Assembled Wires

The ultimate MOSFET structure is a cylindrical wire-like gate all-around device, as show in Figure 2.15. This structure provides the best electrostatic control of the channel by the gate [54]. The natural length, $\lambda$, for such a device is given by Equation 2.16.

$$\lambda = \sqrt{\frac{2\varepsilon_{si} t_{si}^2 \ln(1 + \frac{2t_{ox}}{t_{si}}) + \varepsilon_{ox} t_{si}^2}{16\varepsilon_{ox}}} \qquad (2.16)$$

where $t_{si}$ is the diameter of the wire. To keep short-channel effects under control, $t_{si}$ shrinks with gate length. When $t_{si}$ is below 10nm, the device structure is typically called a "nanowire MOSFET".

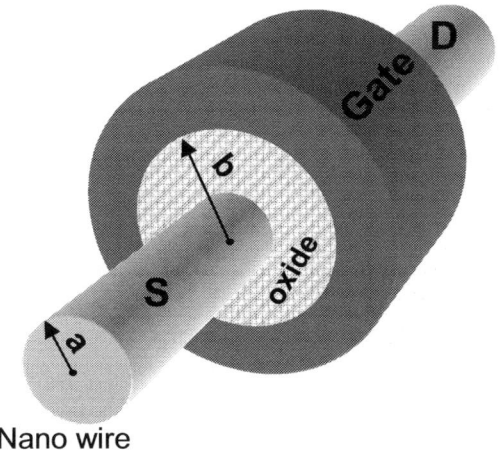

**Fig. 2.15.** Illustration of a silicon nanowire transistor. The gate stack is wrapped around the semiconductor nanowire. $a$ is the radius of the nanowire.

Because the nanowire transistor has a cylindrical cross section, its capacitance can no longer be calculated using a simple parallel-plate model. Considering Figure 2.15, the inversion charge in the nanowire can be approximated by a charge sheet on the outer surface of the wire. The charge on the gate is on the inner wall of the gate electrode. The capacitance can be obtained by calculating the voltage drop across the gate oxide. Applying Gauss's law to an infinitely long cylinder one can find the electrical field outside the cylindrical wire:

$$E = \frac{Q}{2\pi\varepsilon_{ox}r} \qquad (2.17)$$

Where $Q$ is the charge per unit length at the surface of the wire, and $r$ is the radius of the wire. The voltage drop across the gate oxide can be found by integrating the electrical field along a radial line:

$$\Delta V = \frac{Q}{2\pi\varepsilon_0 r}\int_a^b \frac{1}{r}dr = \frac{Q}{2\pi\varepsilon_{ox}r}\ln(\frac{b}{a}) \qquad (2.18)$$

where $a$ is the nanowire radius, and $b = a + t_{ox}$. The capacitance per unit gate length is:

$$\frac{C}{L_g} = \frac{Q}{\Delta V} = \frac{2\pi\varepsilon_{ox}}{\ln(\frac{b}{a})} \qquad (2.19)$$

Converting Equation 2.19 to a capacitance per unit area of the nanowire surface, we obtain:

$$\frac{C}{A} = \frac{2\pi\varepsilon_{ox}}{\pi a \ln(\frac{b}{a})} = \frac{2\varepsilon_{ox}}{a \ln(\frac{b}{a})} \qquad (2.20)$$

Equation 2.20 indicates the effective gate capacitance of the nanowire transistor is a function of the nanowire radius as well as the gate oxide thickness. This relationship can be exploited to reduce the effective gate oxide thickness without using higher $k$ dielectric. As an example, a nanowire MOSFET with $a$=4nm, $t_{ox}$ = 2nm, the effective gate oxide thickness is 1.62nm. The above calculation does not take bulk inversion or quantum-induced volume inversion effects into account. In those cases, the effective nanowire radius is decreased, since the charges are not limited to the nanowire surface.

Silicon nanowires can be made with conventional patterning described in section 2.2.2. A circular cross section can be made by first removing the buried oxide below the silicon, and then rounding of the wires to an approximately circular cross section can be achieved using $H_2$ annealing.[29]

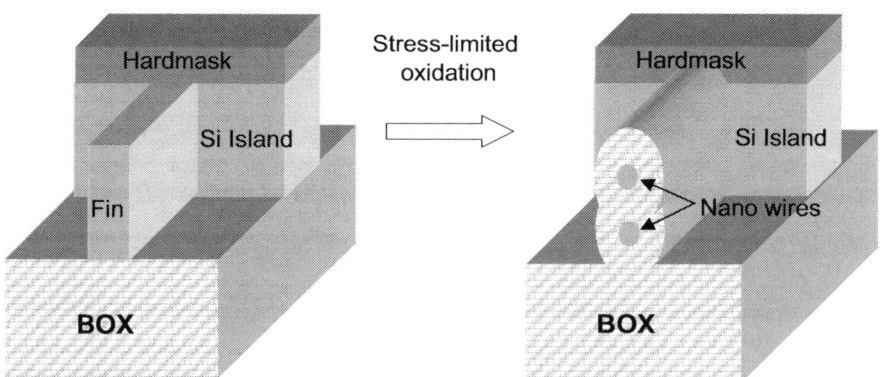

**Fig. 2.16:** Twin Si nanowires formed through stress-limited oxidation. A hard mask is needed to keep the Si island region from oxidizing. The presence of Si island at the end of the wires is needed to anchor the wires.

A twin-wire structure has been made using stress-limited oxidation of tall fins. The stress-limited oxidation oxidizes the wire slower than planar silicon, which produces a final structure comprising two Si wires, one on the top of the fin, and the one other at the bottom of the fin (Figure 2.16). It is important to protect the source and drain region during the oxidation. After removing the oxide, two silicon wires are formed.[55] Twin nanowires can also be formed by oxidation using doping-dependent oxidation rate effects.[126]

The formation of high-density, self-assembled nanowires has been demonstrated using cylindrical-phase diblock copolymers.[56] The polymer has the ability to divide a lithographic pattern ($W$) into integer number of sub-unit with a distance, $l$, between cylinders. $l$ is intrinsic length of the polymer which determine the number of sub-units ($n$) fits in $W$. The relationship between $n$ and $l$ is defined by Equation 2.21.

$$n = \frac{2}{\sqrt{3}}\left(\frac{W}{l}\right) \qquad (2.21)$$

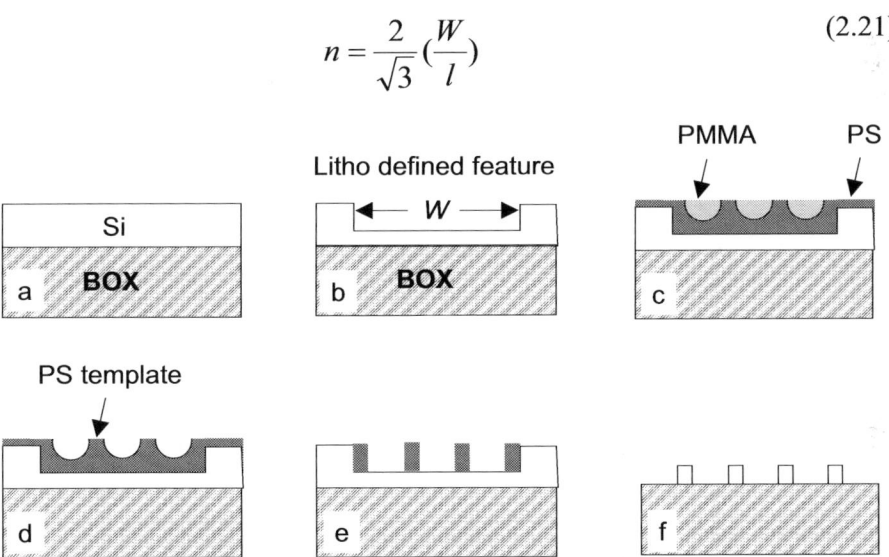

**Fig. 2.17:** Self-assembled nanowire formation. The initial lithothography-defined pattern on an SOI wafer (a,b) is divided into self-aligned patterns by cylindrical-phase diblock copolymers (c). d) PS template formation for plasma etching (e). The remaining PS features are used as mask to etch the silicon nanowires (f).

The self-assembly process is illustrated in Figure 2.17. The sidewalls of the lithography-defined trench initiate the assembly of the polymer domain. In Reference 56, diblock copolymers composed of 70% polystyrene (PS) and 30% poly (methyl-methacrylate) (PMMA) (70:30 PS:PMMA) with total molecular weight Mn=64 kg/mol are shown to form hexagonal lattices of 20 nm diameter cylindrical PMMA domains ($l = 35$ nm) in a matrix of PS. After film assembly, PMMA is removed from the half-cylinder domains by immersing the substrate in acetic acid. Plasma etch is then used to define the Si nanowires.

## 2.3 Gate Stack

### 2.3.1 Gate Patterning

This section presents the process steps used in the gate formation of a multi-gate MOSFET. The topography created by the fins presents many challenges. One of these challenges is illustrated in Figure 2.18: photoresist has a surface planarization effect when deposited on a non-planar topography. The wavy gate material stepping over the fins results in different photoresist thickness along the width of the transistor. The photoresist critical dimension (CD) after exposure is a function of the resist thickness. This effect is known as "swing curves". Without gate planarization, gate CD will vary periodically through the entire width of the device, with a period equal to the fin pitch (Figure 2.18). Therefore, the use of a planarized gate is highly desirable.

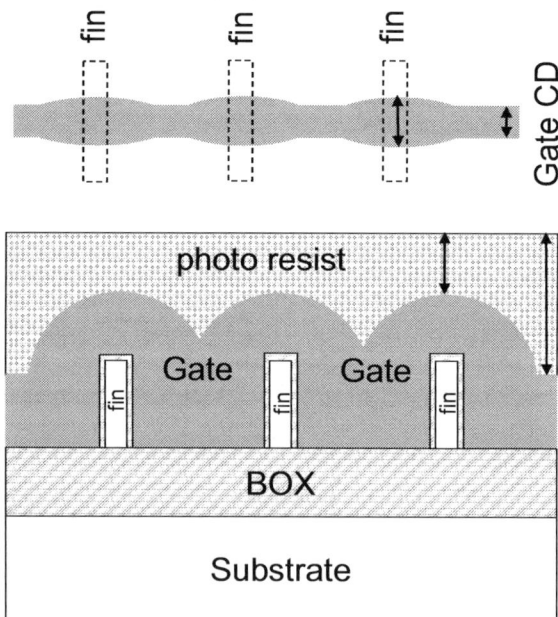

**Fig. 2.18.** Patterning gate on an irregular topography results in wavy gate CD due to the variation in resist thickness. A planarization step is required.

The gate etching process must also be adjusted to account for the fin topography. Gate etch first reaches the fin top surface, then over-etch is required to clear all of the gate material between the fins. The overetch must be selective to the gate dielectric material for tri-gate type devices. A hard mask on top of the fin can help to protect the fins from gate etch. Additional isotropic etch is required for Ω-gate devices in order to clear the gate material in the undercut region of the fin (Figure 2.19).

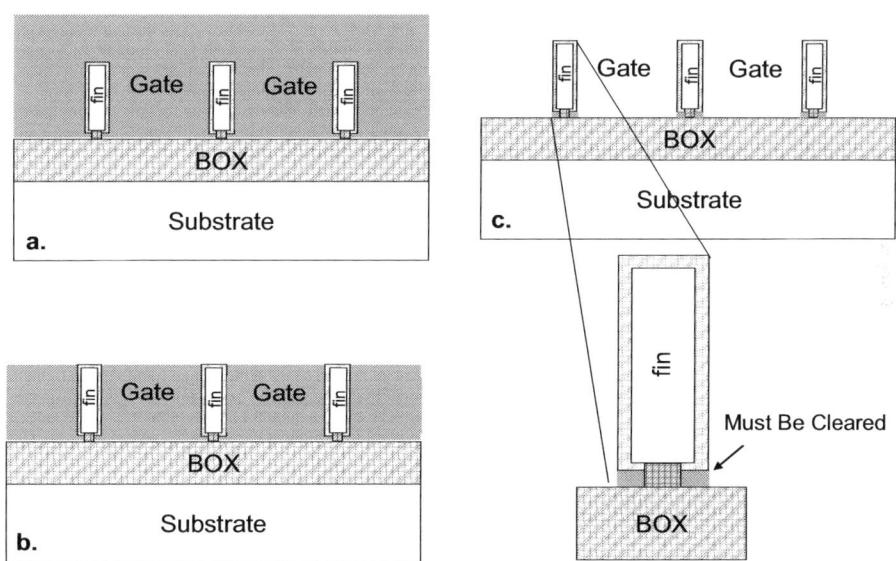

**Fig. 2.19.** The key aspects of the gate etch are the ability to etch the gate material between the fins without damaging the fin (a,b), the removal of the gate material in the undercut region between the fin and buried oxide (c).

### 2.3.2 Threshold Voltage and Gate Workfunction Requirements

The threshold voltage of a long-channel MOSFET, $V_t$, is classically is defined by Equation 2.22.[57]

$$V_t = \phi_{ms} + 2\phi_f + \frac{Q_D}{C_{ox}} - \frac{Q_{ss}}{C_{ox}} \qquad (2.22)$$

where $Q_{ss}$ represents the charges in the gate dielectric, $C_{ox}$ is the gate capacitance, $\Phi_{ms}$ is the workfunction difference between the semiconductor and the gate electrode, $\Phi_f$ is the difference between the semiconductor Fermi level $E_F$ and the intrinsic semiconductor Fermi level

$E_{Fi}$, and $\Phi_f$ is the Fermi potential, given by Equation 2.23 for P-type silicon:

$$\phi_f = \frac{kT}{q}\ln(\frac{N_a}{n_i}) \qquad (2.23)$$

In a planar bulk MOSFET, strong inversion occurs when the band bending (surface potential) at the semiconductor surface reaches $2\Phi_f$. $Q_D$ is the depletion charge in the channel. In a fully depleted double-gate MOSFET $Q_D$ is given by Equation 2.24.

$$Q_D = q\frac{1}{2}Fin_{WIDTH}N_a \qquad (2.24)$$

Let us now examine if this $V_t$ definition is applicable to ultra-thin body devices. Let us assume a metal gate electrode with a workfunction equal to the middle of the band gap of the silicon. The silicon body is doped p-type with $N_a = 2 \times 10^{15}/cm^3$. The fin cross section is 11nm x 60nm and the gate dielectric is 1.8nm-thick $SiO_2$. Let us also keep in mind that for 11nm fin width quantum effects on $V_t$ are negligible. The Fermi potential, $\Phi_f$, can be calculated by Equation 2.23 and is equal to 0.33V at room temperature. Since the gate material has a midgap workfunction $\Phi_{ms} = -\Phi_f$ and Equation 2.22 becomes:

$$V_t = \phi_f + \frac{Q_D}{C_{ox}} - \frac{Q_{ss}}{C_{ox}} \qquad (2.25)$$

Considering the width of the fin, the total depletion charge $Q_D$ contributes 0.05mV to Equation 2.25. If the gate dielectric charge $Q_{ss}$ is under $1 \times 10^{12}/cm^2$, the contribution to Equation 2.25 is less than a few $\mu V$. Thus, $V_t$ from $Q_D$ and $Q_{ss}$ is negligible compared to $\Phi_f$. We can conclude that $V_t$ in our example should be equal to 0.33V. But this is 0.12V less than experimentally observed using maximum $g_m$ method in reference [58].

This discrepancy is due to the estimation of the surface potential at the threshold of inversion in thin-body devices. For undoped ultra-thin body devices, the availability of inversion charges is quite limited when the surface potential is at $2\Phi_f$. The surface potential at the onset strong inversion (and thus at threshold) is higher than $2\Phi_f$. Therefore, Equation 2.22 needs to be modified:

$$V_t = \phi_{ms} + 2\phi_f + \frac{Q_D}{C_{ox}} - \frac{Q_{ss}}{C_{ox}} + V_{inv} \qquad (2.26)$$

where $V_{inv}$ represents the additional surface potential to $2\Phi_f$ that is needed to bring enough inversion charge into the channel for the transistor to reach threshold. Several publications tackle the issue of modeling $V_{inv}$. $V_{inv}$ is a function of fin width and doping concentration. The narrower the fin and the lower the doping level, the higher the value of $V_{inv}$. The use of wide fins and higher doping concentration reduces $V_{inv}$ and brings Equation 2.26 closer to Equation 2.22.[58-60]

The contribution of the depletion charge $Q_D$ to the threshold voltage is negligible unless the doping concentration is very high. Therefore, the threshold voltage of a MuGFET is primarily determined by the effective workfunction of the gate stack (the workfunction of the gate electrode and the charges in the gate dielectric or at the interfaces). Figure 2.20 shows the required gate stack workfunction for various devices.

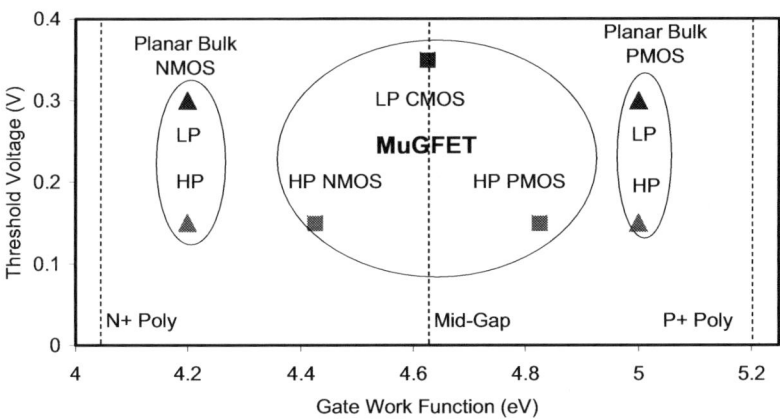

**Fig. 2.20.** Gate workfunctions for different device types. High Performance (HP) and Low Power (LP) planar bulk MOSFETs need workfunctions close to the silicon band edges, while HP and LP multi-gate MOSFETs need workfunction values close to the midgap of the silicon.

For high-performance CMOS, where low threshold voltages are desired, a two-workfunction metal gate system is required. For n-channel FETs, the workfunction needs to be ~200mV below the midgap of the silicon. For p-channel devices, the workfunction needs to be ~200mV above the midgap of silicon. In low-power CMOS, a single midgap metal gate system can be used to obtain a symmetrical threshold voltage for n-channel and p-channel devices at approximately ±350mV.[58]

### 2.3.2.1 Polysilicon Gate

Heavily doped polysilicon gates have an effective workfunction close to the band edges. For depletion-mode operation, if $N^+$ poly is used for the NMOS devices, $V_t$ is less than zero, and if $P^+$ poly is used for PMOS devices, their $V_t$ is greater than zero. These values are not suitable for standard digital applications.

To obtain more acceptable threshold voltage values, attempts have been made to place an $N^+$ poly gate on one side of the fin and a $P^+$ poly gate on the other side.[22,62] The resulting device is commonly referred to as an asymmetrical double-gate FinFET. Figure 2.21 illustrates the process used to make an $N^+/P^+$ poly-gate FinFET. Two high-tilt-angle gate implants were used to dope the gate on one side of the fin $N^+$ and the other side $P^+$. This process requires a thin polysilicon gate and tall fins so that an implant "shadow" effect can be used to prevent sidewall gates from receiving both implants. The threshold voltages observed experimentally in [22] are 0.15V for NMOS transistor sand -0.1V for PMOS devices.

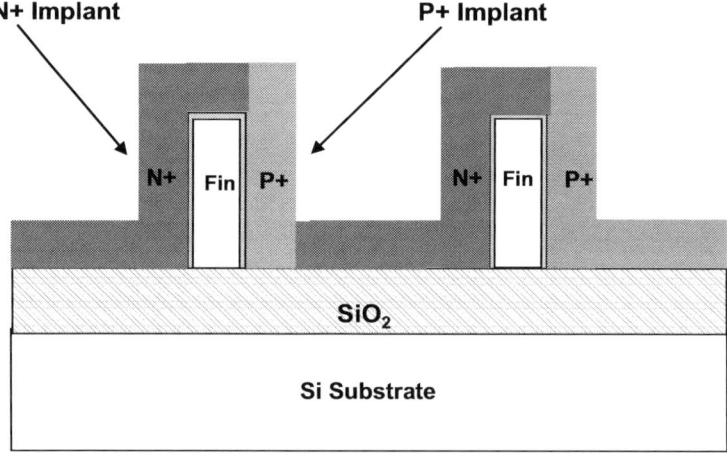

**Fig. 2.21.** $N^+$ and $P^+$ asymmetrical poly-gate formation. Large-tilt-angle gate implant is employed to dope the gates on one side of the fin $N^+$, and the other side $P^+$. The implant shadowing effect created by the tall fin helps to achieve this result.

The advantage of such an approach is that the gate material is familiar to semiconductor manufacturing. The disadvantages are: 1) higher Gate Induced Drain Leakage (GIDL) due to the high offset workfunction on one side of the gate [63,64] (more on GIDL in Section 2.3.1.4); 2) $V_t$ is not tunable through gate stack; 3) $P^+$ poly-silicon gate is not compatible with high-K gate dielectrics due to Fermi level pinning [65,66] and; 4) there is a poly depletion effect which increases the effective oxide thickness.

### 2.3.2.2 Metal Gate

The use of a metal gate eliminates the poly depletion problem. It increases carrier mobility by reducing the transverse electrical field at a given gate overdrive. Metal gate and high-K dielectrics are key scaling enablers for all MOS technologies. Integrating metal gate electrodes into the MOSFET front-end processes is difficult. One of the main reasons is most metals are not compatible with high-temperature (>1000°C) processing.

Some metal nitrides are stable at very high temperatures, on the other hand. Examples are TiN, TiSiN, TaN and TaCN. The workfunctions of these metals after high-temperature annealing are typically between 4.4eV to 4.7eV, depending on: 1) the layer thickness, 2) alloy composition, 3) crystal orientation and 4) on the gate dielectric material. Metal nitrides are not suitable gate materials for planar CMOS, but they do have the right workfunction for low-power, multi-gate CMOS applications, and have thus been used to fabricate multi-gate MOSFETs.[67] The digital circuit performance of MuGFET made with TiN and TiSiN has been demonstrated to exceed the planar technology with similar gate length (see Chapter 7).

The integration of metal nitride with multi-gate devices is straightforward. A thin layer of metal nitride is deposited either by Atomic Layer Deposition (ALD) or by Chemical Vapor Deposition (CVD). ALD and CVD give conformal step coverage of the fin sidewalls, which is important for high aspect ratio fins. The metal layer is then covered with a thick polysilicon capping layer for gate patterning. The polysilicon capping layer serves as a planarization medium and reduces gate resistance after silicidation (Figure 2.18).

### 2.3.2.3 Tunable Workfunction Metal Gate

For high-performance CMOS applications, gate stacks with workfunctions at +/- 200mV away from the silicon midgap are sufficient to set the threshold voltages, but this requires integrating two or more workfunction metal gate systems onto the same chip, one for N-channel transistors and one for P-channel devices.

Several integration schemes have been proposed to produce gates with tunable workfunction materials: Fully Silicide (FUSI) metal gate, metal layer inter-diffusion, the use of multiple metal thicknesses, and nitrogen-implanted molybdenum (Mo) gates.

### 2.3.2.3.1 Fully Silicided Metal Gate (FUSI)

The use of Fully Silicided (FUSI) metal gate on multi-gate device was first reported in [68]. FUSI metal gates are formed through a silicidation process after the dopant activation anneal, which avoids the problem of compatibility issues of the metal gate with front-end thermal cycles. Figure 2.22 illustrates the general principle of FUSI metal gate formation on a planar device.

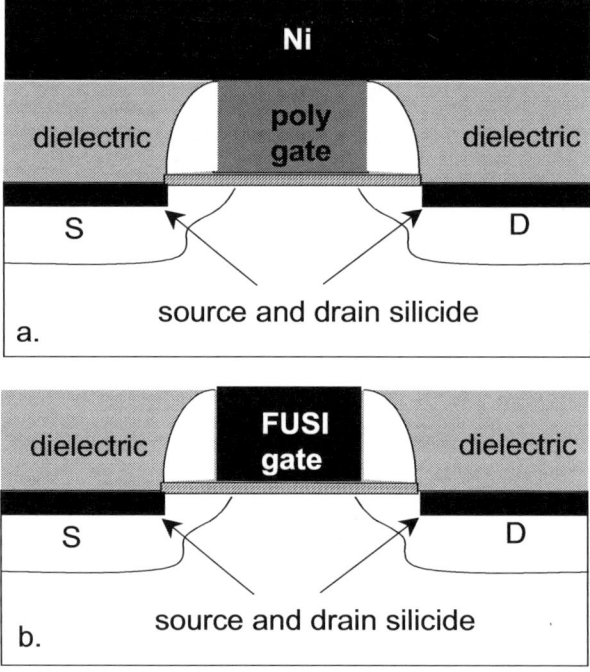

**Fig. 2.22.** Planar MOSFET Fully Silicided (FUSI) metal gate fabrication concept.

The front-end process, up to the gate silicide formation step, is the same as for a standard polysilicon gate device. After source and drain activation and silicidation, a layer of filling dielectric material is deposited and polished until the gate is reached. Then, a thick layer of metal (usually Ni or Co) is deposited on top of the gate. After proper thermal anneal, the gate is completely silicided all the way to the gate dielectric. Any unreacted metal is then stripped off using a wet process.

The FUSI gate used in a multi-gate MOSFET fabrication process is similar to that used for planar devices. Due to the differential gate poly thickness caused by the fin topography, however, the deposited Ni needs to be thick enough to form a silicide layer all the way to the buried oxide. It is important to control the formation a uniform silicide phase throughout the gate and keeping the Ni from diffusing though the gate dielectric (Figure 2.23).

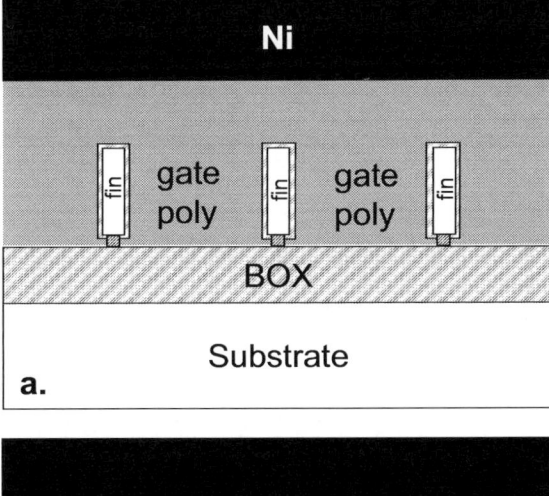

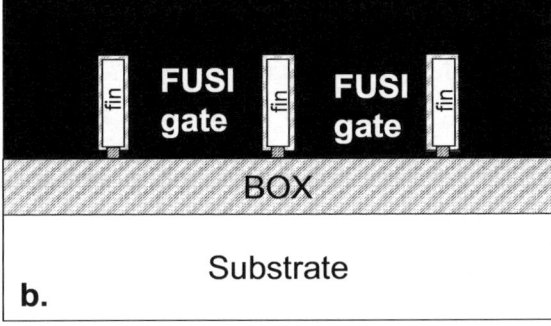

**Fig. 2.23.** Multi-gate FUSI metal gate formation, a) after metal deposition and b) after silicide formation. Fin topography creates differential gate poly thickness between the fin top and fin bottom. The silicide formation front first reaches the fin top, and must continue to the bottom of the fin. The silicide phase uniformity throughout the gate thickness must be controlled and metal diffusion through the gate dielectric through the extended silicidation process must be limited.

The most commonly used silicides, NiSi and $CoSi_2$, have workfunctions near midgap. In order to adjust the workfunction, a high concentration of dopant is implanted into the polysilicon prior to the gate silicide formation. The effect of dopants on the final effective workfunction of FUSI gate depends on the type of gate dielectric used: on a nitrided gate oxide, boron- and aluminum-doped FUSI have been reported to increase the workfunction up to 4.9eV, while phosphorus, arsenic and antimony can reduce the workfunction down to 4.2eV.[66-68]

The origin of such effective workfunction changes with pre-doped poly silicon is a dopant segregation effect that occurs during silicide formation. Doping atoms have much smaller solid solubility in NiSi than in polysilicon. As Ni consumes Si to form NiSi, dopants in the polysilicon start to segregate out and are pushed toward the gate dielectric interface. When the silicon in the gate is completely consumed and transformed in NiSi, an extremely high concentration of dopant is formed at the interface between the silicide and the gate dielectric, which alters the effective workfunction of the gate stack.

The "dopant segregation" effect is less pronounced on high-K dielectrics. Boron doping has little impact on the workfunction due to Fermi level pinning.[72]. The use of nickel-rich silicides can help to reduce the Fermi level pinning. A workfunction of 4.85eV was achieved with $Ni_2Si$.[73] Extremely high concentrations of Ytterbium in NiSi was used to move the workfunction up to 4.22eV.[74] Unfortunately, the simultaneous integration of multi-phase Ni silicides on the same chip is by no means an easy task.

### 2.3.2.3.2 Metal Nitride Thickness and Nitrogen Content

The effective workfunction (EWF) of TiN and TiSiN has been found to change with the thickness of the layer.[75,76] Up to 250mV workfunction shift were observed when TiN thickness was increased from 3nm to 20nm. The workfunction saturates at mid band gap for TiN layers thicker than 20nm. TiSiN has a workfunction that varies from 4.44eV to 4.83eV when the layer thickness is increased from 2.5nm to 20nm.[77,78,127] Similarly, the workfunction of TiSiN saturate at 20nm (Figure 2.24). The origin of the workfunction shift with TiN and TiSiN thickness is the effective nitrogen content in the metal. The top portion of the TiN is oxidized and become nitrogen deficient. For thin TiN layers, a relatively Ti-rich TiN is formed, which yields a lower workfunction value.

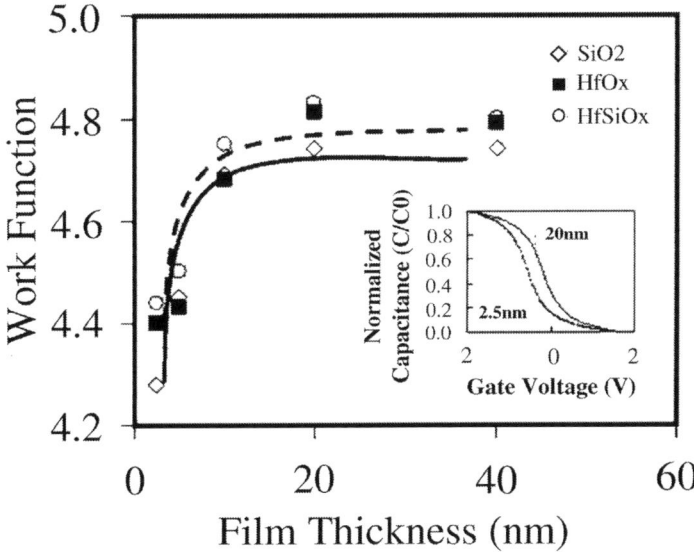

**Fig. 2.24.** TiSiN effective workfunction as a function of the TiSiN film thickness and gate dielectric materials. REPRINTED WITH PERMISSION FROM [17]. COPYRIGHT 2006 AMERICAN INSTITUTE OF PHYSICS.

The effective workfunction of molybdenum (Mo) varies with the nitrogen content in the metal.[79] Molybdenum is a good candidate for gate material because its high melting temperature (2623°C) makes it compatible with CMOS processing. In addition, it has a low resistivity.[80] The workfunction of Mo is 4.9eV. Nitrogen implant can reduce the Mo effective workfunction to 4.5eV. The amount of workfunction shift is a function of the implanted nitrogen dose and the thermal anneal done after the implant. Low nitrogen implant energy should be used to avoid penetration of nitrogen ions in the gate dielectric.

### 2.3.3 Gate EWF and Gate Induced Drain Leakage (GIDL)

Gate Induced Drain Leakage (GIDL) is caused by band-to-band tunneling in the drain region under the gate.[81-84] When there is a large drain-to-gate bias, enough band-bending occurs in the Si at the gate dielectric interface, causing valence band electrons to tunnel into the conduction band. In the case of n-channel devices, the generated electrons flow to the drain while holes flow to the lowest potential electrode, *i.e.* to the source.

The band-to-band tunneling current is a function of the total electrical field in the drain overlap region, and a function of the bandgap of the semiconductor.[85-86] For direct-bandgap materials, $J_{GIDL}$ can be modeled by Equation 2.27.

$$J_{GIDL} = AE_{TOT} \exp(\frac{-B}{E_{TOT}})  \quad (2.27)$$

where $A$ is a pre-exponential parameter given by Equation 2.28 and $B$ is a physically-based exponential parameter given by Equation 2.29. $E_{TOT}$ is the electrical field at point the maximum band-to-band tunneling.

$$A = \frac{2qm_r \pi E_g^2}{h^3} \quad (2.28)$$

$$B = \frac{\pi^2 \sqrt{m_r} E_g^{3/2}}{qh\sqrt{2}} \quad (2.29)$$

$E_g$ is the energy bandgap of the semiconductor channel. $E_g$ has a strong influence on the band-to-band tunneling current. $E_{TOT}$ depends on 1) the doping profile of the drain overlap region, 2) the gate oxide thickness and 3) the applied bias. CMOS scaling requires an abrupt doping profile and a thin gate dielectric. Both of these requirements increase $E_{TOT}$ and the GIDL current. Figure 2.25 illustrates the GIDL current component in the $I_d(V_g)$ curve of a device. GIDL is most pronounced at large negative gate-to-drain biases.

A large component of the $E_{TOT}$ is the surface electrical field $E_S$. $E_S$ is defined by Equation 2.30. If we ignore the doping gradient effect, Equation 2.30 is the $E_{TOT}$.

$$E_S = \frac{V_{DG} - V_{FB} - 1.2}{3T_{OX}} \quad (2.30)$$

where $V_{GD}$ is the drain to gate bias and $V_{FB}$ is the flat-band voltage. Equation 2.30 is not directly applicable to double-gate devices, since $E_s$ depends on the silicon body thickness.[87] The transverse electric field is reduced when the silicon film thickness is decreased. Fully depleted double-gate devices have lower surface electrical field than planar bulk MOSFETs and FDSOI devices.

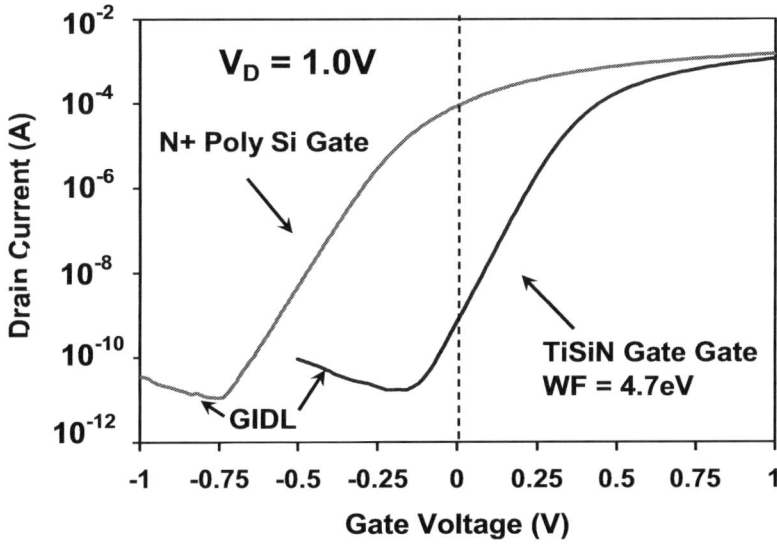

**Fig. 2.25.** GIDL current as a function of gate voltage and gate electrode workfunction. Using a higher gate effective workfunction shifts the $I_d(V_g)$ curve to the right and increases the GIDL current at any given gate voltage.[67]

Equation 2.30 indicates that the flat-band voltage partially determines $E_S$. $V_{FB}$ for mid-band gap metal gate is, by definition, different from heavily doped poly-silicon gate by half a bandgap, $E_g/2$. Thus, the use of a midgap metal gate will cause a higher electric field and a higher GIDL current. A simple way to interpret the change in GIDL caused by the use of different gate workfunctions is shown in Figure 2.25. Using a gate material with a higher workfunction shifts the entire $I_d(V_g)$ curve to the right. As a result, for any given gate-to-drain voltage, the midgap metal gate will give rise to a higher GIDL. In principle, the use of a thin-body, double-gate structure reduces the surface field, while the use of a midgap metal gate increases $E_S$. The final GIDL performance is determined by the overall device design.

The generation of GIDL current is limited to the drain overlap region under the gate. The amount of overlap plays an important role in determining the total current. If the drain is pushed away from the gate edge and if no drain-to-gate overlap exists (underlap), the electrical field is reduced significantly. Several papers have suggested improved $I_{on}$ vs. $I_{off}$ performance can be obtained in MuGFETs with drain underlap.[88-90]. Such a design can reduce GIDL to negligible values compared to subthreshold leakage. Figure 2.26 illustrates the source and drain underlap structure concept.

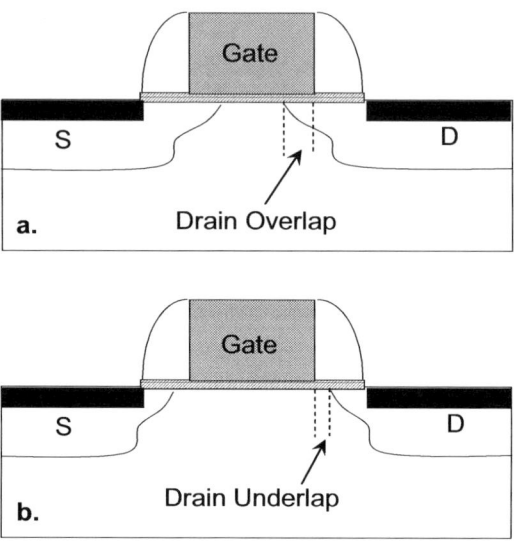

**Fig. 2.26.** a) Drain overlap MOSFET: GIDL occurs in the overlap region; b) Drain underlap MOSFET: Drain is moved away from the gate. As a result, the electrical field and the GIDL current are reduced.

### 2.3.4 Independently Controlled Gates

It is possible to fabricate FinFETs in which the two gates can be biased independently.[91-94] These devices are commonly called Multiple Independent Gate MOSFETs (MIGFETs) or 4 terminally–driven MuGFET (4T-MuGFET). The four terminals are the source, the drain, Gate 1 (G1) and Gate 2 (G2) (see Figure 2.27). 4T-MuGFET enables the threshold voltage of G1 to be raised or lowered by G2 to achieve a higher range of $I_{on}/I_{off}$ performance (Figure 2.28). The 4T-MuGFET structure can be particular useful in power management, where high $I_{on}/I_{off}$ ratio header and footer transistors are used.

The relationship between the threshold voltage of Gate 1, $V_{tG1}$ and G2 bias can be modeled by Equation 2.31.[95] This is a simplified model, since it uses a charge sheet approximation of the channel carrier distribution. It also ignores SCE and quantum mechanical effects. Nevertheless, Equation 2.31 gives a simple way to understand how the G1 threshold voltage changes with G2 bias. $C_{ox1}$ is the G1 gate oxide capacitance and $C_{ox2}$ is the G2 gate oxide capacitance, $C_{Fin} = \varepsilon_{si}/Fin_{WIDTH}$ is the depletion capacitance of the silicon fin.

$$\frac{\Delta Vt_{G1}}{\Delta V_{G2}} = -\frac{C_{Fin}C_{ox2}}{C_{ox1}(C_{Fin}+C_{ox2})} \approx -\frac{3t_{ox1}}{3t_{ox2}+Fin_{WIDTH}} \tag{2.31}$$

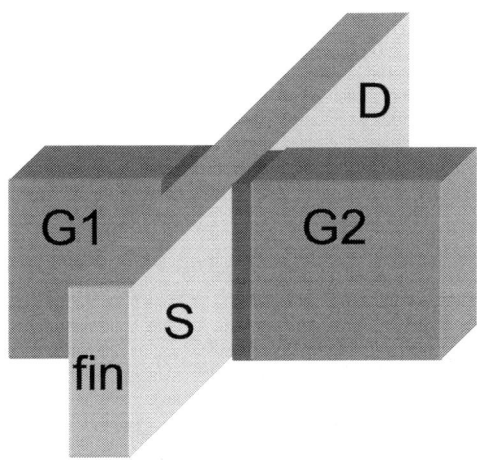

**Fig. 2.27.** 4T-MuGFET, Gate 1 (G1) and Gate 2 (G2) are electrically isolated from one another. This configuration allows for the independent biasing of G1 and G2.

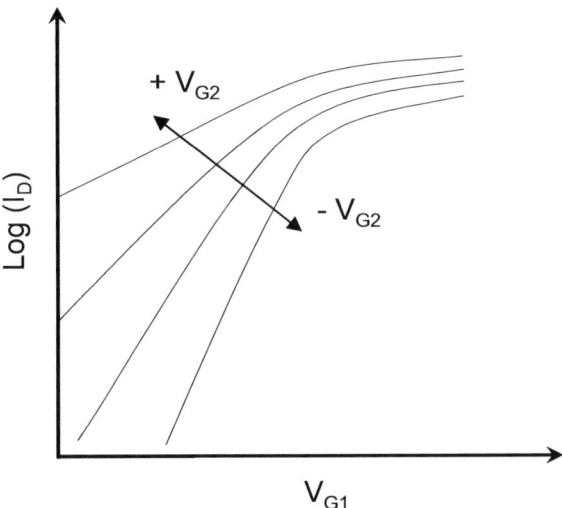

**Fig. 2.28.** 4T-MuGFET drain current as a function of G1 voltage ($V_{G1}$) and G2 voltage ($V_{G2}$). $V_{G2}$ affects the $I_D$ vs. $V_{G1}$ curves.

Further analysis of the $\Delta V_{tG1}/\Delta V_{G2}$ relationship by Zhang et al. [96] reveals that $V_{G1}$ and $V_{G2}$ both control the charge density and the inversion charge centroid position in the fin. The change of inversion carrier centroid position within the fin changes the effective thickness of $t_{ox1(eff)}$ and $Fin_{WIDTH(eff)}$. Equation 2.32 can be modified based on the position of the inversion layer charge centroid in the fin, $x_c$, where $0 < x_c < Fin_{WIDTH}$.

$$\frac{\Delta V t_{G1}}{\Delta V_{G2}} \approx -\frac{3t_{ox1(eff)}}{3t_{ox2} + Fin_{WIDTH(eff)}} = \frac{3(t_{ox1} + \frac{x_c}{3})}{3t_{ox2} + (Fin_{WIDTH} - x_c)} \quad (2.32)$$

To fabricate a 4T-MuGFET, the gate material on top of the fin needs to be removed to provide electrical isolation between G1 and G2. Chemical Mechanical Polishing (CMP) can be used, providing a fin hardmask is in place to stop the CMP before it starts to etch the fin. This approach converts all the devices on the chip to 4T-MuGFETs (Figure 2.29). If only some devices on the chip need to become 4T-MuGFETs, a photo mask step must be employed to selectively open the area just on top of the fin. In this case, plasma etch is used to remove the gate material. The etch end point signal can be provided by the fin hardmask.

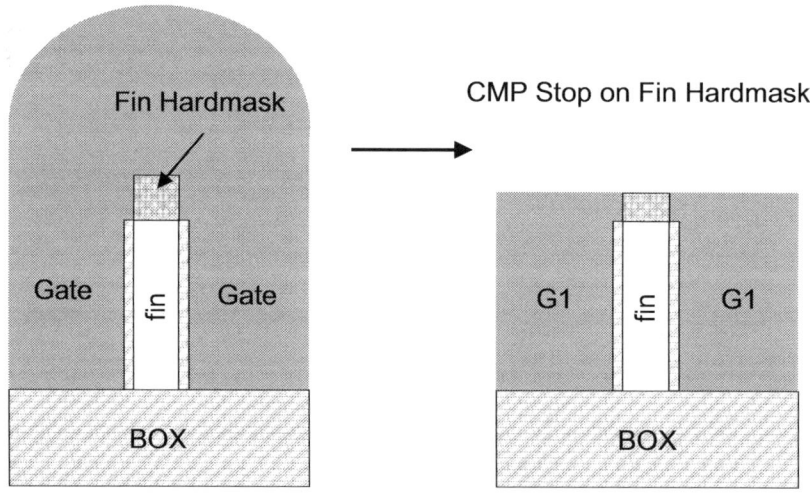

**Fig. 2.29:** Fabrication of a 4T-MuGFET: The unwanted gate electrode material is removed by Chemical Mechanical Polishing (CMP). The fin hardmask serves as CMP etch stop.

## 2.4 Source/Drain Resistance and Capacitance

### 2.4.1 Doping the Thin Fins

Since multi-gate channels are on the sidewall of the fins, the source and drain should be uniformly doped along the sidewalls to avoid variable parasitic resistances across fin height. Large-angle, two-pass, tilted implants are widely used to dope the source and drain extensions and the deep source and drain. Figure 2.30 illustrates the implant process. Two adjacent fins will cast implant shadows on each other. The maximum tilt angle that is allowed while avoiding that the shadow effect blocks the implant from reaching the adjacent fin is limited by the fin height and the fin spacing. It is given by Equation 2.33:

$$Tilt\_Angle \leq \arctan\left(\frac{P_{FIN} - Fin_{WIDTH}}{Fin_{HEIGHT}}\right) \quad (2.33)$$

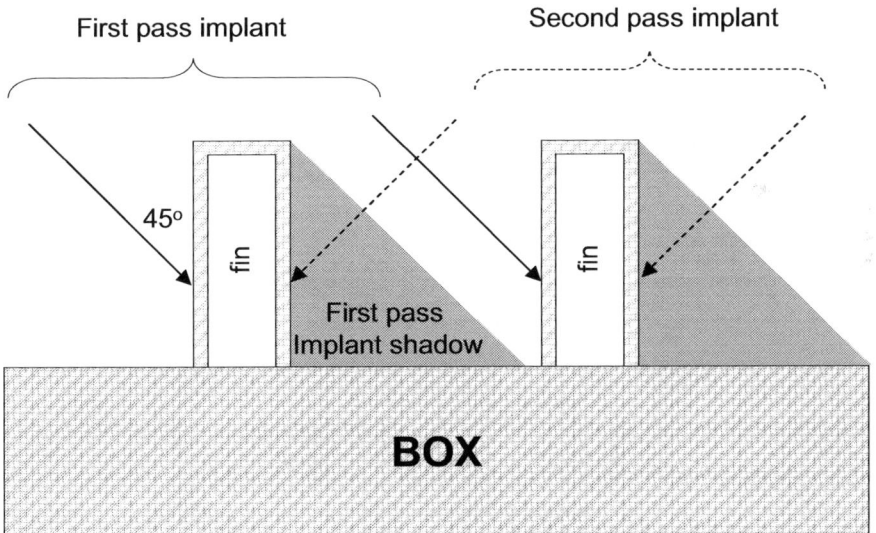

**Fig. 2.30.** Tilted S/D implant.

If the fin top surface is exposed to the implantation, as in Figure 2.30, the top portion of the fin will receive two implants. Furthermore, if the implant tilt is less than 45 degrees, the incident angle of the ion beam for the top surface is higher than for the sidewalls. Thus, the top surface will receive a higher implant dose and the dopants will penetrate deeper. This results in lower resistance for the top portion of the fin and smaller effective channel length for the top portion compared to the bottom portion. Figure 2.31 illustrates this effect.

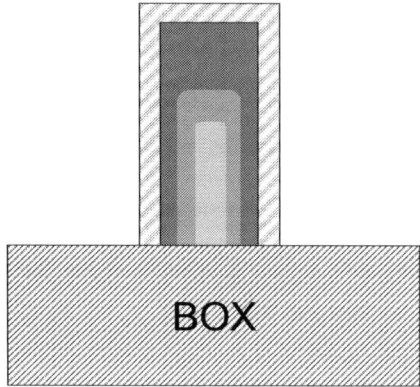

**Fig. 2.31.** Illustration of the impact of tilted implant on the doping profile in the fin. The darker color represents a higher doping concentration. Double implants on the top of the fin result in non-uniform doping concentration and resistance.

Larger tilt implant angles can reduce the problem [97], but the best way to dope the fin uniformly is through isotropic implant. Pulsed Plasma Doping (PLAD) has been developed for high throughput, sub-1keV implants. The ion incident angle distribution can be scattered depending on process parameters to achieve isotopic implantation. Details on the PLAD implant technique can be found in [98].

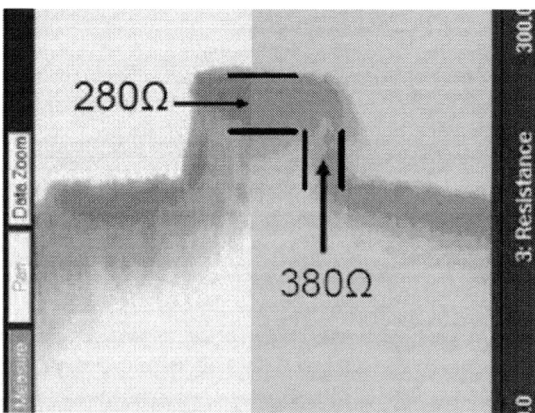

**Fig. 2.32.** SSRM profile of a PLAD implanted fin. The doping uniformity of the fin top and fin sidewall is not perfect, but it is better than what can be achieved by 45 degree tilt implant.[100] Copyright© 2006 IEEE.

Scanning Spreading Resistance Microscopy [99] has been used to study the 2-dimentional resistance distribution of the fin cross section, which can be used to approximate the doping profile. Figure 2.32 shows the SSRM scan of a fin that was implanted with a PLAD process. The resistance ratio between the sidewall and fin top surface is 1.4. This is better than the 2:1 achieved by 45 degree tilted implant. [100]

### 2.4.2 Junction Depth

The physical $Fin_{WIDTH}$ limits the maximum junction depth in a multi-gate MOSFET to one half of the fin width. Therefore, the junction depth is no longer limited by implant energy and thermal cycle. It is still important to control the implant energy and thermal cycle of multi-gate MOSFET, since dopants will move laterally. The lateral movement of the dopants determines the source and drain overlap or underlap, and therefore, the effective channel length. Furthermore, the parasitic resistance is highly dependent on the junction abruptness and doping gradient in the source and drain extension. Smaller gradients and abrupt junctions reduce the extension resistance, but increase short channel effects.

### 2.4.3 Parasitic Resistance/Capacitance and Raised Source and Drain Structure

In Section 2.2.1 we have shown that in order to maintain a good SCE control, fin width needs to be less than the gate length. Continued scaling of the CMOS will inevitably push the fin width below 20nm. Sources and drains made in such thin fins have a high resistance.

One way to reduce this resistance in the source and drain regions is to widen the fin outside the gate area by selective epitaxial growth (SEG) (Figure 2.34).[13,101] The resulting structure is commonly referred to as Raised Source and Drain (RSD). The RSD structure reduces the $R_{SD}$ component of the parasitic resistance, however, $R_{ext}$, the resistance in the source and drain extension regions under the gate spacer (Figure 2.33) remains high due to the thin fin. Placing the raised source and drain closer to the gate reduces $R_{ext}$, but increases the parasitic capacitance $C_{ext}$. To optimize the AC performance, design trade-off between $R_{ext}$ and $C_{ext}$ is required. The advantage of the raised source and drain structure is not just limited to $R_{SD}$ reduction. It also gives more process margin to form silicided source and drain and gives a larger contact landing area.

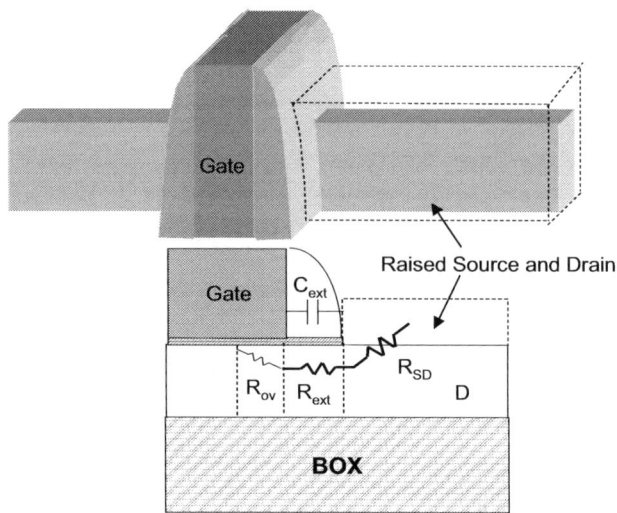

**Fig. 2.33.** The raised source and drain structure reduces the $R_{SD}$ but increasess $C_{ext}$.

The RSD process integration is complicated by the gate spacers left on the sidewalls of the fins (Figure 2.34a). Spacers on the sidewall of the fins must be removed prior to epitaxial growth. One way to remove the fin sidewall spacers is to use a gate hardmask taller than the fins. In this approach, dry etch can remove all the spacers on the sidewall of the fins, while leaving enough hardmask and spacer on the gate (Figure 2.34).[13]

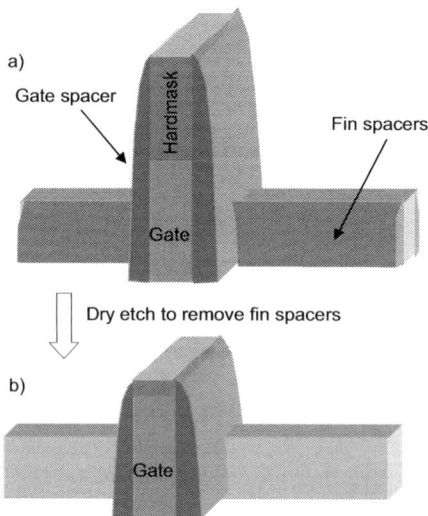

**Fig. 2.34.** Parasitic fin spacers are formed at the same time as gate spacer. The fin spacer must be removed prior to RSD growth process.

It is important to leave enough gate spacers and gate hardmask to prevent "mushroom" and "Mickey Mouse Ears" defects from forming. "Mushroom" formation happens when the gate hardmask is completely removed from exposing the gate to surface to SEG. "Mickey Mouse Ears" happens if the gate top corners are exposed during SEG. Both "mushroom" and "Mickey Mouse Ears" defects can cause shorts between gate and the source and drain (Figure 2.35).

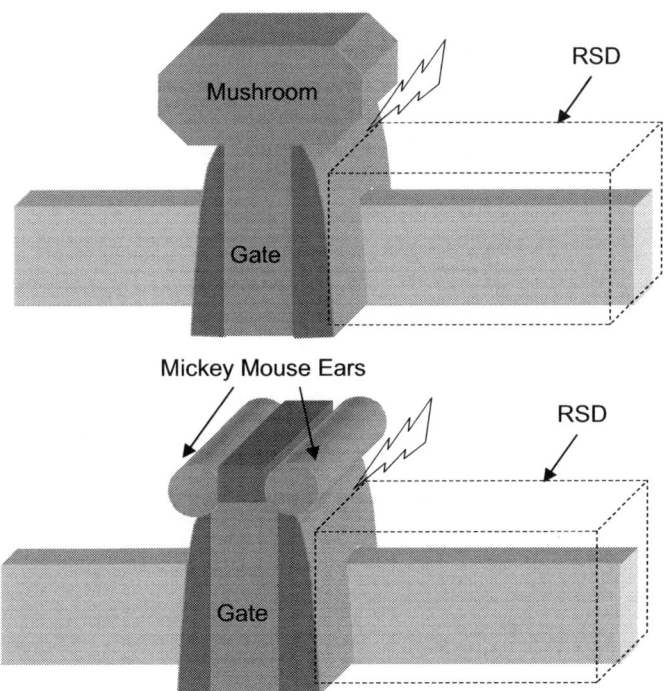

**Fig. 2.35.** If the polysilicon gates are NOT completely covered by protective dielectric, gate "mushroom" or "Mickey Mouse Ears" defects will occur. Both defects can cause short circuit between gate and source and drain.

Another integration method, proposed by Kaneko *et al.* [102] avoids the formation of parasitic fin sidewall spacer altogether. After gate formation, a filling material is deposited and planarized by CMP. An etch-back process recesses the filling material below the gate hardmask. Then spacers are formed on the sides of the exposed gate hardmask. At this time the fins are still embedded in the filling material. After etching the filling material using the newly formed spacer as a hardmask, spacers are only formed along the sidewall of the gate, and not on the sidewalls of the fins (Figure 2.36).

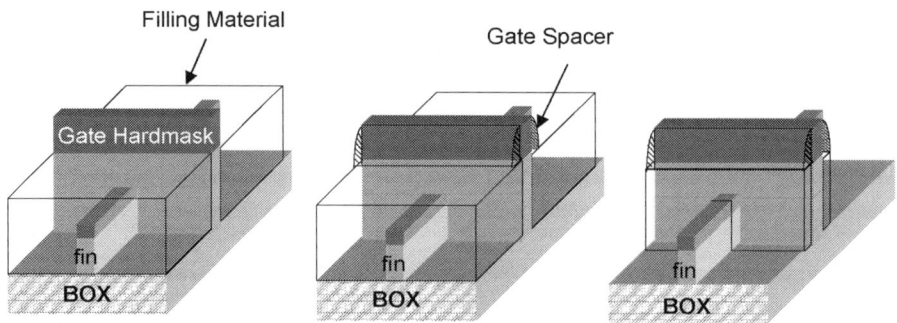

**Fig. 2.36.** Integration scheme to form gate spacer without parasitic fin spacer.

The preparation of the silicon surface before epitaxial growth is important. An in-situ hydrogen pre-bake step at temperature ~ 800°C is typically used before the growth process. Thin silicon fins are susceptible to damage at this temperature at low $H_2$ partial pressure.[47] The deposition of a nitride hardmask on top of the fin can help to protect the fins.[102]

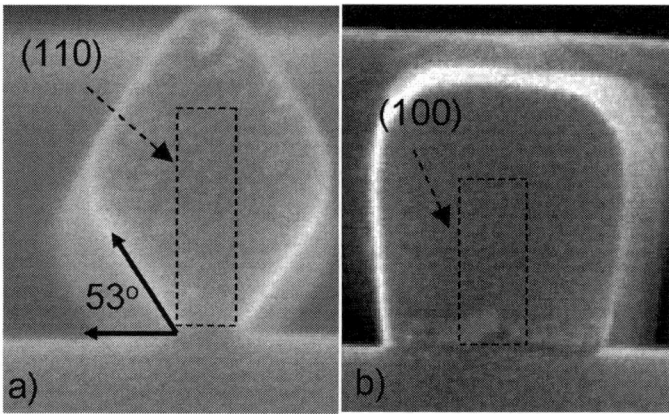

**Fig. 2.37.** Cross sectional SEM images on fins after epitaxial growth. a) (110) side wall surface result in diamond-shaped fin; b) (100) sidewall surface result in square-shaped fin.

The selective epitaxial growth (SEG) surface faceting is different for different surfaces. The faceting determines the final shape of the source and drain region. For (100) sidewall surfaces, a rectangular shaped fin is possible. For (110) sidewall surface, the faceting will result in a diamond shaped surface [102] (Figure 2.37). Diamond-shape fins present challenges to interlayer dielectric fill. Any unfilled region could become a potential defect site in the contact and back-end of the line processes.

## 2.5 Mobility and Strain Engineering

### 2.5.1 Introduction

The carrier mobility, $\mu$, can be defined by Drude's Equation

$$\mu = \frac{q\tau}{m^*} \tag{2.31}$$

where $q$ is the elemental charge, $\tau$ is the mean scattering time and $m^*$ is the effective mass of the carriers. The value of both $\tau$ and $m^*$ can be modified through strain engineering. Under external applied stress, the Si crystal lattice constant (or inter-atomic spacing) is altered, thus changing the band structure, the density of states and the effective mass of the carriers. The strain effects on mobility are anisotropic.[103-104] It highly depends on the surface orientation and the channel direction. A more detailed theoretical discussion of strain engineering can be found in Chapter 5.

Strain effect on carrier mobility was first used to fabricate piezoresistance-based sensors.[105] The application of strain engineering to CMOS can be traced to the 1980s [106-107], where a thin Si layer was first grown on SiGe. The Si lattice takes on the larger lattice constant of SiGe, which produces biaxial tensile strain in the silicon. Since then, uniaxial strain techniques (stressors) have been adopted as one of the main performance boosters for planar CMOS.[108-109]. Drain current improvement in excess of 50% for PMOS transistors and 32% for NMOS devices has been reported.[110]

In order for multi-gate MOSFET technology to be successful, it must be compatible with these powerful performance booster techniques. The following sections describe the response of the multi-gate MOSFET to various strain techniques.

As mentioned in Sections 2.2.1 and 2.5.1 the surface orientation and current flow directions of a FinFET can be (110)/<110>, (100)/<100>, as well as (111)/<111>, respectively. Each surface orientation and channel direction has a different mobility response to a given stress. Since most of the multi-gate devices made to date have been on (100) and (110) surfaces, we will focus our discussion of process strain techniques on these surfaces.

## 2.5.2 Wafer Bending Experiment

The basic responses of multi-gate channel mobility to various stress components were studied in a wafer bending experiment.[111] Figure 2.38 illustrates the wafer bending apparatus.[112] Circular ridges were used to produce a bi-axial stress in Reference 111. The surface stain is measured based on the sample thickness and the radius of the curvature.

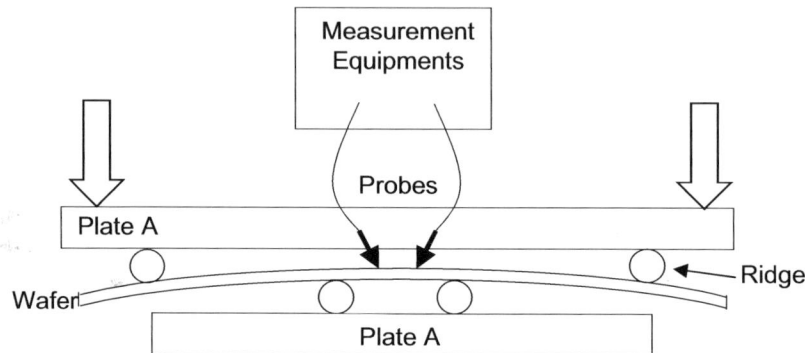

**Fig. 2.38.** Wafer bending apparatus. A wafer or a piece of wafer containing devices is bent by pressing Plate A toward Plate B. Bi-axial stress is applied to the wafer through the circular ridges.

For a 0.11% surface strain (~200MPa), the peak electron mobility was shown to improve by +14% for (100)/<100> fins and +26% for (110)/<110> fins. The peak hole mobility shift with the same surface strain was +8% in (100)/<100> devices and -3% in (110)/<110> PMOS devices. Table 2.1 summarizes the experimental results and the estimated stress levels based on the bulk silicon piezoresistance (PR) coefficients from reference.[111]

There are three stress components: 1) $\sigma_{xx}$ is the stress in the direction parallel to channel current flow; 2) $\sigma_{yy}$ is in the plane of the channel conduction, perpendicular to the current flow; 3) $\sigma_{zz}$ is in the direction vertical to the plane of the channel conduction. (See the figure in the insert of Table 2.1.)

The estimated stress in each direction is similar to the applied stress in the experiment. These wafer bending exercise prove that strain engineering is possible in multi-gate MOSFET, and the response of carrier mobility to stress is similar to planar bulk devices. It is now an engineering undertaking to make the processes work.

**Table 2.1.** Projected stress in each crystal direction using bulk PR coefficients. [111] Copyright© 2006 IEEE.

|  | STRESS (Mpa) | | | MOBILITY IMPROVEMENT (%) | | | |
|---|---|---|---|---|---|---|---|
|  | $\sigma_{xx}$ | $\sigma_{yy}$ | $\sigma_{zz}$ | (100) NMOS | (100) PMOS | (110) NMOS | (110) PMOS |
| Experiment |  |  |  | 14.4 | 8.1 | 25.9 | -2.6 |
| PR Theory | 290 | 350 | -160 | 14.4 | 1.7 | 25.6 | -2.6 |

### 2.5.3 Nitride Stress Liners

Depending on the deposition condition, a nitride film can be either tensile or compressive. Tensile nitride film is a common stressor used to improve NMOS performance. Tensile films tend to shrink, thus when deposited on top of the gate and over the source and drain, the portion of the nitride film on top of the source and drain produces tensile stress in the channel direction ($\sigma_{xx}$). The nitride film on the sidewall of the gate exerts a compressive stress in vertical to the channel ($\sigma_{yy}$) (Figure 2.39). (100)/<110> Electrons mobility is enhanced by tensile stress in the $\sigma_{xx}$ direction and compressive stress in the $\sigma_{yy}$ direction, therefore a net gain in electron mobility is achieved with tensile stressor.

The amount of the strain in the channel is a function of: 1) the stress level in the stressor film, 2) the transistor spacer width, 3) the gate height, 4) the size of the source and drain, and 5) the gate length. Narrow spacers bring the stressor closer to the gate, and therefore improve the stress transfer into the channel. A tall gate allows more stressor volume to induce compressive stress in the $\sigma_{yy}$ direction. For the same reason, long sources and drains produce higher stress levels in the channel.[114] Therefore, in Figure 2.39, the drain side induces higher channel strain than the source side. As the gate length increases, the stressor is located further away from the channel center, therefore the overall channel strain induced by the stressors become smaller. For this reason, the nitride stressors are not effective in improving long-channel device performance.

Hole mobility is enhanced by the opposite channel stress. A compressive liner can be used for PMOS. The difference between PMOS and NMOS is that the hole mobility is not sensitive to the stress in the $\sigma_{yy}$ direction, unlike the electron mobility. Tensile and compressive nitride stressors on multi-gate have been theoretically studied in [115]. The stress induced in the channel by stressors is the same for multi-gate devices as for planar devices. Stress-induced improvement in drive current has been demonstrated experimentally.[116-118]

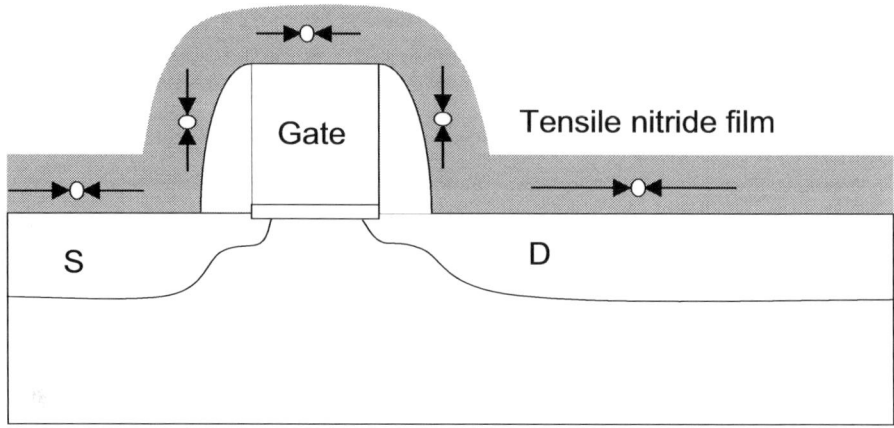

**Fig. 2.39.** A tensile stressor tends to shrink. The stressor on the source and drain pulls apart the ends of the channel and produce a tensile strain $\sigma_{xx}$ in the silicon. The stressor on top of the gate squeezes the gate and produces a compressive stress $\sigma_{zz}$.

### 2.5.4 Embedded SiGe and SiC Source and Drain

Embedded SiGe source and drain was first investigated for the purpose of reducing the contact resistance for in PMOS devices. SiGe has a lower Schottky barrier height for holes than Si. The experimental results have shown a larger than expected drive current improvement when SiGe source and drain structures are used.[106] It was later found that SiGe, with its larger lattice constant than Si, produces a compressive stress in the channel which improves the hole mobility. Figure 2.40 illustrates the compressive stress SiGe source and drain exert in the Si channel. The amount of strain in the channel increases with Ge content (larger SiGe lattice). The SiGe lattice constant $\alpha_{Si_{(1-y)}Ge_y}$ is a function of the Ge content and can be estimated by Equation 2.32. [119]

$$\alpha_{Si_{(1-y)}Ge_y} = (1-y)\alpha_{Si} + y\alpha_{Ge} \qquad (2.32)$$

where $y$ is the Ge fraction (between 0 and 1). $\alpha_{Si}$ is the Si lattice constant equal to 5.431Å and $\alpha_{Ge}$ is the Ge lattice constant, equal to 5.646Å.

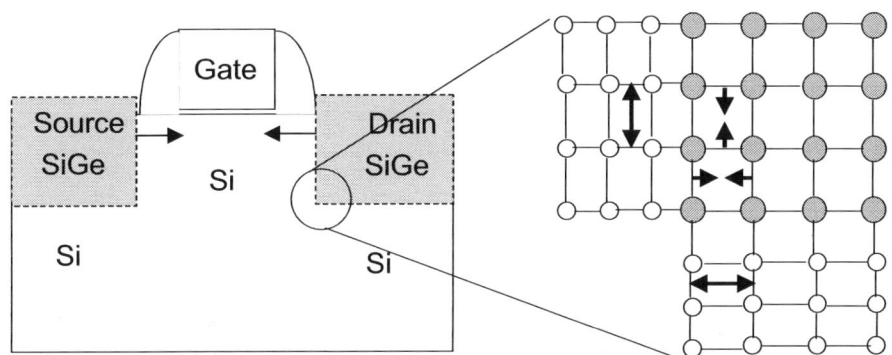

**Fig. 2.40.** The lattice mismatch between SiGe and Si results in channel compressive stress in the $\sigma_{xx}$ direction and tensile stress in the $\sigma_{yy}$ direction. $\sigma_{xx}$ compressive stress improves hole mobility.

The stress in the channel is not just depended on the lattice mismatch; it also depends on the SiGe volume (size) of the source and drain and the proximity of the SiGe to the channel. A larger volume produces more stress. Therefore, the SiGe impact on PMOS device performance depends on the layout, gate spacer width and gate length, similar to the effect created by a nitride film described in the previous section.

SiGe sources and drains were used in multigate MOSFETs in [31,120]. 25% to 40% drive current improvement was seen compared to samples without SiGe. The improvement is less than what has been reported for planar bulk MOSFETs with (001)/<110> orientation. The reduced improvement can be explained by: 1) reduced PR coefficient for (110)/<110> surface/direction and 2) reduced SiGe volume in the source and drain of the multi-gate MOSFETs.

SiC has a smaller lattice constant than silicon. Thus, the lattice mismatch for a SiC source and drain structure produces a tensile stress in the Si channel, which improves the mobility of electrons. SiC has been successfully demonstrated on multi-gate MOSFETs where one percent C in SiC source and drain produced 20% improvement in transistor current. [121]

### 2.5.5 Local Strain from Gate Electrode

The gate electrode of a multigate device wraps around the fin from three sides. If stress can be directly applied from the gate, it will have a large impact on the channel mobility. Furthermore, gate stressor would have reduced gate/active pitch dependence. Some metal deposition processes can produce a large intrinsic stress in the metal film. Tensile stress in the gigapascal range has been reported in metal gates.[27] When such a high tensile metal gate is applied to a multi-gate device, the electron mobility on the (110)/<110> surface/direction can increase by ~100%. Such a tensile stress gate electrode does not change the electron mobility for the (100)/<100> surface/direction, however and it has no impact on hole motility, regardless of the surface orientation and channel direction.[27,75] This approach makes the (110) surface very attractive for MOSFET fabrication.

Metal gate can produce further stress from subsequent processing.[122] Typical metals have thermal expansion coefficients very different from that of silicon. If a SiN capping layer is placed on top of the gate, a limited upward expansion of the metal gate takes place during high temperature processes. The result is a compressive stress in the channel. Since the metal gate is squeezing the silicon fin from both sides, tensile stress is induced in the direction of the current flow, and the electron mobility is increased (Figure 2.41).

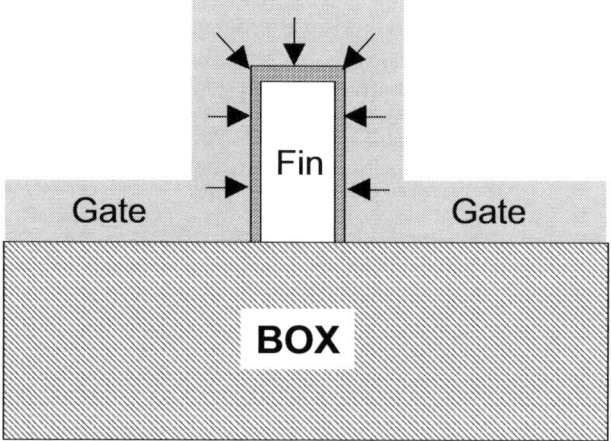

**Fig. 2.41.** Expansion of metal gate electrode during thermal processing exerts a compressive stress in the fin.

It is interesting to note that the mobility improvements brought about by tensile gate stress were reported on long channel devices. This is very different from other process induced local strain, where the improvement in mobility decreases as the channel length is increased. The main reason for this difference is that the stress is applied from the electrode directly to the channel. For this reason, gate induced stress is not layout dependent.

## 2.5.6 Substrate Strain: Strained Silicon on Insulator

This final section on strain engineering focuses on substrate-induced strain. Strain is generated in thin Si layers grown on SiGe. The Si lattice takes on the larger lattice constant of SiGe and is submitted to a biaxial tensile strain. Biaxial tensile strain has been shown to improve electron mobility and, if the strain is large enough, hole mobility can be improved as well.[123] However, the IC industry has not adopted the substrate strain for manufacturing, largely due to the relatively poor quality of the substrates: misfit defects and threading dislocations present in these wafers give rise to unacceptable leakage currents.[124]

Strained silicon on insulator (sSOI) can help to reduce above problems by eliminating the need for a SiGe layer below the strained silicon.[125] Figure 2.42 illustrates the manufacturing process of a sSOI wafer. The process starts with a bulk sSi (strained silicon) wafer, where a thin silicon film is grown on a relaxed SiGe layer. The sSi wafer is first implanted with hydrogen and bonded to a host wafer. Then after applying a layer transfer (ion-cut) process, the SiGe layer is selectively etched away, leaving the strained silicon layer directly on the insulator. The oxide below the strained silicon helps to keep the strain intact. An optional epitaxial process can then be used to increase the strain silicon layer thickness without reducing the strain. This approach eliminates the SiGe buffer underneath strained silicon, therefore reducing defect density and Ge up-diffusion. sSOI films with a thickness in excess of 60nm are commercially available.

Thick sSOI layers have been used to fabricate multi-gate MOSFETs. Since channel conduction is located at the sidewall of the fin, the bi-axial stress of sSi produces tensile stress in $\sigma_{xx}$ direction, compressive stress in $\sigma_{zz}$ direction and tensile stress in $\sigma_{yy}$ direction. For narrow fins, the $\sigma_{yy}$ stress is expected to relax, while the $\sigma_{xx}$ and $\sigma_{yy}$ stresses are intact. Experimental data indicates that the (110)/<110> electron mobility is improved by 60% [39] and the (100)/<100> electron mobility is improved by 30%. The stress from sSOI is additive to those generated by a tensile nitride liner and a metal gate.[27,117-118] Improvements of drive current up to 40% has been observed when sSOI, tensile nitride liner and metal gate electrode are used. The tensile stress in the $\sigma_{xx}$ direction degrades the (110)/<110> p-channel performance, but increases the (100)/<100> hole mobility.

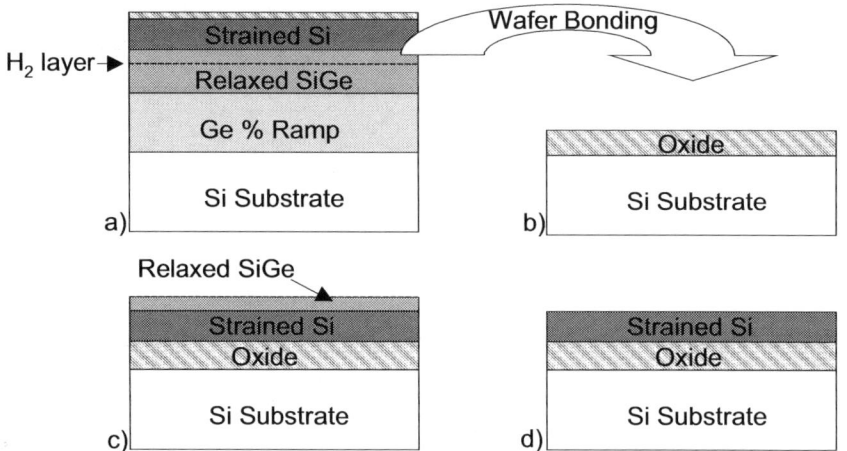

**Fig. 2.42.** Illustration of the fabrication sequence of a sSOI wafer. a) Starting bulk sSi wafer is implanted with $H_2$. b) The sSi wafer is bonded to a host wafer, and c) an ion-cut process is used for layer transfer. d) The remaining SiGe is selectively etched away from the strained Silicon on Insulator material.

## 2.6 Contacts to the Fins

Since the back end of the line process for a MuGFET is the same as for a planar MOSFET, the final section of this chapter reviews the contact formation process for MuGFETs. The unique layout of MuGFET gives three major options for contact: 1) Dumbbell source and drain contact, 2) Saddle contact (individual and slot), and 3) Contact on merged fins. These three options are illustrated in Figure 2.43.

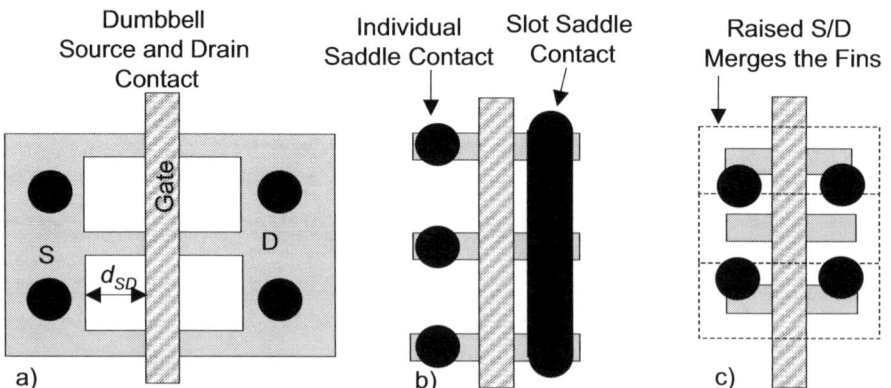

**Fig. 2.43.** Three contact options for MuGFETs: a) contact to the dumbbell source and drain; b) saddle contact to individual fin, or slot contact; c) contact to the merged-fin source and drain.

### 2.6.1 Dumbbell Source and Drain Contact

Dumbbell layout provides the most straightforward contact process since it is identical to a planar MOSFET contact. The dumbbell layout is not area efficient, however. The contact to gate spacing is limited by the dumbbell-to-gate distance $d_{SD}$ (Figure 2.43a).

### 2.6.2 Saddle Contact

The saddle contact is attractive for the following reasons: 1) contact-to-gate spacing can be smaller than for the dumbbell layout, and 2) the saddle contact touches the fin top surface, as well as the fin sidewall surfaces, potentially giving rise to a larger contact area, which can reduce contact resistance (Figure 2.44a). In order to realize a saddle contact, one needs to form a metal silicide on the sidewall of the fin.

The concern with the saddle contact is the alignment of the contact to the active fin. If the contact landing only touches one sidewall of the fin, the contact resistance will be higher than if the contact touches both sides of the fin (Figure 2.44b). Therefore the contact alignment is very tight. This is especially important when the contact size is similar to the fin width in the source and drain area.

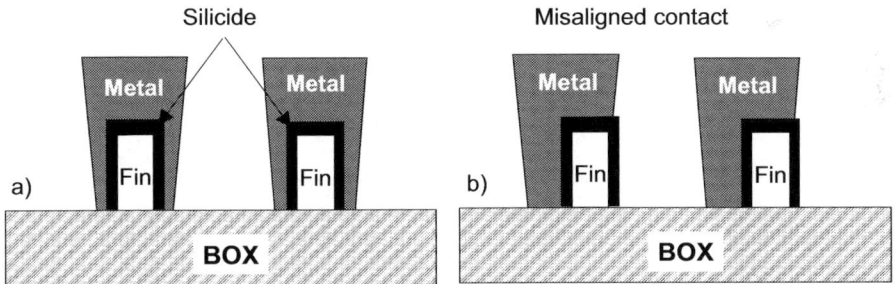

**Fig. 2.44.** Saddle contact: a) perfect contact to fin alignment; b) misaligned contact.

The major issue with saddle the contact is the contact pitch. If the contact pitch cannot be the same as small as the fin pitch, then making a saddle contact to individual fins is not possible. Slot contacts must be used instead. The alignment constraints on slot contacts are less strict than on saddle contacts (Figure 2.44b). However, slot contacts increase the gate-to-contact capacitance due to the extra contact metal between the fins.

### 2.6.3 Contact to Merged Fins

To eliminate the contact to fin pitch matching requirement, one can also use epitaxial growth to merge adjacent fins. This is possible since a tight fin pitch is desirable in order to increase drive current per footprint of a transistor. In addition, raised source and drain structure (RSD) is needed to reduce parasitic source and drain resistance. If the increase in fin width from RSD formation is equal to 50% of the fin spacing, adjacent fins will touch one another. In this configuration, the contact to the source and drain will not touch the fin sidewalls. The topography of the source and drain surface will depend on the source and drain epitaxial faceting and the fin conditions (Section 2.4.3, Figure 2.37). For (110) sidewalls the final fin cross section is either hexagonal or diamond-shaped. The contact will land on a non-planar surface (Figure 2.45a). Some contact overetch is needed to ensure that the contact lands on silicide when the landing area is in the groove. For (100) sidewalls the cross section of the epitaxially grown silicon is rectangular and contacts will land on a flat surface (Figure 2.45b). Contacts to merged fins have been shown to reduce parasitic capacitance, compared to other types of contacts described in previous sections.[15]

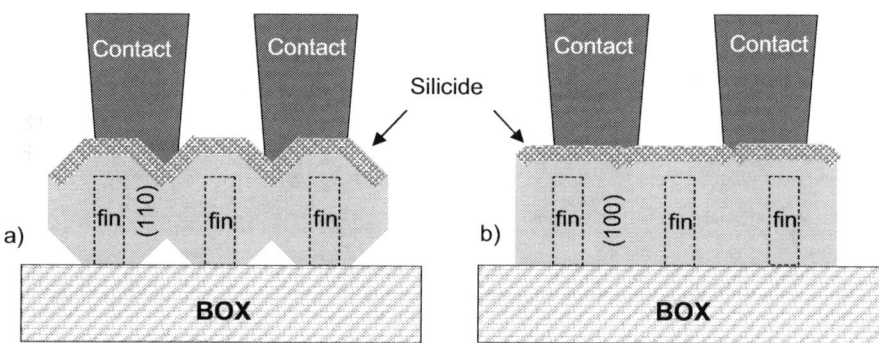

**Fig. 2.45.** Contact to merged fin: a) if the fin sidewall surface is (110), the contact lands on a wavy topography; b) if the fin sidewall surface is (100), the contact lands on a flat surface.

# Acknowledgments

I would like to express my sincere appreciation to Dr. Rinn Cleavelin, Dr. Andrew Marshall, Dr. Allen Bowling, Dr. Linton Salmon and Dr. Hans Stork at Texas Instruments Inc. for their support of this work.

I would also like to extend my appreciation to my colleagues at Infineon AG, Dr. Klaus Schrüfer, Dr. Thomas Schulz, Dr. Klaus von Armin and the authors of chapter 7 of this book: Dr. Gerhard Knobinger, Dr. Christian Pacha and Dr. Michael Fulda.

I would like acknowledge the contributions made by Dr. Guanghua Ma, Dr. Tony Lin and Mr. Che-Hua Hsu from the United Micro-electronic Company; Mr. Gabriel Gebara, Mrs. Jyoti Zaman and Mr. Jason Saulters from the Advanced Technology Development Facility Inc.; and Mr. Paul Patruno, Dr. Ian Cayrefourcq and Dr. Carlos Mazure from SOITEC SA.

And finally, as a non-native English speaker, I cannot thank enough my lovely wife Heidi Ratzlaff Xiong for proofreading this chapter and the never-ending laughter from my daughters Ling and LiLi who kept me motivated.

## References

1   T. Sekigawa, Y. Hayashi: Calculated threshold-voltage characteristics of an XMOS transistor having an additional bottom gate. Solid-State Electronics **27**, 827 (1984)
2   D. Hisamoto, T. Kaga, Y. Kawamoto, E. Takeda: A fully depleted lean-channel transistor (DELTA)-a novel vertical ultra thin SOI MOSFET. *Technical Digest of IEDM*, 833 (1989)
3   D. Hisamoto, Wen-Chin Lee, J. Kedzierski, H. Takeuchi, K. Asano, C. Kuo, E. Anderson, Tsu-Jae King, J. Bokor, Chenming Hu: FinFET-a self-aligned double-gate MOSFET scalable to 20 nm. IEEE Transaction on Electron Devices **47-12**, 2320 (2000)
4   X. Baie, J.P. Colinge, V. Bayot, E. Grivei: Quantum-wire effects in thin and narrow SOI MOSFETs. *Proceedings of IEEE International SOI Conference*, 66 (1995)
5   B. Doyle, B. Boyanov, S. Datta, M. Doczy, S. Hareland, B. Jin, J. Kavalieros, T. Linton, R. Rios, R. Chau: Tri-gate fully depleted CMOS transistors: Fabrication, design and layout. *Symposium on VLSI Technology* 133 (2003)
6   B.S. Doyle, S. Datta, M. Doczy, S. Hareland, B. Jin , J. Kavalieros, T. Linton, A. Murthy, R. Rios, R. Chau: High performance fully-depleted tri-gate CMOS transistors. IEEE Electron Device Letters **24-4**, 263 (2003)
7   J.T. Park, J.P. Colinge, C. H. Diaz: Pi-gate SOI MOSFET. IEEE Electron Device Letters **22-8**, 405 (2001)
8   F.L. Yang, H.Y. Chen, F.C. Cheng, C.C. Huang, C.Y. Chang, H.K. Chiu, C.C. Lee, C.C. Chen H.T. Huang, C.J. Chen, H.J. Tao, Y.C. Yeo, M.S. Liang, C. Hu: 25 nm CMOS Omega FETs, *Technical Digest of IEDM*, 255 (2002)

9. Y.-K. Choi, T.-J. King, C. Hu: Nanoscale CMOS spacer FinFET for the terabit era. IEEE Electron Device Letters **23-1**, 25 (2002).
10. C. Johnson, S. Ogura, J. Riseman, N. Rovedo, J. Shepard: Method for making submicron dimensions in structures using sidewall image transfer techniques. IBM Technical Disclosure Bulletin **26**, 4587 (1984).
11. Yang-Kyu Choi, Leland Chang, P. Ranade, Jeong-Soo Lee, Daewon Ha, S. Balasubramanian, A. Agarwal, M. Ameen, Tsu-Jae King, J. Bokor: FinFET process refinements for improved mobility and gate workfunction engineering. *Technical Digest of the IEDM*, 259 (2002)
12. W. Xiong, G. Gebara, J. Zaman, M. Gostkowski, B. Nguyen, G. Smith, D. Lewis, C. R.Cleavelin, R. Wise, S.Yu, M. Pas, T.J. King, J.P. Colinge: Improvement of FinFET electrical characteristics by hydrogen annealing. IEEE Electron Devices Letters **25-8**, 541 (2004)
13. J. Kedzierski, M. Ieong, E. Nowak, T. S. Kanarsky, Y. Zhang, R. Roy, D. Boyd, D. Fried, H.-S.Wong: Extension and source/drain design for high-performance FinFET devices. IEEE Transaction on Electron Devices, **50-4**, 952 (2003)
14. A. Dixit, A. Kottantharayil, N. Collaert, M. Goodwin, M. Jurczak, K. De Meyer: Analysis of the Parasitic S/D Resistance in Multiple-Gate FETs. IEEE Transactions on Electron Devices **52-6**, 1132 (2005)
15. H. Shang, L. Chang, X. Wang, M. Rooks, Y. Zhang, B. To, K. Babich, G. Totir, Y. Sun, E. Kiewra, M. Ieong, W. Haensch: Investigation of FinFET device for 32nm technologies and beyond. *Symposium on VLSI Technology*, 66 (2006)
16. S.M. Sze *Physics of semiconductor devices*. J. Wiley & Sons (1981)
17. R.-H. Yan, A. Ourmazd, K. F. Lee: Scaling the Si MOSFET: From Bulk to SOI to Bulk. IEEE Transaction on Electron Devices **39-7**, 1704 (1992)
18. S.A. Parke, J.E. Moon, H.C. Wann, P.K. Ko, C. Hu: Design for Suppression of Gate-Induced Drain Leakage in LDD MOSFET's Using a Quasi-Two-Dimensional Analytical Model. IEEE Transactions on Electron Devices **39-7**, 1694 (1992)
19. Y. Liu, K. Ishii, T. Tsutsumi, M. Masahara, E. Suzuki: Ideal Rectangular Cross-Section Si-Fin Channel Double-Gate MOSFETs Fabricated Using Orientation-Dependent Wet Etching. IEEE Electron Device Letters **24-7**, 484 (2003)
20. K. Suzuki, T. Tanaka, Y. Tosaka, H. Horie, Y. Arimoto: Scaling theory for double-gate SOI MOSFETs. IEEE Transaction Electron Devices **40-12**, 2326 (1993)
21. K.K. Young: Short-channel effect in fully-depleted SOI MOSFETs. IEEE Transaction on Electron Devices **36-2**, 399 (1989)
22. J. Kedzierski, D.M. Fried, E.J. Nowak, T. Kanarsky, J.H. Rankin, H. Hanafi, W. Natzle, D. Boyd, Y. Zhang, R.A. Roy, J. Newbury, C. Yu, Q. Yang, P. Saunders, C.P. Willets, A. Johnson, S.P. Cole, H. E. Young, N. Carpenter, D. Rakowski, B.A. Rainey, P.E. Cottrell, M. Ieong, H.-S. P. Wong: High-performance symmetric-gate and CMOS-compatible Vt asymmetric-gate FinFET devices. *Technical Digest of IEDM*, 437 (2001)

23  J.P. Colinge: Multiple-gate SOI MOSFETs. Solid-State Electronics **48-6**, 897 (2004)
24  K.G. Anil, K. Henson, S. Biesemans, N. Collaert: Short-channel effect in fully-depleted SOI MOSFET's. *Proceedings of ESSDERC*, 139 (2003)
25  T. Ludwig, I. Aller, V. Gernhoefer, J. Keinert, E. Nowak, R.V. Joshi, A. Mueller, S. Tomaschko: FinFET technology for future microprocessors. *Proceedings of IEEE International SOI Conference*, 33 (2003)
26  S.H. Kim, J.G. Fossum, V.P. Trivedi: Bulk Inversion in FinFETs and Implied Insights on Effective Gate Width. IEEE Transactions on Electron Devices **52-9**, 1904 (2005)
27  W. Xiong, K. Shin, C.R. Cleavelin, T. Schulz, K. Schruefer, I. Cayrefourcq, M. Kennard, C. Mazure, P. Patruno, T.J. King: FinFET Performance Enhancement with Tensile Stress Metal Gate Electrode and Strained Silicon on Insulator Substrate. *Proceedings of Device Research Conference*, 39 (2006)
28  J. Kedzierski, E. Nowak, T. Kanarsky, Y. Zhang, D. Boyd, R. Carruthers, C. Cabral, R. Amos, C. Lavoie, R. Roy, J. Newbury, E. Sullivan, J. Benedict, P. Saunders, Ke. Wong, D. Canaperi, M. Krishnan, K.L. Lee, B.A. Rainey, D. Fried, P. Cottrell, H.S.P. Wong, M. Ieong, W. Haensch: Metal-gate FinFET and fully-depleted SOI devices using total gate silicidation. *Technical Digest of IEDM,* 247 (2002)
29  Fu-Liang Yang, Di-Hong Lee, Hou-Yu Chen, Chang-Yun Chang, Sheng-Da Liu, Cheng-Chuan Huang, Tang-Xuan Chung, Hung-Wei Chen, Chien-Chao Huang, Yi-Hsuan Liu, Chung-Cheng Wu, Chi-Chun Chen, Shih-Chang Chen, Ying-Tsung Chen, Ying-Ho Chen, Chih-Jian Chen, Bor-Wen Chan, Peng-Fu Hsu, Jyu-Horng Shieh, Han-Jan Tao, Yee-Chia Yeo, Yiming Li, Jam-Wem Lee, Pu Chen, Mong-Song Liang, Chenming Hu: 5nm-Gate Nanowire FinFET. *Symposium on VLSI Technology*, 196 (2004)
30  N. Collaert, M. Demand, I. Ferain, J. Lisoni, R. Singanamalla, P. Zimmerman, Y. Yim, T. Schram, G. Mannaert, M. Goodwin, J. Hooker, F. Neuilly, M. Kim, K. De Meyer, S. De Gendt, W. Boullart, M. Jurczak, S. Biesemans, "Tall triple-gate device with TiN/HfO2 gate stack. *Symposium on VLSI Technology*, 108 (2005)
31  J. Kavalieros, B. Doyle, S. Datta, G. Dewey, M. Doczy, B. Jin, D. Lionberger, M. Metz, W. Rachmady, M. Radosavljevic, U. Shah, N. Zelick, R. Chau: Tri-Gate Transistor Architecture with High-k Gate Dielectrics, Metal Gates and Strain Engineering. *Symposium on VLSI Technology*, 62 (2006)
32  Yang-Kyu Choi, Tsu-Jae King, Chenming Hu: A Spacer Patterning Technology for Nanoscale CMOS. IEEE Transactions on Electron Devices **49-3**, 436 (2002)
33  Y.K. Choi, J.S. Lee, J. Zhu, G.A. Somorjai, L.P. Lee, J. Bokor: Sublithographic nanofabrication technology for nanocatalysts and DNA chips. Journal of Vacuum Science Technology B **21-6**, 2951 (2003)

34  S. Takagi, A. Toriumi, M. Iwase, H. Tango: On the Universality of Inversion Layer Mobility in Si MOSFET's: Part II-Effects of Surface Orientation. IEEE Transactions on Electron Devices **41-12**, 2362 (1994)
35  T. Sato, Y. Takeishi, H. Hara: Effects of Crystalgraphical Orientation on Mobility, Surface State Density and Noise in p-Type Inversion Layer on oxidized Si Surfaces. Japanese Journal of Applied Physics **8-5**, 588 (1969)
36  T. Sato, Y. Takeishi, H. Hara: Mobility Anisotropic of Electron in Inversion Layers on Oxidized Silicon Surfaces. Physical Review B **4-6**, 1950 (1971)
37  M. Yang, M. Ieong, L. Shi, K. Chan, V. Chan, A. Chou, E. Gusev, K. Jenkins, D. Boyd, Y. Ninomiya, D. Pendleton, Y. Surpris, D. Heenan, J. Ott, K. Guarini, C. D'Emic, M. Cobb, P. Mooney, B. To, N. Rovedo, J. Benedict, R. Mo, H. Ng: High Performance CMOS Fabricated on Hybrid Substrate with Different Crystal Orientations. *Technical Digest of IEDM*, 18.7.1 (2003)
38  Leland Chang, Yang-Kyu Choi, Daewon Ha, Pushkar Ranade, Shiying Xiong, Jeffrey Bokor, Chenming Hu, Tsu-Jae King: Extremely Scaled Silicon Nano-CMOS Devices. Proceedings of the IEEE, **91-11**, 1860 (2003)
39  W. Xiong, C.R. Cleavelin, P. Kohli, C. Huffman, T. Schulz, K. Schruefer, G. Gebara, K. Mathews, P. Patruno, I. Cayrefourcq, M. Kennard, C. Mazure, K. Shin, T.-J. King Liu: Impact of Strained Silicon on Insulator (sSOI) Substrate on FinFET Mobility. IEEE Electron Device Letters **27-7**, 612 (2006)
40  E. Landgraf, W. Rosner, M. Staadele, L. Dreeskornfeld, J. Hartwich, F. Hofmann, J. Kretz, T. Lutz, R.J. Luyken, T. Schulz, M. Specht, L. Risch: Influence of crystal orientation and body doping on trigate transistor performance. Solid State Electronics **50-1**, 38 (2006)
41  Z. Guo, S. Balasubramanian, R. Zlatanovici, T.-J. King, B. Nikolic: FinFET-Based SRAM Design. *International Symposium on Low Power Electronics and Design ISLPED*, 2 (2005)
42  Y.X. Liu, E. Sugimata, M. Masahara, K. Endo, K. Ishii, T. Matsukawa, H. Takashima, H. Yamauchi, E. Suzuki: Comparative study on effective electron mobility in FinFETs with a (111) channel surface fabricated by wet and dry etching processes. Microelectronic Engineering, **80-17**, 390 (2005)
43  Weize Xiong, J.W. Park, J.P. Colinge: Corner Effect in Multiple-Gate SOI MOSFETs. *IEEE International SOI Conference Proceedings*, 111 (2003)
44  J. Preteta, S. Cristoloveanu, A. Vandooren, L. Mathew, J. Jomaah, B. -Y. Nguyen: Coupling effects and channels separation in FinFETs. Solid State Electronics **48-4**, 535 (2004)
45  S. Kuppurao, H.S. Joo, G. Miner: In Situ Steam Generation: A New Rapid Thermal Oxidation Techniques. Solid-State Technology **43-7**, 233 (2000)
46  T.Y. Luo, M. Laughery, G.A. Brown, H.N. Al-Shareef, V.H.C. Watt, A. Karamcheti, M.D. Jackson, H.R. Huff: Effect of $H_2$ Content on Reliability of Ultrathin In-Situ Steam Generated (ISSG) $SiO_2$. IEEE Electron Device Letters **21- 9**, 430 (2000)
47  R.J. Zaman, W. Xiong, A. Franke, R. Quintanilla, T. Schulz, N. Chaudhary, C.R. Cleavelin, R. Wise, M. Pas, S. Yu, K. Schruefer, S. Banerjee: A

Detailed Study of Hydrogen Annealing Process Effects on silicon Nano-structures. *Proceedings of Spring MRS Conference,* J3.1 (2005)

48 T. Park, S. Choi, D. H. Lee, J.R. Yoo, B.C. Lee, J.Y. Kim, C.G. Lee, K K. Chi, S.H. Hong, S.J. Hyun, Y.G. Shin, J.N. Han, I.S. Park, U.I. Chung, J.T. Moon, E. Yoon, J.H. Lee: Fabrication of body-tied FinFETs (Omega MOSFETs) using bulk Si wafers. *Symposium on VLSI Technology,* 135 (2003)

49 R. Katsumata, N. Tsuda, J. Idebuchi, M. Kondo, N. Aoki, S. Ito, K. Yahashi, T. Satonaka, M. Morikado, M. Kim, M. Kido, T. Tanaka, H. Aochi, T. Hamannoto: Fin-Array-FET on bulk silicon for sub-100nm fin trench capacitor DRAM. *Symposium on VLSI Technology,* 61, 2003

50 C.H. Lee, J.M. Yoon, C.Lee, H.M. Yang, K.N. Kim, T.Y. Kim, H.S. Kang, Y.J. Ahn, D. Park, K. Kim: Novel body tied FinFET cell array transistor DRAM with negative word line operation for sub 60 nm technology and beyond. *Symposium on VLSI Technology,* 130 (2004)

51 T. Park, D. Park, J.H. Chung, E.J. Yoon, S.M. Kim, H.J. Cho, J.D. Choe, J.H. Choi, B.M. Yoon, J.J. Han, B.H. Kim, S. Choi, K. Kim, E. Yoon, J.H. Lee: PMOS body-tied FinFET (Omega MOSFET) characteristics. *Proceedings of Device Research Conference,* 33 (2003)

52 Tai-Su Park, Hye Jin Cho, Jeong Dong Choe, Il Hwan Cho, Donggun Park, Euijoon Yoon, Jong Ho Lee: Characteristics of Body-Tied Triple-Gate pMOSFETs. IEEE Electron Device Letters **25-12**, 798 (2004)

53 Tai-Su Park, Siyoung Choi, Deok-Hyung Lee, U-In Chung, Joo Tae Moon, Euijoon Yoon, Jong-Ho Lee: Body-tied triple-gate NMOSFET fabrication using bulk Si wafer. Solid-State Electronics **49-3**, 377 (2005)

54 C.P. Auth, J.D. Plummer: Scaling theory for cylindrical, fully-depleted, surrounding-gate MOSFETs. IEEE Electron Device Letters **18-2**, 74 (1997)

55 N. Singh, A. Agarwal, L.K. Bera, T.Y. Liow, R.Yang, S.C. Rustagi, C.H. Tung, R. Kumar, G.Q. Lo, N. Balasubramanian, D.L. Kwong: High-Performance Fully Depleted Silicon Nanowire (Diameter ≤ 5 nm) Gate-All-Around CMOS Devices. IEEE Electron Device Letters, **27-5**, 383 (2006)

56 C. T. Black: Self-aligned self assembly of multi-nanowire silicon field effect transistors. Applied Physics Letters **87**, 163116 (2005)

57 W. Xiong, C. R. Cleavelin, T. Schulz, K. Schrüfer, P. Patruno,J.P. Colinge: MuGFET CMOS process with midgap gate material. *Abstracts of NATO International Advanced Research Workshop, Nanoscaled Semiconductor-on-Insulator Structures and Devices,* 95 (2006)

58 P. Francis, A. Terao, D. Flandre, F. Van de Wiele: Modeling of Ultrathin Double-Gate nMOS/SOI Transistors. IEEE Transactions on Electron Devices **41-5**, 715 (1994)

59 Y. Tahara, Y. Omura: Empirical Quantitative Modeling of Threshold Voltage of Sub-50-nm Double-Gate SOI MOSFETs. *Proceedings of International Conference on Solid State Device and Materials,* 618 (2005)

60 Y. Tahara, Y. Omura: Empirical Quantitative Modeling of Threshold Voltage of Sub-50-nm Double-Gate Silicon-on-Insulator Metal–Oxide–

Semiconductor Field-Effect Transistor. *Japanese Journal of Applied Physics* **45-4**, 3074 (2006)

61 L. Chang, S. Tang, T.J. King, J. Bokor, C. Hu: Gate length scaling and threshold voltage control of double-gate MOSFETs. *Technical Digest of IEDM*, 719 (2000)

62 L. Mathew, M. Sadd, B.E. White, A. Vandooren, S. Dakshina-Murthy*, J. Cobb, T. Stephens, R.. Mora, D. Pham, J. Conner, T. White, Z. Shi, A.V-Y Thean, A. Barr, M. Zavala, J. Schaeffer, M.J. Rendon, D. Sing, M. Orlowski, B.-Y. Nguyen, J. Mogab: Finfet with isolated n+ and p+ gate regions strapped with metal and polysilicon. *Proceedings of IEEE International SOI Conference,* 109 (2003)

63 Nick Lindert, Makoto Yoshida, Clement Wann, Chenming Hu: Comparison of GIDL in p+-poly PMOS and ns-poly PMOS Devices. IEEE Electron Device Letters **17- 6**, 285 (1996)

64 Yang-Kyu Choi Daewon Ha, Tsu-Jae King, J. Bokor: Investigation of gate-induced drain leakage (GIDL) current in thin body devices: single-gate ultra-thin body, symmetrical double-gate, and asymmetrical double-gate MOSFETs. Japanese Journal of Applied Phys. **42-4**, 2073 (2003)

65 C.C. Hobbs, L.R.C. Fonseca, A. Knizhnik, V. Dhandapani, S.B. Samavedam, W.J. Taylor, J.M. Grant, L.G. Dip, D.M. Triyoso, R.I. Hegde, D.C. Gilmer, R. Garcia, D. Roan, M. Luke Lovejoy, R.S. Rai, E.A. Hebert, Hsing-Huang Tseng, S.G.H. Anderson, B.E. White, P.J. Tobin: Fermi-level pinning at the polysilicon/metal oxide interface—Part I. IEEE Transaction on Electron Devices **51-6**, 971 (2004)

66 C.C. Hobbs, L.R.C. Fonseca, A. Knizhnik, V. Dhandapani, S.B. Samavedam, W.J. Taylor, J.M. Grant, L.G. Dip, D.M. Triyoso, R.I. Hegde, D.C. Gilmer, R. Garcia, D. Roan, M. Luke Lovejoy, R.S. Rai, E.A. Hebert, Hsing-Huang Tseng, S.G.H. Anderson, B.E. White, P.J. Tobin: Fermi-level pinning at the polysilicon/metal-oxide interface—Part II. IEEE Transaction on Electron Devices **51-6**, 978 (2004)

67 W. Xiong, C.R. Cleavelin, Che-Hua Hsu, M. Ma, K. Schruefer, K. Von Armin, T. Schulz, I. Cayrefourcq, C. Mazure, P. Patruno, M. Kennard, K. Shin, X. Sun, T.J. King Liu, K. Cherkaoui, J.P Colinge: Intrinsic advantages of SOI multiple-gate MOSFET (MuGFET) for low-power applications. Electrochemical Society Transactions **6-4**, 59 (2007)

68 J. Kedzierski, E. Nowak, T. Kanarsky, Y. Zhang, D. Boyd, R. Carruthers, C. Cabral, R. Amos, C. Lavoie, R. Roy, J. Newbury, E. Sullivan, J. Benedict, P. Saunders, K. Wong, D. Canaperi, M. Krishnan, K.L. Lee, B.A. Rainey, D. Fried, P. Cottrell, H.S. Philip Wong, M. Ieong, W. Haensch: Metal-gate FinFET and fully-depleted SOI devices using total gate silicidation. *Technical Digest of IEDM,* 247 (2002)

69 D. Aimé, B. Froment, F. Cacho, V. Carron, S. Descombes, Y. Morand, N. Emonet, F. Wacquant, T. Farjot, S. Jullian, C. Laviron, M. Juhel, R. Pantel, R. Molins, D. Delille, A. Halimaoui, D. Bensahel, A. Souifi: Workfunction tuning through dopant scanning and related effects in Ni fully silicided gate for sub-45nm nodes CMOS. *Technical Digest of IEDM,* 87 (2004)

70  W.P. Maszara, Z. Krivokapic, P. King, J. Goo, M. Lin, "Transistors with dual workfunction metal gates by single full silicidation (FUSI) of polysilicon gates. *Technical Digest of IEDM*, 367 (2002)

71  S. Yu, J.P. Lu, F. Mehrad, H. Bu, A. Shanware, M. Ramin, M. Pas, M.R. Visokay, S. Vitale, S.-H. Yang, P. Jiang, L. Hall, C. Montgomery, Y. Obeng, C. Bowen, H. Hong, J. Tran, R. Chapman, S. Bushman, C. Machala, J. Blatchford, R. Kraft, L. Colombo, S. Johnson, B. McKee: 45-nm node NiSi FUSI on nitrided oxide bulk CMOS fabricated by a novel integration process. *Technical Digest of IEDM*, 221 (2005)

72  T. Nabatame, M. Kadoshima, K. Iwamoto, N. Mise, S. Migita, M. Ohno, H. Ota, N. Yasuda, A. Ogawa, K. Tominaga, H. Satake, A. Toriumi: Partial silicides technology for tunable workfunction electrodes on high-k gate dielectrics—Fermi level pinning controlled PtSix for HfOx(N) pMOSFET. *Technical Digest of IEDM*, 83 (2004)

73  A. Lauwers, A. Veloso, T. Hoffmann, M.J.H. van Dal, C. Vrancken, S. Brus, S. Locorotondo, J.-F. de Marneffe, B. Sijmus, S. Kubicek, T. Chiarella, M.A. Pawlak, K. Opsomer, M. Niwa, R. Mitsuhashi, K.G. Anil, H.Y. Yu, C. Demeurisse, R. Verbeeck, M. de Potter, P. Absil, K. Maex, M. Jurczak, S. Biesemans, J.A. Kittl: CMOS integration of dual workfunction phase controlled Ni FUSI with simultaneous silicidation of NMOS (NiSi) and PMOS (Ni-rich silicide) gates on HfSiON. *Technical Digest of IEDM*, 400 (2005).

74  J.D. Chen, H.Y. Yu, M.F. Li, D.L. Kwong, M.J.H. van Dal, J.A. Kittl, A. Lauwers, P. Absil, M. Jurczak, S. Biesemans: Yb-Doped Ni FUSI for the n-MOSFETs Gate Electrode Application. IEEE Electron Device Letters **27-3** 160 (2006)

75  C.Y. Kang, R. Choi, S.C. Song, K. Choi, B.S. Ju, M.M. Hussain, B.H. Lee1, G. Bersuker, C. Young, D. Heh, P. Kirsch, J. Barnet, J.W. Yang, P. Zeitzoff, W. Xiong, H.H Tseng, R. Jammy: A novel electrode induced strain engineering for high performance SOI FinFET utilizing Si (110) channel for both nMOS and pMOS. *Technical Digest of IEDM*, 1-4 (2006)

76  C.Y. Kang, R. Choi, S.C. Song, B.S. Ju, M.M. Hussain, B.H. Lee, J.W. Yang, P. Zeitzoff, D. Pham, W. Xiong, H.H Tseng: Effects of ALD TiN Metal Gate Thickness on Metal Gate /High-k Dielectric SOI FinFET Characteristics. *Proceedings of IEEE International SOI Conference*, 135 (2006)

77  D. H. Luan, H. N. Alshareef, H. R. Harris, H. C. Wen, K. Choi, Y. Senzaki, P. Majhi, and B-H. Leed: Evaluation of titanium silicon nitride as gate electrodes for complementary metal-oxide semiconductor. Applied Physics Letters **88-4**, 142113 (2006)

78  Z.B. Zhang, S.C. Song, K. Choi, J.H. Sim, P. Majhi, B.H. Lee: An Integratable Dual Metal Gate/High-k CMOS Solution for FD-SOI and MuGFET Technologies. *Proceedings of IEEE International SOI Conference*, 157 (2005)

79 Daewon Ha, H. Takeuchi, Y.K. Choi, T.J. King, W.P. Bai, D.L. Kwong, A. Aganval, M. Ameen: Molybdenum-Gate $HfO_2$ CMOS FinFET Technology. *Technical Digest of IEDM*, 643 (2004)

80 P. Ranade; Yang-Kyu Choi; Daewon Ha; A. Agarwal, M. Ameen, T.J. King: Tunable workfunction molybdenum gate technology for FDSOI-CMOS. *Tech. Digest of IEDM* 363 (2002)

81 J. Chen, T.Y. Chan, I.C. Chen, P.K. Ko, C. Hu: Subbreakdown Drain Leakage Current in MOSFET. IEEE Electron Device Letters **8-11**, 515 (1987)

82 S. Banerjee, J. Coleman, W. Richardson, A. Shah: A band-to-band tunneling effect in the trench transistor cell. *Proceedings Symposium on VLSI Technology*, 97 (1987)

83 M. Takayanagi, S. Iwabuchi: Theory of band-to-band tunneling under nonuniform electric fields for subbreakdown leakage currents. IEEE Transaction on Electron Devices **38-6** 1425 (1991)

84 S.A. Parke, J.E. Moon, H.J.C. Wann, P.K.Ko, C. Hu: Design for Suppression of Gate-Induced Drain Leakage in LDD MOSFET's Using a Quasi-Two-Dimensional Analytical Model. IEEE Transaction on Electron Devices **39-7**, 1694 (1992)

85 T.Y. Chan, J. Chen, P.K. Ko, C. Hu: The impact of gate-induced drain-leakage current on MOSFET scaling. *Technical Digest of IEDM*, 718 (1987)

86 H.J. Wann, P.K. Ko, C. Hu: Gate-induced band-to-band tunneling leakage current in LDD MOSFET's. *Technical Digest of IEDM*, 147 (1992)

87 Yang-Kyu Choi, Daewon Ha, T.J. King, J. Bokor: Investigation of Gate-Induced Drain Leakage (GIDL) Current in Thin Body Devices: Single-Gate Ultra-Thin Body, Symmetrical Double-Gate, and Asymmetrical Double-Gate MOSFETs. Japanese Journal of Applied Physics **42-4**, 2073 (2003)

88 J.G. Fossum, M.M. Chowdhury, V.P. Trivedi, T.J. King, Y.K. Choi, J. An, B. Yu: Physical insights on design and modeling of nanoscale FinFETs. *Technical Digest of IEDM*, 679 (2003)

89 V. Trivedi, J.G. Fossum, M.M. Chowdhury: Nanoscale FinFETs with gate-source/drain underlap. IEEE Trans. Electron Devices **52-1**, 56 (2005)

90 Ji-Woon Yang, P.M. Zeitzoff, Hsing-Huang Tseng: Highly manufacturable double-gate FinFET with gate-source/drain underlap. IEEE Transactions on Electron Devices **54-6**, 1464 (2007)

91 D.M. Fried, J.S. Duster, K.T. Kornegay: High-performance p-type independent-gate FinFET. IEEE Electron Device Letters, **25-4**, 199 (2004)

92 Y.X. Liu, M. Masahara, K. Ishii, T. Tsutsumi, T. Sekigawa, H. Takashima, H. Yamauchi, E. Suzuki: Flexible threshold voltage FinFETs with independent double gates and an ideal rectangular cross-section Si-Fin channel. *Technical Digest of IEDM*, 986 (2003)

93 L. Mathew: Multiple independent gates FET (MIGFET) - Multi-Fin RF mixer architecture, three independent gates (MIGFET-T) operation and temperature characteristics. *Symposium on VLSI Technology*, 200 (2005)

94 M. Masahara, R. Surdeanu, L. Witters, G. Doornbos, V.H. Nguyen, G. Van den bosch, C. Vrancken, M. Jurczak, S. Biesemans: Experimental

Investigation of Optimum Gate Workfunction for CMOS Four-Terminal Multigate MOSFETs (MUGFETs). IEEE Transactions on Electron Devices **54-6**, 1431 (2007)
95  H. Lim, J.G. Fossum: Threshold voltage of thin-film silicon-on-insulator MOSFETs. IEEE Transaction Electron Devices **30-10**, 1244 (1983)
96  Weimin Zhang, J.G. Fossum, L. Mathew, Yang Du: Physical Insights Regarding Design and Performance of Independent-Gate FinFETs. IEEE Transactions on Electron Devices **52-10**, 2198 (2005)
97  D. Pham, L. Larson, Ji-Woon Yang: FINFET Device Junction Formation Challenges. *Proceedings of International workshop on Junction technology*, 73 (2006)
98  S. Radovanov, L. Godet, R. Dorai, Z. Fang, B.W. Koo, C. Cardinaud, G. Cartry, D. Lenoble, A. Grouillet: Ion energy distributions in a pulsed plasma doping system. Journal of Applied Physics **98** 113307 (2005)
99  D. Alvarez, J. Hartwich, M. Fouchier, P. Eyben, W. Vandervorst: Sub-5-nm-spatial resolution in scanning spreading resistance microscopy using full-diamond tips. Applied Physics Letters **82-11**, 1724 (2003)
100 D. Lenoble   K. Anil   A. De Keersgieter   P. Eybens   N. Collaert   R. Rooyackers   S. Brus   P. Zimmerman   M. Goodwin   D. Vanhaeren   W. Vandervorst   S. Radovanov   L. Godet   C. Cardinaud   S. Biesemans   T. Skotnicki   M. Jurczak: Enhanced Performance of PMOS MUGFET via integration of conformal plasma-doped source/drain extensions. *Symposium on VLSI Technology*, 168 (2006)
101 A. Dixit, K.G. Anil, R. Rooyackers, F. Leys, M. Kaiser, N. Collaert, K. De Meyer, M. Jurczak, S. Biesemans: Minimization of specific contact resistance in multiple gate NFETs by selective epitaxial growth of Si in the HDD regions. Solid State Electronics **50-4**, 587 (2006)
102 A. Kaneko, A. Yagishita, K. Yahashi, T. Kubota, M. Omura, K. Matsuo, I. Mizushima, K. Okano, H. Kawasaki, S. Inaba, T. Izumida, T. Kanemura, N. Aoki, K. Ishimaru, H. Ishiuchi, K. Suguro, K. Eguchi, Y. Tsunashima: Sidewall Transfer Process and Selective Gate Sidewall Spacer Formation Technology for Sub-15nm FinFET with Elevated Source/Drain Extension. *Technical Digest of IEDM*, 844 (2005)
103 D. Colman, R.T. Bate, J.P. Mize: Mobility anisotropy and piezoresistance in silicon p-type inversion layers. Journal of Applied Physics **39-4**, 1923 (1968)
104 G. Dorda, I. Eisele, H. Gesch: Many-valley interactions in n-type silicon inversion layers. Physical Review B (Solid State) **17-4**, 1785 (1978)
105 R. Craddock: Sensors based on silicon strain gauges. *IEE Colloq. Sens. via Strain*, 5/1 (1993)
106 H.M. Manasevit, I.S. Gergis, A.B. Jones: Electron mobility enhancement in epitaxial multilayer Si–Si$1-x$Ge$x$ alloy films on (100) Si. Applied Physics Letters **41-5**, 464 (1982).
107 R. People, J.C. Bean, D.V. Lang, A.M. Sergent, H.L. Stormer, K.W. Wecht, R.T. Lynch, K. Baldwin: Modulation doping in Ge$_x$Si$_{1-x}$/Si strained layer heterostructures. Applied Physics Letters **45-11**, 1231 (1984)

108 P. R. Chidambaram, C. Bowen, S. Chakravarthi, C. Machala, R. Wise: Fundamentals of Silicon Material Properties for Successful Exploitation of Strain Engineering in Modern CMOS Manufacturing. IEEE Transactions on Electron Devices **53-5**, 944 (2006)

109 S.E. Thompson, Guangyu Sun, Youn Sung Choi, T. Nishida: Uniaxial-Process-Induced Strained-Si: Extending the CMOS Roadmap: IEEE Transactions on Electron Devices **53-5**, 1010 (2006)

110 M. Horstmann, A. Wei, T. Kammler, J. Höntschel, H. Bierstedt, T. Feudel, K. Frohberg, M. Gerhardt, A. Hellmich, K. Hempel, J. Hohage, P. Javorka, J. Klais, G. Koerner, M. Lenski, A. Neu, R. Otterbach, P. Press, C. Reichel, M.Trentsch, B. Trui, H. Salz, M. Schaller, H.J. Engelmann, O. Herzog, H. Ruelke, P. Hübler, R. Stephan, D. Greenlaw, M. Raab, N. Kepler: Integration and Optimization of Embedded-SiGe, Compressive and Tensile Stressed Liner Films, and Stress Memorization in Advanced SOI CMOS Technologies. *Technical Digest of IEDM*, 233 (2005)

111 Kyoungsub Shin, W. Xiong, C.Y. Cho, C.R. Cleavelin, T. Schulz, K. Schruefer, P. Patruno, L. Smith, T.J. King Liu: Study of Bending-Induced Strain Effects on MuGFET Performance. IEEE Electron Device Letters **27-8**, 671 (2006)

112 K. Uchida, R. Zednik, C.H. Lu, H. Jagannathan, J. McVittie, P. C. McIntyre, Y. Nishi, "Experimental study of biaxial and uniaxial strain effects on carrier mobility in bulk and ultrathin-body SOI MOSFETs: *Technical Digest of IEDM*, 229 (2004)

113 C. Smith: Piezoresistance effect in germanium and silicon. Phys. Rev. **94-1** 42 (1954)

114 G. Eneman, P. Verheyen, A. De Keersgieter, M. Jurczak, K. De Meyer: Scalability of Stress Induced by Contact-Etch-Stop Layers: A Simulation Study. IEEE Transactions on Electron Devices **54-6**, 1446 (2007)

115 K. Shin, C.O. Chui, T.J. King: Dual stress capping layer enhancement study for hybrid orientation finFET CMOS technology. *Technical Digest of IEDM*, 988 (2005)

116 N. Collaert, A. De Keersgieter, K.G. Anil, R. Rooyackers, G. Eneman, M. Goodwin, B. Eyckens, E. Sleeckx, J.F. de Marneffe, K. De Meyer, P. Absil, M. Jurczak, S. Biesemans: Performance improvement of tall triple gate devices with strained SiN layers. IEEE Electron Device Letters **26-11**, 820 (2005)

117 N. Collaert, R. Rooyackers, F. Clemente, P. Zimmerman, I. Cayrefourcq, B. Ghyselen, K. San, B. Eyckens, M. Jurczak, S. Biesemans: Performance Enhancement of MUGFET Devices Using Super Critical Strained-SOI (SC-SSOI) and CESL. *Symposium on VLSI Technology,* 52 (2006)

118 Che-Hua Hsu, Weize Xiong, Chien-Ting Lin, Yao-Tsung Huang, M. Ma, C.R Cleavelin, P. Patruno, M. Kennard, I. Cayrefourcq, Kyoungsub Shin, Tsu-Jae King Liu: Multi-Gate MOSFETs with Dual Contact Etch Stop Liner Stressors on Tensile Metal Gate and Strained Silicon on Insulator (sSOI). *Symposium on VLSI Technology, Systems and Applications,* T115 (2007)

119 Yee-Chia Yeo, Jisong Sun, Eng Hong Ong: Strained Channel Transistor Using Strain Field Induced By Source and Drain Stressors. *Proceedings of MRS*, B10.4.1 (2004)
120 N. Verheyen, N. Collaert, R. Rooyackers, R. Loo, D. Shamiryan, A. De Keersgieter, G. Eneman, F. Leys, A Dixit. M.Goodwin, 4 Y.S.Yim, M. Caymax, 1 K. De Meyer, P. Absil, M. Jurczak, S. Biesemans: 25% Drive Current improvement for p-type Multiple Gate FET (MuGFET) Devices by the Introduction of Recessed Si0.8Ge0.2 in the Source and Drain Regions. *Symposium on VLSI Technology,* 194 (2005)
121 Tsung-Yang Liow, Kian-Ming Tan, R.T.P. Lee, Anyan Du, Chih-Hang Tung, G.S. Samudra, Won-Jong Yoo, N. Balasubramanian, Yee-Chia Yeo: Strained N-Channel FinFETs with 25 nm Gate Length and Silicon-Carbon Source/Drain Regions for Performance Enhancement. *Symposium on VLSI Technology*, 56 (2006)
122 Kian-Ming Tan, Tsung-Yang Liow, Rinus T. P. Lee, Chih-Hang Tung, Ganesh S. Samudra, Won-Jong Yoo: Drive-Current Enhancement in FinFETs Using Gate-Induced Stress. IEEE Electron Device Letters **27-9**, 769 (2006)
123 F. M. Bufler, W. Fichtner: Hole and electron transport in strained Si: Orthorhombic versus biaxial tensile strain. Applied Physics Letters, **81-1**, 82 (2002)
124 H.C. Wang, Y. Wang, S. Chen, C. Ge, S. M. Ting, J. Kung, R. Hwang, H. Chiu, L.C. Sheu, P. Tsai, L. Yao, S. Chen, I. Tao, Y. Yeo, W. Lee, C. Hu: Substrate-strained silicon technology: process integration. *Technical Digest of IEDM*, 61 (2003)
125 C. Mazuré, I. Cayrefourcq: Status of Device Mobility Enhancement through Strained Silicon Engineering. *Proceeding of IEEE International SOI Conference*, 1 (2005)
126 X. Baie, X. Tang, J.P. Colinge: Fabrication of twin nano silicon wires based on arsenic dopant effect. Japanese Journal of Applied Physics, **37/1-3B**, 1591 (1998)

# 3 BSIM-CMG: A Compact Model for Multi-Gate Transistors

Mohan V. Dunga, Chung-Hsun Lin, Ali M. Niknejad and Chenming Hu

## 3.1 Introduction

The scaling of conventional planar CMOS is expected to become increasingly difficult due to increasing gate leakage and subthreshold leakage.[1-2] Multi-gate FETs such as FinFETs have emerged as the most promising candidates to extend the CMOS scaling into the sub-25nm regime.[3-4] The strong electrostatic control over the channel originating from the use of multiple gates reduces the coupling between source and drain in the subthreshold region and it enables the Multigate transistor to be scaled beyond bulk planar CMOS for a given dielectric thickness. Numerous efforts are underway to enable large scale manufacturing of multi-gate FETs. At the same time, circuit designers are beginning to design and evaluate multi-gate FET circuits.

A compact model serves as a link between process technology and circuit design. It is a concise mathematical description of the complex device physics in the transistor. A compact model maintains a fine balance between accuracy and simplicity. An accurate model stemming from physics basis allows the process engineer and circuit designer to make projections beyond the available silicon data (scalability) for scaled dimensions and also enables fast circuit/device co-optimization. The simplifications in the physics enable very fast analysis of device/circuit behavior when compared to the much slower numerical based TCAD simulations. It is thus necessary to develop a compact model of multi-gate FETs for technology/circuit development in the short term and for product design in the longer term.

## 3.2 Framework for Multigate FET Modeling

One of the biggest challenges in modeling multi-gate FETs is the need to model the several flavors of multi-gate FETs. The silicon body can be controlled by either two gates, three gates or four gates. The gates can all be electrically interconnected or they can be biased independently. Multi-gate FETs can be built on SOI or bulk silicon.

Figure 3.1 illustrates some of the different architectures of Multi-gate FETs which need to be accounted in the compact model. It is important to obtain a versatile model that can model all the different types of Multi-gate FETs without making the model computationally intensive.

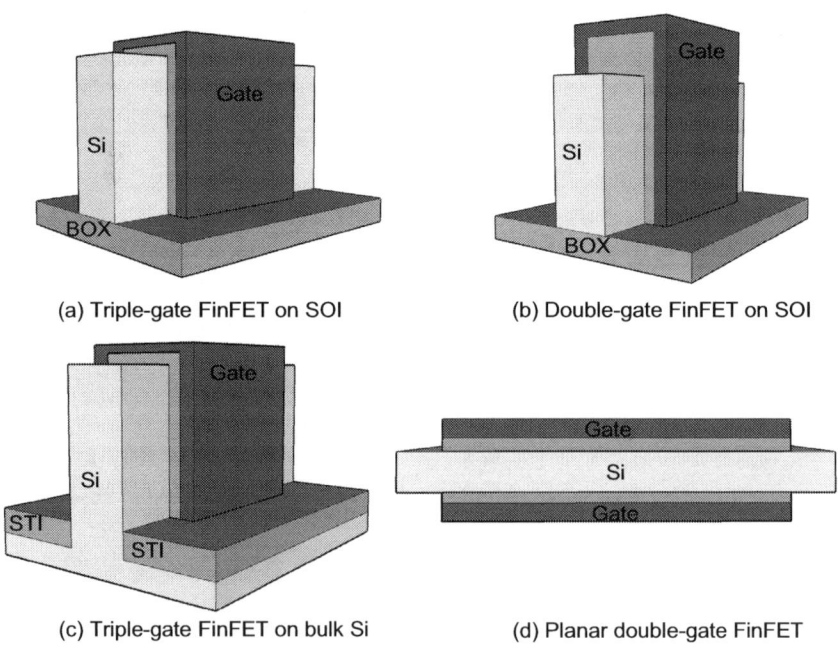

**Fig. 3.1.** Different Multigate FET architectures.

One possible technique to handle the different Multi-gate FET architectures is to classify them into two categories and introduce a separate model for each category: a common gate model and an asymmetric/independent gate model. The term "common-gate" means that all the gates in the multi-gate FET (double-gate or triple-gate or quadruple-gate FinFET) are electrically interconnected and are biased at the same electrical gate voltage. The common-gate model further assumes that the gate workfunctions and the dielectric thicknesses on the two, three or four active sides of the fin are the same. However, the carrier mobilities in the inversion layers on the horizontal and vertical active sides of the fin can be different due to different crystal orientations and/or strain. The asymmetric/independent gate model allows different workfunctions and dielectric thicknesses on the top and bottom of the fin. The asymmetric/independent gate model also permits that the two gates can be biased independently.

## 3.3 Multigate Models: BSIM-CMG and BSIM-IMG

Compact models for multi-gate FETs: BSIM-CMG and BSIM-IMG, have been developed based on the framework described in previous section. BSIM-CMG (Berkeley Short-channel IGFET Model – Common MultiGate) models the common-gate multi-gate FETs and BSIM-IMG (BSIM-Independent Multi-Gate) models the independent/asymmetric multi-gate FETs.

BSIM-CMG and BSIM-IMG are surface potential based models. Surface potential models have an inherent advantage of yielding continuous and smooth expressions for terminal currents and charges over the different regions of operations: subthreshold, linear and saturation regions. Starting with core long-channel double-gate models, full-scale compact models are developed by inclusion of numerous physical effects such as quantum mechanical effect (QME), poly-depletion effect (PDE), short-channel effect (SCE), mobility degradation and carrier velocity saturation. The expressions for terminal currents and charges are $C_\infty$-continuous, which makes the two models suitable for mixed-signal design. Both the models are accurate, predictive and scalable as demonstrated through extensive 2-D and 3-D TCAD simulations.

### 3.3.1 The BSIM-CMG Model

The existing modeling efforts for the multi-gate FETs are limited to undoped or lightly doped silicon body for double-gate (DG) FETs.[5-7] A multiple threshold voltage technology based on common-gate multi-gate FETs will likely require a significant concentration of body dopant for threshold voltage tuning. As a result, BSIM-CMG models the effect of finite body doping on the electrical characteristics of a multi-gate FET in Poisson's equation.[8]

Starting from a core long-channel symmetric DG-FET framework, the model is extended to triple-gate FinFETs and quadruple-gate FinFETs through 3-D modeling of SCE. A unique field penetration length model is developed to capture the stronger electrostatic control over the channel due to multiple gates, which leads to reduced subthreshold leakage in multi-gate FETs.

Common-gate multi-gate FETs can be built on SOI or bulk silicon. BSIM-CMG allows the user to select SOI mode or bulk SOI mode through the addition of the body node for bulk multi-gate FET. The BSIM-CMG model has been successfully used to describe the measured electrical characteristics of SOI FinFETs and bulk FinFETs.[9] The model formulation will be described in further detail in this chapter.

### 3.3.2 The BSIM-IMG Model

BSIM-IMG models the independent/asymmetric multi-gate FET. Unlike the BSIM-CMG model, BSIM-IMG assumes a lightly doped body in the Poisson equation for simplicity. For an independent/asymmetric multi-gate FET, the threshold voltage of the transistor can be tuned by adjusting the back gate voltage. As a result, a lightly doped body is expected to be used even for a multiple-threshold voltage technology and heavy body doping can be avoided in the thin body. Many of the physical effects models are borrowed from BSIM-CMG model with appropriate changes for an independent gate operation. As a result, only a brief description of BSIM-IMG will be provided in this chapter.

## 3.4 BSIM-CMG

BSIM-CMG is a surface potential based model. All electrical variables such as terminal currents, charges and capacitances are derived from the surface potentials at the source and the drain end. The calculation of the surface potentials forms the basis of the model. The core model for BSIM-CMG is a long-channel Double-Gate FET model. Numerous physical phenomena observed in an advanced Multigate FET technology are added to the core model to yield the final model.

This section describes the BSIM-CMG model formulation. The core model is first developed by analyzing the physics of a long-channel DG-FET. Next, modeling of some of the important physical effects such as QME, SCE and PDE are described. The section finally concludes with the verification of the model against experimental data.

### 3.4.1 Core Model

#### 3.4.1.1 Surface Potential Model

The long-channel, symmetric double-gate (DG) FET forms the core of BSIM-CMG. Figure 3.2 shows a schematic of the symmetric, common-gate DG-FET under study. The convention for the axes and the symbols used in this section are indicated in Figure 3.2. The electronic potential in the body is obtained by solving Poisson's equation. For a long-channel transistor, the gradual channel approximation is used which states that the horizontal electric field is much smaller than the vertical electric field. The use of gradual channel approximation results in a 1-D Poisson's equation (in the vertical dimension).

The 1-D Poisson equation including both inversion carriers and bulk charge in the body can be written as:

$$\frac{\partial^2 \psi(x,y)}{\partial x^2} = \frac{qn_i}{\varepsilon_{Si}} \cdot e^{\frac{q(\psi(x,y) - \phi_B - V_{ch}(y))}{kT}} + \frac{qN_A}{\varepsilon_{Si}} \tag{3.1}$$

where $\psi(x,y)$ is the electronic potential in the body, $V_{ch}(y)$ is the channel potential ($V_{ch}(0) = 0$ and $V_{ch}(L) = V_{ds}$), $N_A$ is the body doping and

$$\phi_B = \frac{kT}{q} \cdot \ln\left(\frac{N_A}{n_i}\right) \quad (3.2)$$

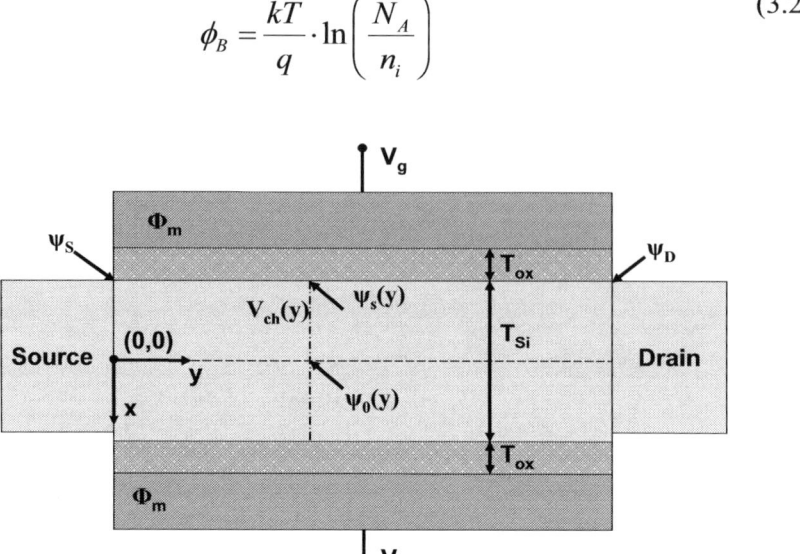

**Fig. 3.2.** Schematic of the symmetric common-gate DG-FET under study.

For a lightly doped body, the body doping can be neglected and Eq. (3.1) can be solved easily by integrating it twice and using Gauss's law as the boundary condition.[5] However, for moderate to heavy body doping conditions, the doping term cannot be neglected, which complicates the calculation of surface potential as Eq. (3.1) cannot be integrated analytically twice. To overcome this limitation, perturbation approach is used to solve Poisson's equation in presence of significant body doping.[8] Under the perturbation approach, the potential in the body can be written as sum of two terms:

$$\psi(x,y) = \psi_1(x,y) + \psi_2(x,y) \quad (3.3)$$

The first term, $\psi_1(x,y)$, is the potential due to the inversion carriers term in Eq. (3.1). The second term, $\psi_2(x,y)$, is the perturbation in potential due to body doping term. The body can be fully depleted or partially depleted depending on applied gate bias ($V_{gs}$), body doping ($N_A$) and body thickness ($T_{Si}$). The perturbation method yields surface potential in both full-depletion and partial-depletion regimes.

In the fully depleted regime, the inversion carriers are spread through the entire body. The contribution of inversion carriers to the potential, $\psi_1(x,y)$, is calculated by neglecting the bulk charge term in Eq. (3.1).

$$\frac{\partial^2 \psi_1(x,y)}{\partial x^2} = \frac{qn_i}{\varepsilon_{Si}} \cdot e^{\frac{q(\psi_1(x,y)-\phi_B-V_{ch}(y))}{kT}} \quad (3.4)$$

Using the fact that the electric field at the mid-plane is zero for a symmetric common gate FET, Eq. (3.4) can be integrated twice to obtain $\psi_1(x,y)$.

$$\psi_1(x,y) = \psi_0(y) - \frac{2kT}{q} \cdot \ln\left(\cos\left(\sqrt{\frac{q^2}{2\varepsilon_{Si}kT}\frac{n_i^2}{N_A}e^{\frac{q(\psi_0(y)-V_{ch}(y))}{kT}}} \cdot \frac{x}{2}\right)\right) \quad (3.5)$$

where $\psi_0(y)$ is the potential at the center of the body as shown in Figure 3.2. Substituting Eq. (3.4) in Eq. (3.3) yields a second order differential equation in $\psi_2(x,y)$.

$$\frac{\partial^2 \psi_2(x,y)}{\partial x^2} = \frac{qn_i}{\varepsilon_{Si}} \cdot e^{\frac{q(\psi_1(x,y)-\phi_B-V_{ch}(y))}{kT}} \cdot \left(e^{\frac{q\psi_2(x,y)}{kT}}-1\right) + \frac{qN_A}{\varepsilon_{Si}} \quad (3.6)$$

Since the contribution of the bulk charge to potential and electric field at mid-plane ($x = 0$) is zero

$$\psi_2(x=0,y) = 0 \quad \text{and} \quad \frac{d\psi_2}{dx}\bigg|(x=0,y) = 0 \quad (3.7)$$

Using the boundary conditions on $\psi_2(x,y)$ in Eq (3.7) and the expression for $\psi_1(x,y)$ in Eq. (3.5), the differential equation Eq. (3.6) can be solved to obtain the expression for $\psi_2(x,y)$.

$$\psi_2(x,y) = \frac{2qn_i}{\varepsilon_{Si}} \cdot \frac{e^{\frac{q\phi_B}{kT}}}{a(x,y)} \cdot \left(\frac{e^{x\frac{\sqrt{a(x,y)}}{2}}-1}{2e^{x\frac{\sqrt{a(x,y)}}{2}}}\right)^2 \quad (3.8)$$

where

$$a(x,y) = \frac{q^2 n_i}{\varepsilon_{Si}kT} \cdot e^{\frac{q(\psi_1(x,y)-V_{ch}(y)-\phi_B)}{kT}} \quad (3.9)$$

The surface potential at a point "y" along the channel is the sum of $\psi_1(x,y)$ and $\psi_2(x,y)$ evaluated at the surface:

$$\psi_s(y) = \psi_1\left(\frac{T_{Si}}{2}, y\right) + \psi_2\left(\frac{T_{Si}}{2}, y\right) \qquad (3.10)$$

Note that $\psi_2(x,y)$ is a function of $\psi_1(x,y)$ and $\psi_1(x,y)$ is a function of $\psi_0(y)$. As a result, $\psi_s(y)$ is a function of only one variable, $\psi_0(y)$, the potential at the center of the body. The electric field at the surface can be easily obtained by integrating Eq. (3.1) once. Gauss's Law at the surface can then be expressed as

$$V_{gs} = V_{fb} + \psi_s(y) + \frac{\varepsilon_{Si}}{C_{ox}} \cdot \sqrt{\frac{2qn_i}{\varepsilon_{Si}} \cdot \left( \frac{e^{\frac{q\psi_s(y)}{kT}} - e^{\frac{q\psi_0(y)}{kT}}}{q/kT} \cdot e^{\frac{-q(V_{ch}(y)+\phi_B)}{kT}} + e^{\frac{q\phi_B}{kT}} \cdot (\psi_s(y) - \psi_0(y)) \right)} \qquad (3.11)$$

Eq. (3.11) can be expressed in terms of only one unknown quantity $\psi_0(y)$. Solving Eq. (3.11) yields $\psi_0(y)$ and hence $\psi_s(y)$ in the fully depleted regime for a given DG-FET structure and a set of external bias voltages.

In the partially depleted regime, the depletion width is bias dependent. At the edge of the depletion width, $x_{dep}$, the electronic potential is zero and hence $\psi_1|_{(x=x_{dep})} = 0$. With these changes, the surface potential can be re-derived for the partially depleted regime similar to the fully depleted regime. The contribution to the surface potential from inversion carriers is

$$\psi_1\left(\frac{T_{Si}}{2}, y\right) = -\frac{2kT}{q} \cdot \ln\left(\cos\left(\sqrt{\frac{q^2}{2\varepsilon_{Si}kT} \frac{n_i^2}{N_A} e^{\frac{-qV_{ch}(y)}{kT}}} \cdot \frac{x_{dep}}{2}\right)\right) \qquad (3.12)$$

and the correction term in the surface potential due to bulk charge is

$$\psi_2\left(\frac{T_{Si}}{2}, y\right) = \frac{2qn_i}{\varepsilon_{Si}} \cdot \frac{e^{\frac{q\phi_B}{kT}}}{a} \cdot \left(\frac{e^{x_{dep}\frac{\sqrt{a}}{2}} - 1}{2e^{x_{dep}\frac{\sqrt{a}}{2}}}\right)^2 \qquad (3.13)$$

where

$$a = \frac{q^2 n_i}{\varepsilon_{Si}kT} \cdot e^{\frac{q\left(\psi_1(T_{Si}/2, y) - V_{ch}(y) - \phi_B\right)}{kT}} \qquad (3.14)$$

Applying Gauss's law in the partially depleted regime yields:

$$V_{gs} = V_{fb} + \psi_s(y) + \frac{\varepsilon_{Si}}{C_{ox}} \cdot \sqrt{\frac{2qn_i}{\varepsilon_{Si}} \cdot \left( \frac{e^{\frac{q\psi_s(y)}{kT}} - 1}{q/kT} \cdot e^{\frac{-q(V_{ch}(y)+\phi_B)}{kT}} + e^{\frac{q\phi_B}{kT}} \cdot \psi_s(y) \right)} \quad (3.15)$$

The unknown variable in the calculation the surface potential of a partially depleted device is $x_{dep}$. It is obtained by solving Eq. (3.15) numerically. Once $x_{dep}$ is determined, the surface potential is calculated using Eq. (3.10) and Eqs. (3.12-3.14).

In order to obtain continuous expressions for the terminal currents and charges, it is necessary to capture the transition between the fully depleted and partially depleted regimes in a smooth manner. Also, the solution of Eq. (3.12) and Eq. (3.15) is computationally expensive due to the complex $\psi_2(x,y)$ perturbation term. To overcome these problems, a simplified expression for $\psi_2(x,y)$ is derived ($\psi_{pert}$).which is continuous between partially depleted and fully depleted regimes. By using $\psi_{pert}$, the surface potential in both the regimes is calculated through a single continuous equation. Using the following transformation of variables:

$$\beta = \frac{T_{Si}}{2} \sqrt{\frac{q^2}{2\varepsilon_{Si} kT} \frac{n_i^2}{N_A} e^{\frac{q(\psi_0(y)-V_{ch}(y))}{kT}}} \quad (3.16)$$

the unified surface potential ($\psi_s$) equation used in the core model for BSIM-CMG can be written as

$$f_0(\beta) = \ln(\beta) - \ln(\cos(\beta)) - \frac{V_{gs} - V_{fb} - V_{ch}}{2\frac{kT}{q}} + \ln\left(\frac{2}{T_{Si}} \sqrt{\frac{2\varepsilon_{Si} kT N_A}{qn_i^2}}\right) + \quad (3.17)$$

$$\frac{2\varepsilon_{Si}}{T_{Si} C_{ox}} \sqrt{\beta^2 \left( \frac{e^{\frac{q\psi_{pert}}{kT}}}{\cos^2(\beta)} - 1 \right) + \frac{\psi_{pert}}{(kT/q)^2} \left( \psi_{pert} - 2\frac{kT}{q} \ln(\cos(\beta)) \right)} = 0$$

where $\beta$ is the only unknown variable. In BSIM-CMG, the transcendental $\psi_s$ equation (Eq. (3.17)) is solved for $\beta$ using an analytical approximation instead of iterative methods to make the model numerically robust. This is another significant advantage of the simplified perturbation term $\psi_{pert}$.

The surface potential is a function of $\beta$ and is given by

$$\psi_s = 2\frac{kT}{q}\left(\ln(\beta) - \ln(\cos(\beta)) + \ln\left(\frac{2}{T_{Si}}\sqrt{\frac{2\varepsilon_{Si}kTN_A}{qn_i^2}}\right)\right) + \psi_{pert} \quad (3.18)$$

Surface potential at source terminal ($\psi_S$) is obtained by solving Eq. (3.17) at the source end, i.e $V_{ch} = 0$. Similarly, the surface potential at the drain end ($\psi_D$) is calculated by solving Eq. (3.17) with $V_{ch} = V_{ds}$.

Figure 3.3(a) compares the surface potential calculated using the model against TCAD. All TCAD simulations for verification of the core model use gate material with mid-gap workfunction and assume constant carrier mobility. The surface potential is calculated as a function of gate voltage for a wide range of body doping ranging from a light doping of $10^{15} cm^{-3}$ to heavy doping of $5\times10^{18} cm^{-3}$. Very good agreement is observed between the model and TCAD for all cases. The transition from partially depleted regime to fully depleted regime with increasing gate voltage is clearly visible in the heavily doped DG-FET. The error in the analytical approximation of $\psi_s$ is limited to only a few nanovolts as shown in Figure 3.3(b).

Eq. (3.17) yields surface potential for both light and heavy body doping as shown in Figure 3.3(a). However, the inclusion of bulk charge in the analysis of a lightly doped body is redundant and it leads to significant overhead in model runtime. For a lightly doped DG-FET, Eq. (3.17) can be simplified into:

$$\ln\beta - \ln(\cos\beta) - \frac{V_{gs} - V_{fb} - V_{ch}}{2\frac{kT}{q}} + \ln\left(\frac{2}{T_{Si}}\sqrt{\frac{2\varepsilon_{Si}kTN_A}{qn_i^2}}\right) + \frac{2\varepsilon_{Si}}{T_{Si}C_{ox}}\beta\tan\beta = 0 \quad (3.19)$$

Based on this insight, BSIM-CMG offers two options to improve runtime for lightly doped DG-FETs:

1. Setting model parameter PHISMOD =0 activates framework for heavily-doped DG-FET
2. Setting model parameter PHISMOD =1 activates framework for lightly-doped DG-FET

Physical effects such as QME, SCE and PDE are added to the core surface potential equation as discussed in Section 3.4.2.

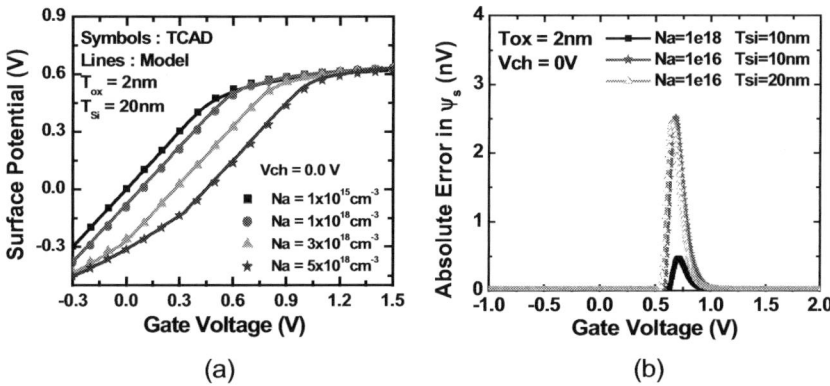

**Fig. 3.3.** Verification of core model surface potential equation (a) Comparison against TCAD for different body doping (b) Error in the analytical approximation for $\psi_s$.

### 3.4.1.2 I-V Model

The I-V model is obtained using drift-diffusion formulation without using any charge-sheet approximation [10]. The current flowing through the body of a DG-FET can be written as:

$$I_d = 2 \cdot \mu \cdot W \cdot Q_{inv}(y) \frac{dV_{ch}}{dy} \qquad (3.20)$$

where $\mu$ is the carrier mobility (assumed position independent), $W$ is the channel width, $Q_{inv}(y)$ is the inversion charge in one half of the body and the factor of two accounts for the front and back channel currents in a symmetric common-gate DG-FET.

The inversion charge ($Q_{inv}$) is simply the difference between the total charge in the body and the bulk charge:

$$Q_{inv}(y) = Q_{total}(y) - Q_{bulk}(y) \qquad (3.21)$$

The bulk charge, $Q_{bulk}$, is obtained from the perturbation potential $\psi_{pert}$:

$$Q_{bulk} = \sqrt{2q\varepsilon_{Si}N_A\psi_{pert}} \qquad (3.22)$$

Gauss's law can be used to determine the total charge in the body:

$$Q_{total}(y) = C_{ox} \cdot \left( V_{gs} - V_{fb} - \psi_s(y) \right) \qquad (3.23)$$

An analytical expression for drain current can be obtained by using Eqs. (3.21-3.23) and Eq. (3.20), and integrating the resulting expression from source to drain. The drain current can be expressed as difference of two terms evaluated at the source and drain end:

$$I_d = 2 \cdot \mu \cdot \frac{W}{L} \cdot \left( f(\psi_S) - f(\psi_D) \right) \tag{3.24}$$

where the function $f(\psi_s(y))$ is given by:

$$f(\psi_s) = \frac{Q_{inv}^2}{2C_{ox}} + 2\frac{kT}{q}Q_{inv} - \frac{kT}{q}\left(5\frac{\varepsilon_{Si}kT}{qT_{Si}} + Q_{bulk}\right) \cdot \ln\left(5\frac{\varepsilon_{Si}kT}{qT_{Si}} + Q_{bulk} + Q_{inv}\right) \tag{3.25}$$

Eqs. (3.24-3.25) predict the drain current for a symmetric DG-FET and they together constitute the I-V model for BSIM-CMG. For the case of a lightly doped body, the drain current expression transforms into a very similar expression as derived in [5] and [7] for an undoped body.

The accuracy and predictivity of the I-V model is verified against TCAD simulations without using any fitting parameters. Figure 3.4 shows the model predicted and TCAD simulated values of $I_d$ for a heavily doped DG-FET ($N_A$ = 3e18cm$^{-3}$). BSIM-CMG can predict very accurate drain current in all the regimes of transistor operation: sub-threshold, linear and saturation. Figure 3.5 tests the accuracy of the I-V model over a wide range of body doping. The model predicts the correct drain current in both fully depleted and partially depleted regimes. This can be seen from the $I_d$-$V_g$ characteristics of the heavily doped DG-FET in Figure 3.5.

A unique behavior of lightly doped DG-FET with a thin body is volume inversion. Due to the absence of bulk charge in the body and presence of only few inversion carriers in the subthreshold regime, there is negligible potential drop between the surface and center of the body. Figure 3.6(a) shows the virtually flat potential profile in lightly doped DG-FETs in sub-threshold. Furthermore, the potential in the body has a very weak dependence on body thickness. Any small increase in gate voltage in subthreshold increases the potential through the entire body causing inversion in the entire body. This phenomenon is called bulk inversion or volume inversion. Since the electronic potential is virtually independent of body thickness, the amount of inversion carriers in the body is linearly proportional to the body thickness for equal area DG-FETs. As a result, the subthreshold current in lightly doped DG-FETs is a linear function of the body thickness.

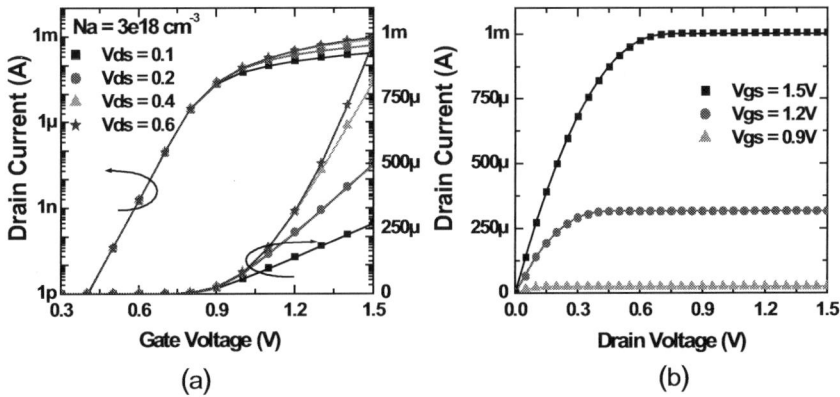

**Fig. 3.4.** (a) Id-Vg and (b) Id-Vd characteristics calculated from the I-V model (lines) and TCAD (symbols). Device data: $N_A = 3\text{e}18\text{cm}^{-3}$, $T_{Si} = 20\text{nm}$, $T_{ox} = 2\text{nm}$.

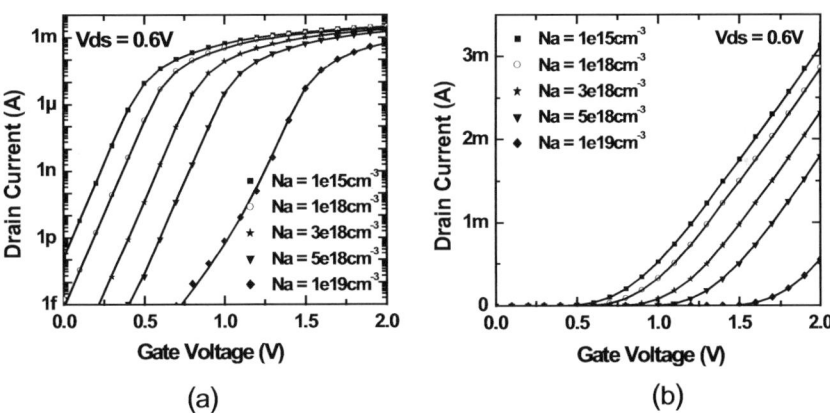

**Fig. 3.5.** Id-Vg characteristics for different body doping concentrations calculated from the I-V model (lines) and TCAD (symbols). Device data: $T_{Si} = 20\text{nm}$, $T_{ox} = 2\text{nm}$.

The I-V model is able to predict this trend correctly as shown in Figure 3.6(b) where the $I_d$ for a 20nm thick body is ~4x of current flowing in a 5nm thick body in the sub-threshold regime.

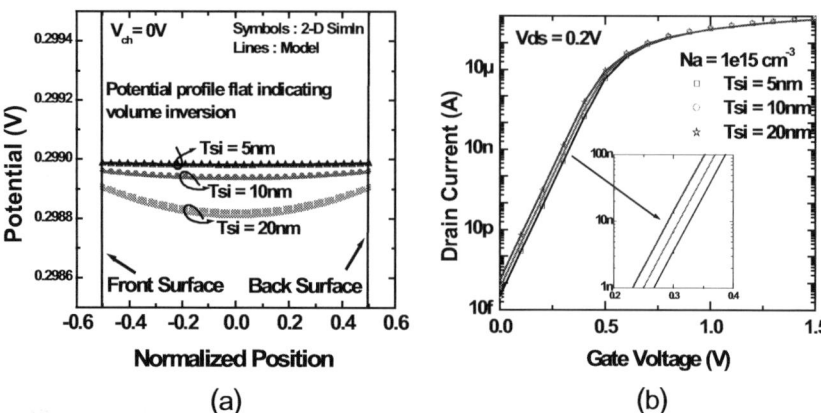

**Fig. 3.6.** Volume inversion in lightly doped DG-FETs (a) Potential profile in the body (b) Sub-threshold $I_d$-$V_g$ characteristics for different body thickness showing volume inversion. In both (a) and (b) lines represent model and symbols represent TCAD. Device data: $N_A = 1\mathrm{e}15\mathrm{cm}^{-3}$, $T_{ox} = 2\mathrm{nm}$.

### 3.4.1.3 C-V Model

No compact model is complete without a capacitance model. The I-V model is adequate only for describing the DC behavior but for transient description the capacitances are absolutely essential. The intrinsic capacitances of the transistor are derived from the terminal charges.

The charge on the top and bottom gate electrodes is equal to total charge in the body. The total charge is computed by integrating the charge along the channel. Since the two gates are electrically interconnected, we have:

$$Q_g = 2WC_{ox} \int_0^L \left( V_g - V_{fb} - \psi_s(y) \right) \cdot dy \qquad (3.26)$$

where $Q_g$ denotes the charge on the electrically interconnected gate. The inversion charge in the body is divided between the source and drain terminals using the Ward-Dutton charge partition approach.[11] The charge on source terminal ($Q_s$) is:

$$Q_s = -2WC_{ox} \int_0^L \left(1 - \frac{y}{L}\right) \cdot \left( V_g - V_{fb} - \psi_s(y) - \frac{Q_{bulk}}{C_{ox}} \right) \cdot dy \qquad (3.27)$$

Using charge conservation, the charge on drain terminal ($Q_d$) can be expressed as:

$$Q_d = -2WC_{ox} \int_0^L \frac{y}{L} \cdot \left( V_g - V_{fb} - \psi_s(y) - \frac{Q_{bulk}}{C_{ox}} \right) \cdot dy \qquad (3.28)$$

The surface potential as a function of the position $y$ along the length of the transistor ($\psi_s(y)$) is obtained using current continuity. Current continuity states that the current is conserved all along the length of the transistor.

$$I_d(L) = I_d(y) \quad \text{where} \quad 0 \leq y \leq L \qquad (3.29)$$

The expression for the drain current in Eqs. (3.24-3.25) is very complex and is not practical for applying current continuity. For the purpose of determining $\psi_s(y)$, a simplified version of I-V model as shown below is used:

$$I_d(y) = 2 \cdot \mu \cdot \frac{W}{y} \cdot \left( g(\psi_S) - g(\psi_s(y)) \right) \qquad (3.30)$$

where the function $g(\psi_s(y))$ is defined as

$$g(\psi_s) = \frac{Q_{inv}^2}{2C_{ox}} + 2\frac{kT}{q} Q_{inv} \qquad (3.31)$$

Eqs. (3.30-3.31) retain good accuracy in the strong inversion regime but overestimate the drain current in subthreshold regime. The advantage of mathematically simple analytical expressions for terminal charges due to this approximation outweighs the resulting loss in accuracy in the C-V model in subthreshold regime, however. Using Eqs. (3.30-3.31), $\psi_s(y)$ can be related to $\psi_S$ and $\psi_D$ by:

$$\frac{y}{L} \cdot (B - \psi_S - \psi_D)(\psi_D - \psi_S) = (B - \psi_S - \psi_s(y))(\psi_s(y) - \psi_S) \qquad (3.32)$$

where

$$B = 2\left( V_g - V_{fb} - \frac{Q_{bulk}}{C_{ox}} + 2\frac{kT}{q} \right) \qquad (3.33)$$

The terminal charges are obtained by substituting $\psi_s(y)$ in Eqs. (3.26-3.28) and evaluating the integrals.

$$Q_g = 2WLC_{ox}\left(V_{gs} - V_{fb} - \frac{\psi_S + \psi_D}{2} + \frac{(\psi_D - \psi_S)^2}{6(B - \psi_D - \psi_S)}\right) \quad (3.34)$$

$$Q_d = -2WLC_{ox}\left(\begin{array}{l}\dfrac{V_{gs} - V_{fb} - Q_{bulk}/C_{ox}}{2} - \dfrac{\psi_S + \psi_D}{4} + \dfrac{(\psi_D - \psi_S)^2}{60(B - \psi_D - \psi_S)} \\ + \dfrac{(5B - 4\psi_D - 6\psi_S)(B - 2\psi_D)(\psi_S - \psi_D)}{60(B - \psi_D - \psi_S)^2}\end{array}\right)$$

$$Q_s = -\left(Q_{fg} + Q_{bg} + Q_{bulk} + Q_d\right)$$

The expressions for terminal charges are continuous and are valid over sub-threshold, linear and saturation regimes of operation. Figure 3.7 shows the terminal charges calculated using Eq. (3.34) as a function of $V_{ds}$ and $V_{gs}$. The ratio of the drain charge to source ratio is 40/60 in the saturation region as seen in Figure 3.7. This is due to Ward-Dutton charge partition which is physically correct under the quasi-static condition.

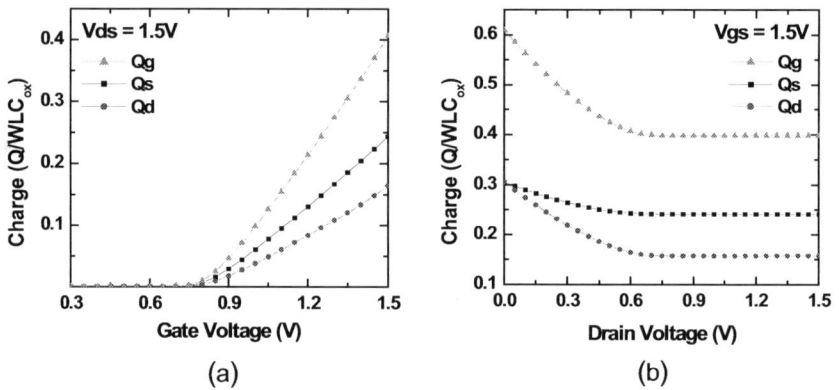

**Fig. 3.7.** Terminal charges using Eq. (3.33) as a function of (a) $V_{gs}$ and (b) $V_{ds}$ Device data: $N_A = 3\text{e}18\text{cm}^{-3}$, $T_{ox} = 2$nm, $T_{Si} = 20$nm.

Eq. (3.34) forms the C-V model for BSIM-CMG. The terminal charges are used as state variables in the circuit simulation. All the capacitances are derived from the terminal charges to ensure charge conservation. The capacitances are defined as:

$$C_{ij} = \frac{\partial Q_i}{\partial V_j} \quad (3.35)$$

where $i$ and $j$ denote the multi-gate FET terminals. Note that $C_{ij}$ satisfies

$$\sum_i C_{ij} = \sum_j C_{ij} = 0 \qquad (3.36)$$

due to charge conservation.

The C-V model is verified against TCAD simulations without the use of any fitting parameters. The capacitances from the C-V model are plotted as a function of gate voltage and drain voltage in Figure 3.8 and Figure 3.9 respectively.

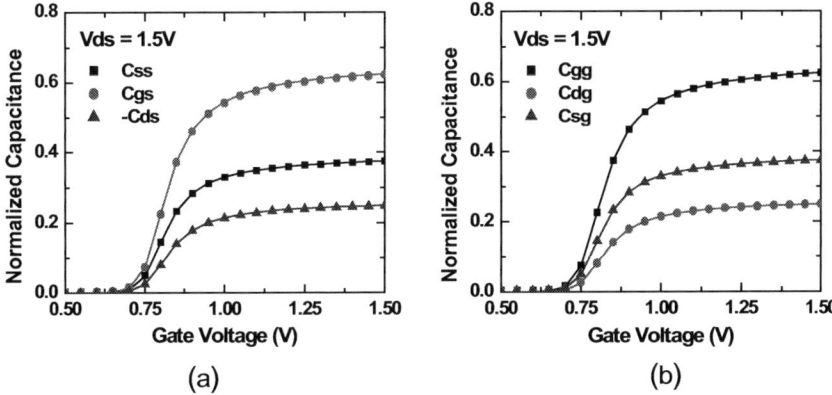

**Fig. 3.8.** Capacitances (normalized to $2WLC_{ox}$) calculated from the C-V model (lines) and TCAD (symbols) as a function of $V_{gs}$. Device data: $N_A = 3\text{e}18\text{cm}^{-3}$, $T_{ox} = 2\text{nm}$, $T_{Si} = 20\text{nm}$.

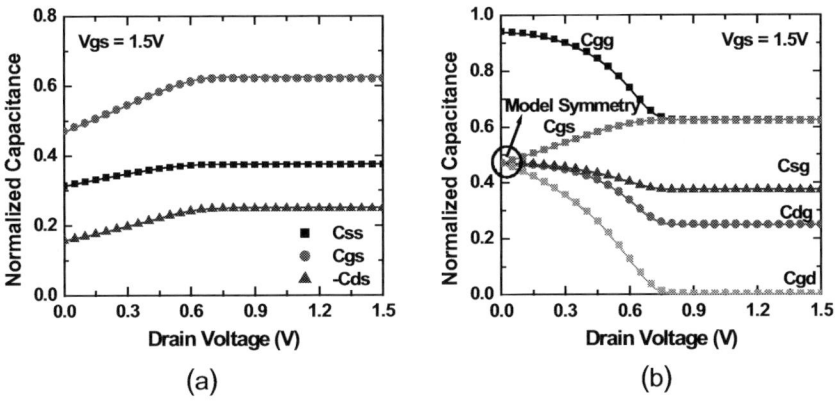

**Fig. 3.9.** Capacitances (normalized to $2WLC_{ox}$) calculated from the C-V model (lines) and TCAD (symbols) as a function of $V_{ds}$. Device data: $N_A = 3\text{e}18\text{cm}^{-3}$, $T_{ox} = 2\text{nm}$, $T_{Si} = 20\text{nm}$.

The capacitance values from the model are in excellent agreement with TCAD simulated values in all regimes of transistor operation. At $V_{ds} = 0$V, Figure 3.9(b) shows that $C_{sg}=C_{dg}$ and $C_{gs}=C_{gd}$. This equality in capacitances at $V_{ds} = 0$V demonstrates the symmetry of the core model. Model symmetry is important for predicting correct distortion metrics for circuits switching about $V_{ds} = 0$V especially in the analog and RF domain.

The surface potential model together with the I-V and C-V model for DG-FET form the core model for BSIM-CMG. The core model is highly predictive and has a high degree of accuracy. The model agrees with TCAD simulations without the use of any fitting parameters.

### 3.4.2 Modeling Physical Effects of Real Devices

The core model is only the beginning of any compact model, which is also the case with BSIM-CMG. BSIM-CMG, in the tradition of BSIM3 and BSIM4, models numerous physical phenomena that are expected to be important to accurately represent advanced multi-gate FET technologies. The thin body in multi-gate FETs experiences significant QME through structural and electrical confinement which are modeled by modifying the core surface potential equation and the C-V model. SCE such as drain-induced barrier lowering (DIBL), threshold voltage ($V_{th}$) roll-off and sub-threshold slope degradation are modeled. BSIM-CMG models the PDE as well since polysilicon-gated FinFETs may be used in low-cost memories to enable continued cell size reduction.

**Table 3.1.** List of physical effects modeled in BSIM-CMG.

1) Quantum Mechanical Effects
2) Short-channel Effects
    a) $V_{th}$ roll-off
    b) DIBL
    c) Sub-threshold slope
    d) Channel length modulation
3) Polysilicon-gate Depletion Effects (PDE)
4) Series resistance
5) Mobility degradation
6) Velocity Saturation
7) Velocity Overshoot/Source-End Velocity Limit
8) Gate Induced Drain (Source) Leakage (GIDL, GISL)
9) Impact Ionization
10) S/D Junction leakage
11) Gate tunneling
12) Parasitic capacitance

All the physical effects included in BSIM-CMG are listed in Table 3.1. Some of the important physical effects are discussed next.

As examples of the physical effect models, the following subsections present the quantum mechanical effect model and the short-channel effect model in some details.

### 3.4.2.1 Quantum Mechanical Effects (QME)

Carrier energy quantization has become significant in state-of-the-art MOSFETs due to increased vertical E-field. The energy quantization and the shift of the inversion charge centroid will delay the formation of inversion charge (threshold voltage ($V_{th}$) shift) and reduce the current driving capability (increase the effective oxide thickness). The quantization effect is more complicated in double-gate (DG) MOSFETs than in bulk MOSFETs due to the extra structural confinement by the thin body. Quantum mechanical (QM) effects have been included in compact models by introducing the effective oxide thickness for a bias-dependent reduction of the gate capacitance, and bias-independent correction for the $V_{th}$ shift, separately.[12] However, the lack of predictability of this approach is particularly undesirable for the DG MOSFETs compact model since less data are available on DG MOSFETs than on single-gate MOSFETs. Several groups have investigated the influence of $T_{si}$ on the quantized carrier distribution and threshold voltage shift numerically [13] and analytically [14-16] in DG MOSFETs. However, these approaches are limited to the carrier distribution and electrostatic potential profile, but not extended to device characteristics, such as I-V and C-V. In this section, a new explicit description of the surface potential, accounting for the QM effects, is introduced. A bias-dependent QM correction for surface potential calculation is derived for all regimes of operation in DG MOSFETs. The QM model agrees with 2-D TCAD simulation results very well.

While the channel carriers are confined in one-dimension, the subband splitting due to the field-induced electrical confinement (EC, Fig. 3.10(a)) has significant impact on device characteristics. The energy quantization is more complicated in the DG MOSFETs due to the extra structural confinement (SC) in the subthreshold region as shown in Fig. 3.10 (b).

The quantum confinement in the DG MOSFETs is affected by the gate work function, gate dielectric thickness, body thickness, substrate doping and gate bias. In order to study the dependence of QM effects on these device parameters, a self-consistent 1-D Schrödinger solver, SCHRED 2.1, is used for simulating DG FETs [17]. Fig. 3.11 shows the simulation results of the ratio of QM-corrected and classical charge density over different substrate doping concentrations for n-type symmetric/common-gate DG MOSFETs.

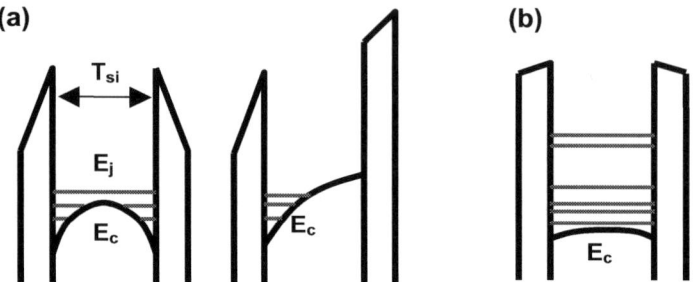

**Fig. 3.10.** Energy-band diagrams of DG-MOSFET under different operation modes illustrate two mechanisms of carrier energy quantization in DG MOSFETs: (a) electrical confinement (EC); (b) structural confinement (SC).

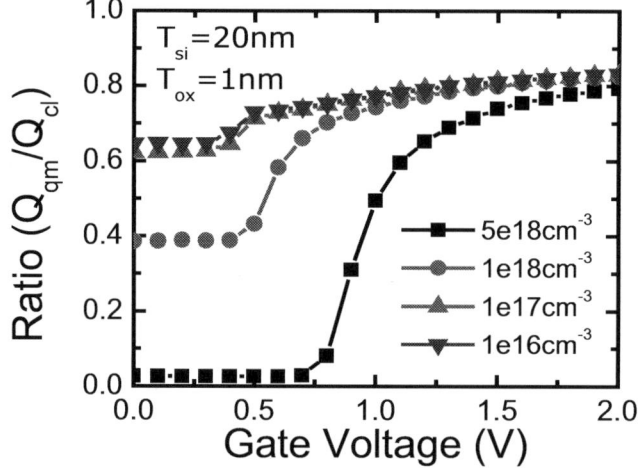

**Fig. 3.11.** The doping concentration dependence of QM effects. QM effects is very significant in heavily-doped device due to larger vertical E-field.

In the subthreshold region, the ratio of charge reflects the $V_{th}$ shift, while in the strong inversion region, the ratio of charge indicates the amount of the gate capacitance degradation. In the undoped DG MOSFETs, QM effect is less significant due to weaker EC (smaller E-field). The SC is weak in this structure due to relatively thick $T_{si}$. The dependence of body thickness on QM effect is shown in Fig. 3.12. The QM effect is less significant in the heavily doped thinner body device due to weaker EC (electrical coupling from two gates).

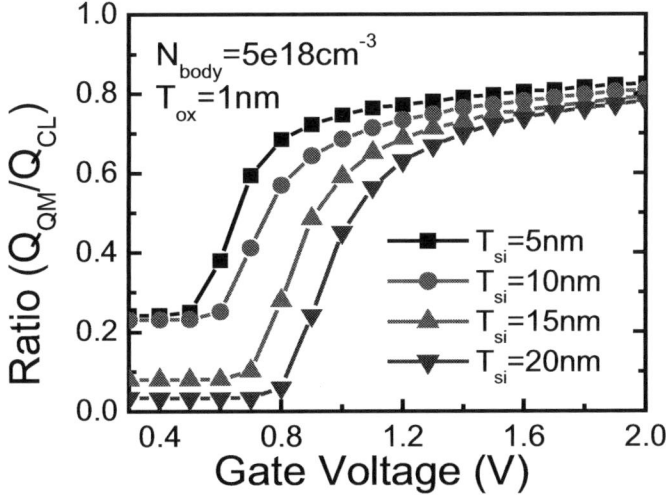

**Fig. 3.12.** The body thickness dependence of QM effects. QM effect is weak in the heavily doped thinner body device due to smaller vertical E-field.

As described in the previous sections, the surface potential is obtained by solving the Poisson's equation and Gauss's law. A bias-dependent ground-state subband energy in the unprimed valley ($E_0$) is added in the surface potential calculation. Trivedi *et al.* have calculated $E_0$ using the vibrational approach to include both structural and electrical confinement in DG MOSFETs.[14] However, the model only considers the threshold voltage shift in intrinsic device with an undoped body. A modification was made to the $E_0$ calculation to extend the model to the strong inversion region for a doped body. The drift-diffusion equation [10] is then employed to obtain a model for drain current in terms of surface potential. The $V_{th}$ shift and gate capacitance degradation are inherently captured without the need of any explicit individual modifications as long as the surface potential is modeled correctly as shown later. A surface potential correction term is added in the inversion carrier term when solving the Poisson's equation:

$$V_g = V_{fb} + \psi_s + \frac{\varepsilon_{si}}{C_{ox}} \qquad (3.37)$$

$$\cdot \sqrt{\frac{2qn_i}{\varepsilon_{si}} \cdot \left( \frac{e^{\frac{q(\psi_s - \Delta\psi_{QM})}{kT}} - e^{\frac{q(\psi_0 - \Delta\psi_{QM})}{kT}}}{q/kT} \cdot e^{\frac{-q\phi_B}{kT}} \cdot e^{\frac{-qV_{ch}}{kT}} + e^{\frac{q\phi_B}{kT}} \cdot (\psi_s - \psi_0) \right)}$$

where $\Delta\psi_{QM} = E_0/q$. $E_0$ is the ground state subband energy obtained by solving Schrödinger equation via the vibrational approach: [14]

$$E_0 \approx \frac{\hbar^2}{2m_x} \left[ \left(\frac{\pi}{T_{si}}\right)^2 + b_0^2 \left( 3 - \frac{4}{3} \frac{1}{\left[\left(\frac{b_0 T_{si}}{\pi}\right)^2 + 1\right]} \right) \right] \qquad (3.38)$$

where $b_0$ represents the vertical electric field dependence. Both structural and electrical confinements are included in the ground state subband energy.

Fig. 3.13 shows the surface potential versus the gate voltage for a long-channel n-type symmetric/ common-gate DG MOSFET where source/drain is grounded. A uniformly doped body ($5e18cm^{-3}$) is assumed. The model agrees with TAURUS 2-D device simulation very well from subthreshold region to strong inversion region. The increase of the surface potential due to the QM effect is well predicted by the model. Larger surface potential is needed to achieve the same level of inversion.

The model can accurately predict surface potential over a wide range of body doping concentration. Fig. 3.14 shows the surface potential versus gate voltage for a symmetric/common-gate DG MOSFET with lightly doped body ($1e16cm^{-3}$).

Fig. 3.15 shows the $I_d$-$V_g$ characteristics of an n-type symmetrical/common-gate DG MOSFET for both classical and QM models. The oxide thickness of 1nm and mid-gap workfunction gate material are used in the simulation. The body doping is $5e18cm^{-3}$. As seen from Fig. 3.15, the QM model predicts the $V_{th}$ shift accurately when compared against the 2-D TCAD simulation results.

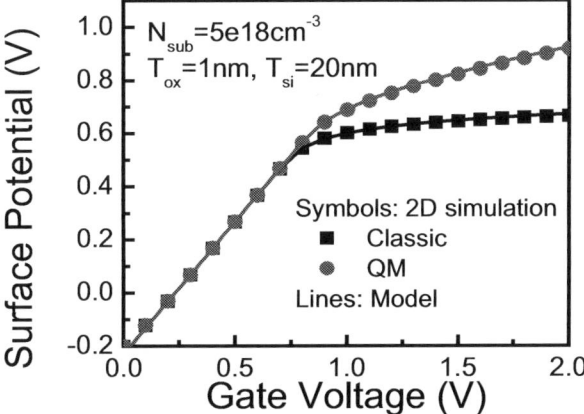

**Fig. 3.13.** QM-corrected surface potential solution agrees with 2-D simulation well. The QM-induced $V_{th}$ shift in subthreshold region and $C_{ox}$ degradation in strong inversion region are simultaneously predicted as long as the QM-corrected surface potential is accurate.

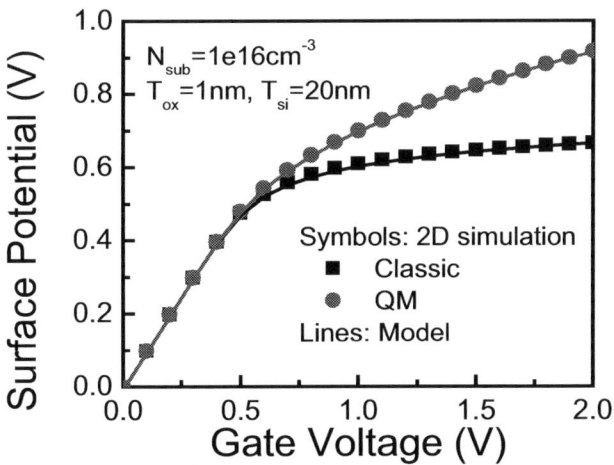

**Fig. 3.14.** Surface potential *vs.* gate voltage for a device with undoped body. The model predicts the doping dependence accurately.

The ratio of QM-corrected and classical drain current obtained from the QM model and 2-D TCAD simulations are shown in Fig. 3.16. It clearly shows that the QM model predicts both the $V_{th}$ shift and gate capacitance degradation accurately. As a result, there is no need to introduce a definition of an effective oxide thickness. The model predicts QM effects accurately in all regimes of operation since the QM-corrected surface potential is modeled accurately.

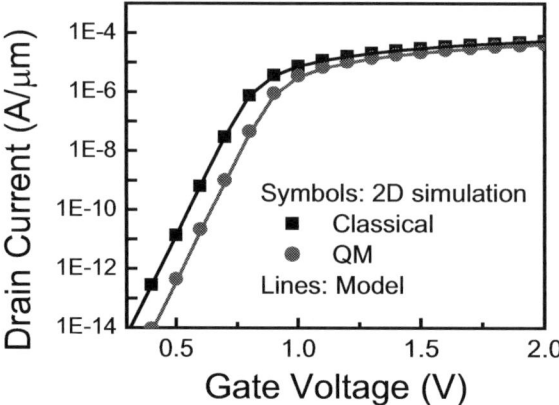

**Fig. 3.15.** The classical and QM-corrected I-V characteristics for an n-type symmetrical/common-gate DG MOSFET.

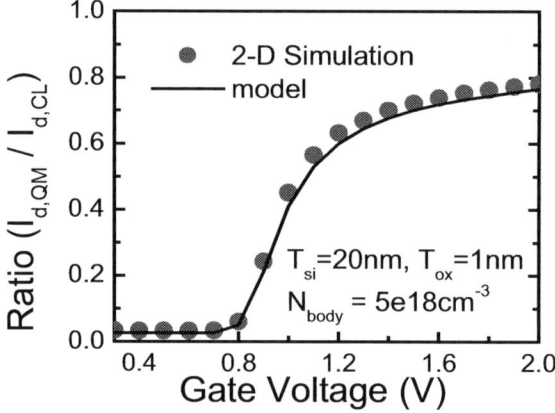

**Fig. 3.16.** The ratio of QM-corrected and classical drain current. The model captures the threshold voltage shift and gate capacitance degradation simultaneously through accurate surface potential calculations.

The predictability of the model is important for multi-gate FET technology/circuits development. The QM model can predict the dependences of QM effects on various device parameters, such as $N_A$, $T_{si}$, $T_{ox}$, etc. Fig. 3.17 shows that the predicted $V_{th}$ shift matches the 2-D simulation results very well over a wide range of body doping concentration. Electrical confinement of carriers dominates in higher doping concentration region, while the structural confinement of carriers dominates in lower doping concentration region.

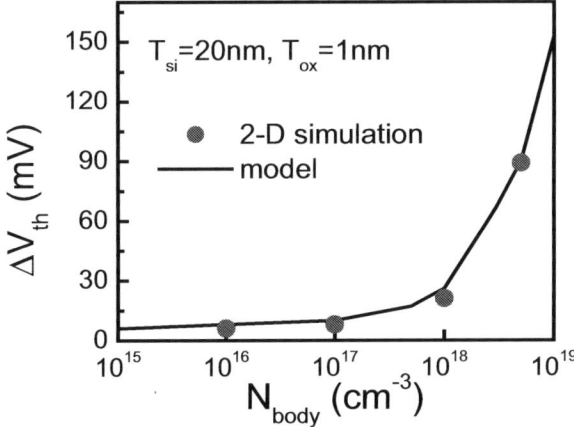

**Fig. 3.17.** $V_{th}$ shift can be modeled accurately over a wide range of body doping. EC dominates in higher doping region, while SC dominates in lower doping region.

### 3.4.2.2 Short-channel Effects (SCE)

The FinFET structure is becoming the most promising candidate to be scaled to the shortest channel length for a given oxide thickness due to the enhanced electrostatic control from the multiple gates. Critical geometry parameters which determine device short-channel behaviors include gate length, fin thickness, fin height, oxide thickness and channel doping. It is very important to include all these parameters in the short-channel model to give correct scalability over a wide range of device parameters.

In BSIM-CMG, the degree of SCE ($V_{th}$ roll-off, drain-induced-barrier-lowering (DIBL), and subthreshold slope degradation) depends on strength of gate control which is modeled by a characteristic field penetration length ($\lambda = f(T_{ox}, T_{si})$) derived from quasi 2-D Poisson's equation. The SCE model shows excellent agreements with 2-D TCAD simulation results without the use of any fitting parameters. Good scalability over $T_{ox}$ and $T_{si}$ down to 30nm channel length ($L_g$) is clearly visible. The SCE model is extended for considering the triple or more gates structures by making $\lambda = f(T_{ox}, T_{si}, H_{fin})$. The SCE model implementation captures the $V_{th}$ roll-off, DIBL and subthreshold slope degradation for short-channel multi-gate FETs simultaneously.

The short-channel behavior is determined by the change in the minmum potential barrier ($\Delta\Psi_m$) inside the conduction channel due to the potential coupling from the drain terminal. Suzuki et al. [18] reported the scaling theory of double-gate MOSFETs by solving the 2-D Poisson's equation of potential inside the conduction channel. $\Delta\Psi_m$ is modeled through a characteristic field penetration length. By linking $\Delta\Psi_m$ to the effective gate bias, this approach is computationally efficient and easily extended to consider QM effect-induced finite inversion charge thickness. Taur et al. [19] reported a full 2-D analytical solution of the potential and the subthreshold current which is accurate and predictive. However, the scale length $\lambda$ in this approach is obtained iteratively. An empirical fitting of $\lambda$ is needed for compact modeling. The flexibility of extending the model in [19] to consider the QM effects remains questionable.

For the BSIM-CMG and BSIM-IMG models, a sophisticated SCE model based on Suzuki's approach is developed. Fig. 3.18 shows the schematic diagram of the symmetric/common gate DG-FET. The 2-D Poisson's equation in the subthreshold region can be written as:

$$\frac{d^2\psi(x,y)}{dx^2} + \frac{d^2\psi(x,y)}{dy^2} = \frac{qN_A}{\varepsilon_{Si}} \tag{3.39}$$

where $N_A$ is body doping. Since the transistor is in subthreshold regime, the inversion carriers are ignored.

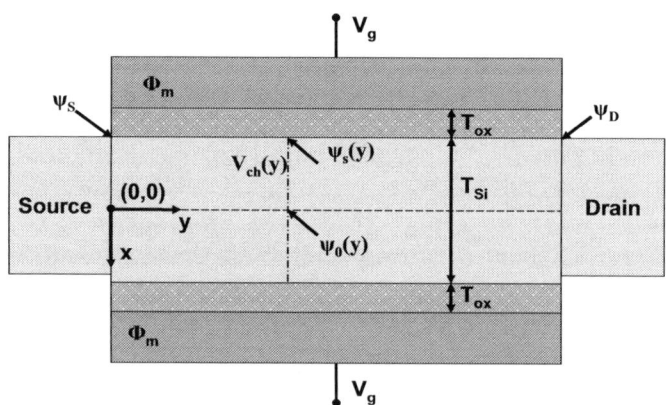

Fig. 3.18. Schematic of the symmetric common-gate DG-FET under study.

In the subthreshold region, the parabolic potential profile is assumed in the vertical x-axis direction: [20]

$$\psi(x,y) = C_0(y) + C_1(y) \cdot x + C_2(y) \cdot x^2 \qquad (3.40)$$

Combined with two boundary conditions at middle of channel ($x=0$) and channel/dielectric interface ($x=T_{si}/2$ and $x=-T_{si}/2$):

$$\left.\frac{d\psi(x,y)}{dx}\right|_{x=0} = 0 \qquad (3.41)$$

$$\left.\frac{d\psi(x,y)}{dx}\right|_{x=\pm\frac{T_{si}}{2}} = \pm\frac{V_g - V_{fb} - \psi_s}{t_{ox}} \cdot \frac{\varepsilon_{ox}}{\varepsilon_{si}} \qquad (3.42)$$

where $V_g$ is the gate voltage, $V_{fb}$ the flat band voltage, and $\Psi_s$ is the surface potential. The potential profile $\Psi(x,y)$ is given by:

$$\psi(x,y) = \psi_s(y) - \frac{V_g - V_{fb} - \psi_s(y)}{t_{ox}} \cdot \frac{\varepsilon_{ox}}{\varepsilon_{si}} \cdot (x + \frac{T_{si}}{2}) \qquad (3.43)$$
$$+ \frac{V_g - V_{fb} - \psi_s(y)}{t_{ox} t_{si}} \cdot \frac{\varepsilon_{ox}}{\varepsilon_{si}} \cdot (x + \frac{T_{si}}{2})^2$$

The SCE is determined by the change of minimum potential barrier. In the DG MOSFETs, the minimum potential barrier, which determines the leakage path, is located in the center plane of the channel. The potential at the center plane of the channel ($\Psi_c$) is obtained by evaluating equation 3.43 at $x=0$:

$$\psi_c(y) = \psi_s(y) - \frac{V_g - V_{fb} - \psi_s(y)}{t_{ox}} \cdot \frac{\varepsilon_{ox}}{\varepsilon_{si}} \cdot \frac{T_{si}}{4} \qquad (3.44)$$

The potential profile $\Psi(x,y)$ is then expressed in terms of $\Psi_c(y)$. The resulting expression is substituted in the 2-D Poisson's equation of potential. We can formulate the differential equation of potential at the center plane of channel in terms of characteristic field penetration length ($\lambda$).

$$\frac{d^2\psi_c(y)}{dx^2} + \frac{V_g - V_{fb} - \psi_c(y)}{\lambda^2} = \frac{qN_A}{\varepsilon_{Si}} \qquad (3.45)$$

where:

$$\lambda = \sqrt{\frac{\varepsilon_{Si}}{2\varepsilon_{ox}} \cdot \left(1 + \frac{\varepsilon_{ox} T_{si}}{4\varepsilon_{Si} t_{ox}}\right) \cdot T_{si} \cdot t_{ox}} \qquad (3.46)$$

Applying two boundary conditions for $\Psi_c(y)$ where $\Psi_c(y=0)=V_{bi}$ and $\Psi_c(y=L)=V_{bi}+V_{ds}$, one can solve the above Poisson's equation for $\Psi_c(y)$:

$$\psi_c(y) = V_{SL} + (V_{bi} - V_{SL})\frac{\sinh\left(\frac{L-y}{\lambda}\right)}{\sinh\left(\frac{L}{\lambda}\right)} + (V_{bi} + V_{ds} - V_{SL})\frac{\sinh\left(\frac{y}{\lambda}\right)}{\sinh\left(\frac{L}{\lambda}\right)} \qquad (3.47)$$

where

$$V_{SL} = V_g - V_{fb} - \frac{qN_A}{\varepsilon_{Si}}\lambda^2 \quad \text{and} \quad V_{bi} = 0.56 + kT/q \cdot \ln\left(\frac{N_A}{n_i}\right) \qquad (3.48)$$

The minimum point of $\phi_c(y)$ will determine the short-channel behavior and is formulated as:

$$\psi_c(\min) = V_{SL} - \frac{V_{ds}^2 \cdot e^{-L/2\lambda}}{\left(e^{L/\lambda} - e^{-L/\lambda}\right)\sqrt{Z_0 Z_L}} + 2\sqrt{Z_0 Z_L}\frac{\sinh(L/2\lambda)}{\sinh(L/\lambda)} \qquad (3.49)$$

where

$$Z_0 = V_{bi} - V_{SL} \quad \text{and} \quad Z_L = V_{bi} - V_{SL} + V_{ds} \qquad (3.50)$$

The minimum potential barrier $\Psi_c(min)$ is controlled by device geometry, channel doping and drain potential. One can use an effective $V_g$ shift ($\Delta V_g$) in the long-channel model to obtain the same potential barrier for short-channel devices:

$$\Delta V_g = \psi_c(\min) - V_{SL} \qquad (3.51)$$

Since the voltage shift is function of gate bias, it captures the change in subthreshold slope simultaneously.

Fig. 3.19 compares the model-predicted $I_d$-$V_g$ characteristics against 2-D TCAD results for difference channel lengths. The model predicts both $V_{th}$ roll-off and subthreshold degradation simultaneously without the use of any fitting parameters.

Fig. 3.20 (a) and (b) compares the threshold voltage ($V_{th}$) roll-off extracted from the model against 2-D TCAD results without the use of any fitting parameter. Good scalability over $T_{ox}$ and $T_{si}$ down to 30nm channel length ($L_g$) is clearly visible.

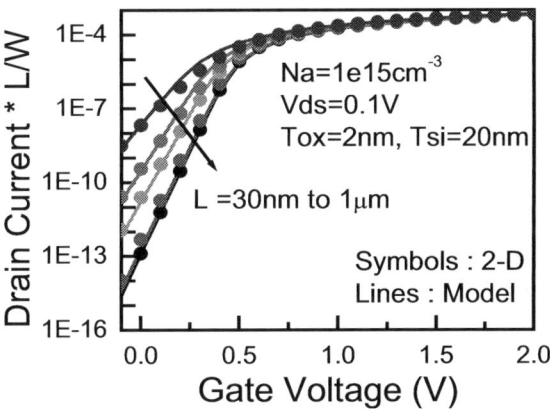

**Fig. 3.19.** Model-predicted L dependence of Id-Vg characteristics against TCAD simulation results.

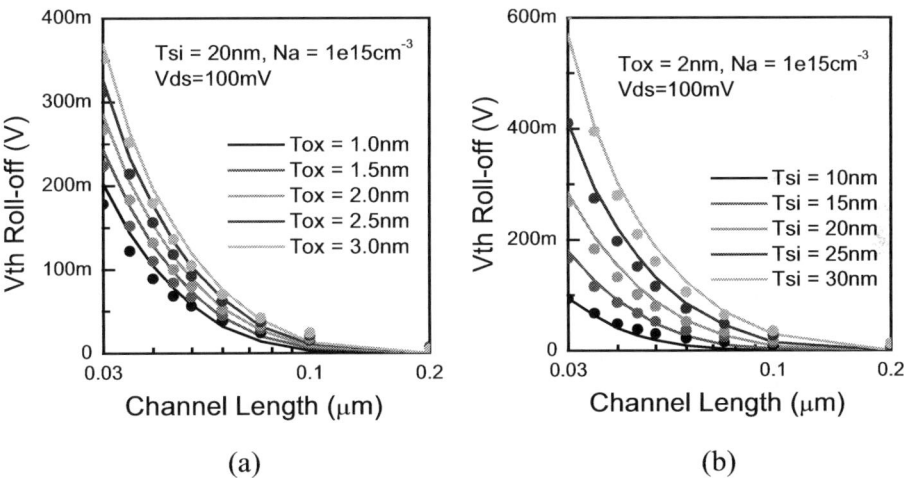

**Fig. 3.20.** Scalability of SCE model has been demonstrated through threshold voltage roll-off for different (a) oxide thickness and (b) body thickness for DG-FETs.

For a FinFET with more than two gates, the physical location of the minimum potential barrier (or the path for maximum drain leakage current) is different from that in a DG-FET. The extra electrostatic control from vertical ends (top gate or bottom gate) reduces short-channel effects. The $V_{th}$ roll-off decreases as fin height ($H_{fin}$) decreases. Fig. 3.21 shows the current density distribution of triple-gate FinFET biased in the subthreshold region. The most leaky channel path is located at the center bottom of the fin where the electrostatic control from the gate is the weakest. The potential barrier at this most leaky path decreases as fin height increases, resulting in an $H_{fin}$ dependence of short-channel effects.

Pei *et al.* have proposed a 3-D analytical electrostatic potential based on the solution of the 3-D Laplace equation in subthreshold region to predict the short-channel behavior of FinFETs.[21] In Eq. 3.46, the DG-FET short-channel effect are modeled by a characteristic field penetration length ($\lambda = f(T_{ox}, T_{si})$). To model the fin height dependence on short-channel effects, a new characteristic field penetration length $\lambda_{Hfin}$ is introduced:

$$\lambda_{Hfin} = \sqrt{\frac{\varepsilon_{Si}}{4\varepsilon_{ox}} \cdot \left(1 + \frac{\varepsilon_{ox} H_{fin}}{2\varepsilon_{Si} T_{ox}}\right) \cdot H_{fin} \cdot T_{ox}} \qquad (3.52)$$

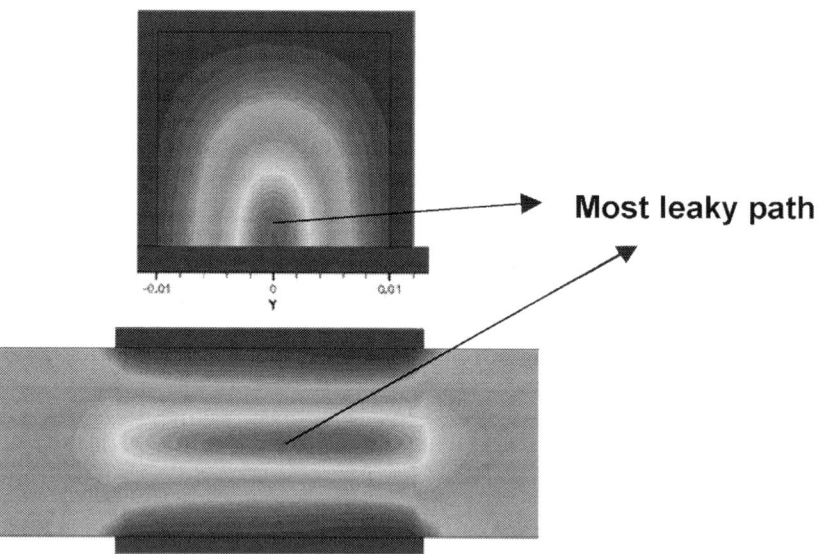

**Fig. 3.21.** Leakage current path is different in triple-gate FETs due to 3D effects. The most leaky path located at the bottom center of the fin.

The final characteristic length used in the short-channel model will be the average of the two scaling lengths:

$$\lambda_{eff} = \frac{1}{\sqrt{\left(\frac{1}{\lambda}\right)^2 + \left(\frac{a}{\lambda_{Hfin}}\right)^2}} \quad (3.53)$$

where $a = 0$ for DG-FET, $a = 0.5$ for triple-gate FET, $a = 1$ for surrounding-gate FET. Note that in the case of triple-gate FET, one can also use different oxide thickness in $\lambda$ and $\lambda_{Hfin}$ to model the thick $SiO_2$ layer (hard mask) on top of the fin. Fig. 3.22 shows the comparison of predicted $V_{th}$ roll-off between compact model and the TCAD simulations. Fin height dependence for the short-channel effect is verified using 3-D TCAD simulation. As shown in Fig. 3.22, the $V_{th}$ roll-off increases as fin height increases for a given fin thickness. The model agrees with the TCAD simulation results very well. By making the characteristic field penetration length as a function of $H_{fin}$, the DG short-channel model is extended to triple-gate and surrounding-gate FETs.

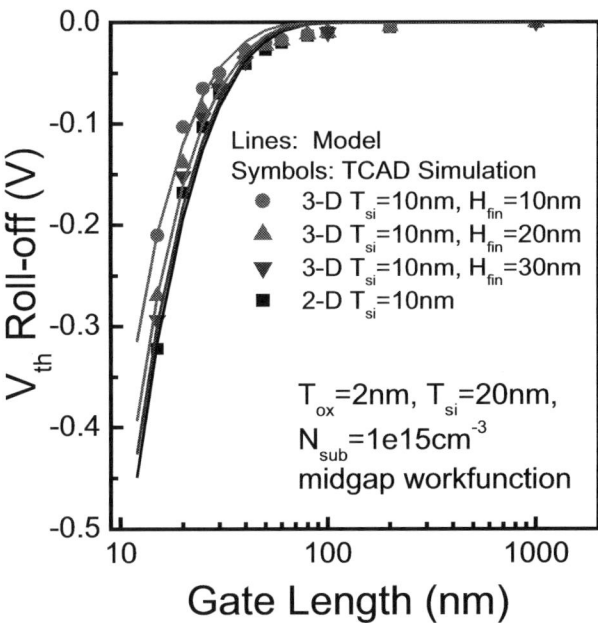

**Fig. 3.22.** SCE model exhibits fin height scalability for tri-gate FETs (Symbols: TCAD, Lines: Model).

### 3.4.3 Experimental Verification

The BSIM-CMG model has been used to describe the measurements of two different FinFET technologies – SOI FinFETs and bulk FinFETs.[9] BSIM-CMG is used to represent the measured drain current and its derivatives, transconductance ($g_m$) and output conductance ($g_{ds}$), for both long-channel and short-channel Multigate FETs.

The SOI FinFETs were fabricated on a lightly doped, 60nm-thick film with 2nm $SiO_2$ dielectric and a strained TiSiN gate. The strained gate strains the channel to enhance the electron mobility, hence increasing the current drive. Measured devices had 20 parallel fins, where each fin is 22nm thick. Figure 3.23 shows the model fitting to $I_d$ and its derivatives for short-channel $L_g$ = 90nm device. Precise modeling of physical phenomenon such as DIBL, mobility degradation, channel length modulation and GIDL is clearly visible from Figure 3.23. To further illustrate the strength of model for use in analog design, Figure 3.24 shows the model description of transconductance efficiency ($g_m/I_d$) and output conductance ($g_{ds}$) of a long-channel device SOI FinFET ($L_g$ = 1μm).

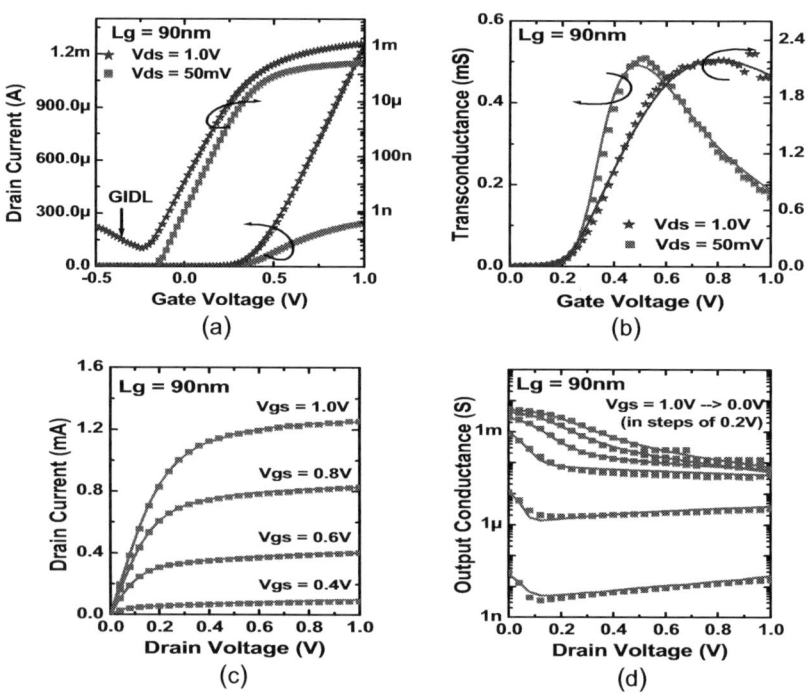

**Fig. 3.23.** BSIM-CMG model fitting to short-channel SOI FinFETs ($L_g$ = 90nm). Symbols represent the measured data and lines indicate model fitting results.

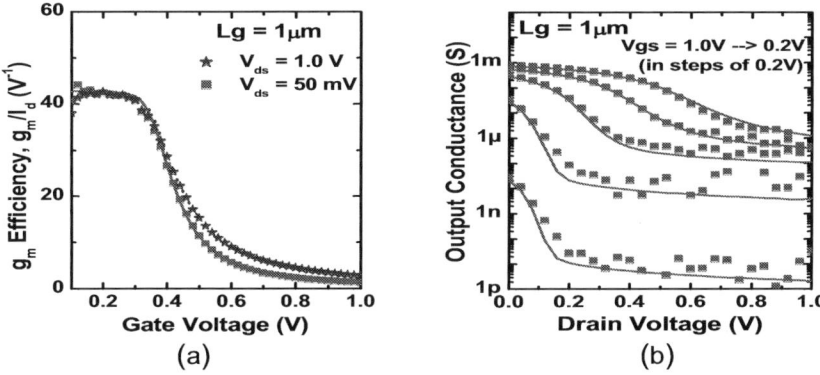

**Fig. 3.24.** BSIM-CMG model fitting to analog design metrics (a) $g_m/I_d$ and (b) $g_{ds}$ for long-channel SOI FinFET ($L_g$ =1μm). Symbols represent the measured data and lines indicate model fitting results.

The BSIM-CMG model has also been verified against bulk FinFET measurements. Bulk FinFETs with moderate doping were fabricated with a TiN gate. Measured devices have 25nm thick fins and an EOT of 1.95nm. Figure 3.25 shows the measured short-channel ($L_g$ = 50nm) characteristics and the corresponding BSIM-CMG fitting results. As explained earlier in Section 3.3.1, BSIM-CMG is extended to bulk FinFETs by the addition of "bulk" node and substrate current model. The measured bulk current for the short-channel bulk FinFET together with model fitting in shown in Figure 3.25(b). Good agreement is observed between the model and the measured data for the long-channel ($L_g$ = 0.97μm) transistor as well. Derivatives of the drain current, $g_m$ and $g_{ds}$, are shown for the long-channel bulk FinFET in Figure 3.26.

The experimental verification shows that BSIM-CMG accurately captures the characteristics of advanced multi-gate FETs. Triple-gate multi-gate FETs were used for model verification demonstrating the ability of the model to capture phenomena such as corner effects which are unique to tri-gate and quadruple-gate FETs. The model is able to describe both SOI and bulk silicon based multi-gate FET technologies. Accurate description of the drain current and its derivatives warrants the use of BSIM-CMG for both digital and analog designs.

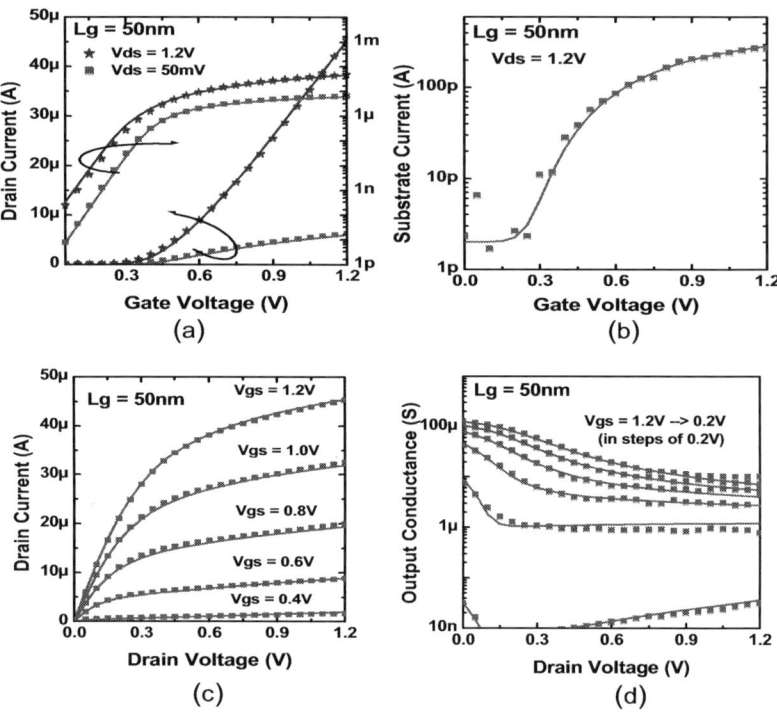

**Fig. 3.25.** BSIM-CMG model fitting to short-channel bulk FinFETs ($L_g$ =50nm). Symbols represent the measured data and lines indicate model fitting results.

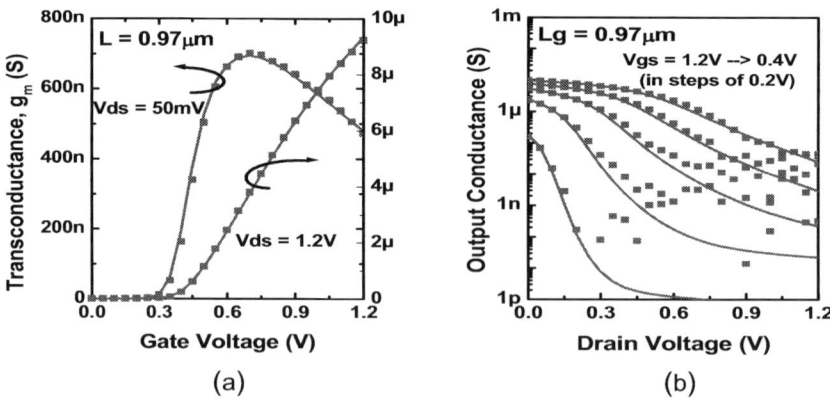

**Fig. 3.26.** BSIM-CMG model fitting to analog design metrics (a) transconductance $g_m$ and (b) output conductance $g_{ds}$ for a long-channel bulk FinFET ($L_g$ =0.97μm). Symbols represent the measured data and lines indicate model fitting results.

## 3.5 The BSIM-IMG Model

BSIM-IMG models multi-gate FETs with independent gate control and asymmetry in the transistor structure. BSIM-IMG model is also surface potential based. A separate core DG-FET model is developed for independent double-gate FETs with a lightly-doped body. The core model consists of a long-channel surface potential model, I-V model and C-V model. Following the core model, several physical effects are added.

This section discusses the surface potential solution of the core long-channel DG-FET with independent gate operation and ends with a brief note on BSIM-IMG implementation and features.

### 3.5.1 Surface Potential of Independent DG-FET

Figure 3.27 shows the schematic of the DG-FET under study. The asymmetric DG-FET can have different front and the back dielectric thickness and different gate workfunctions. Since the threshold voltage of an independent DG-FET can be tuned by adjusting the back gate bias ($V_{gs2}$), there is no need for significant doping in the body. The core model is developed assuming a lightly-doped body.

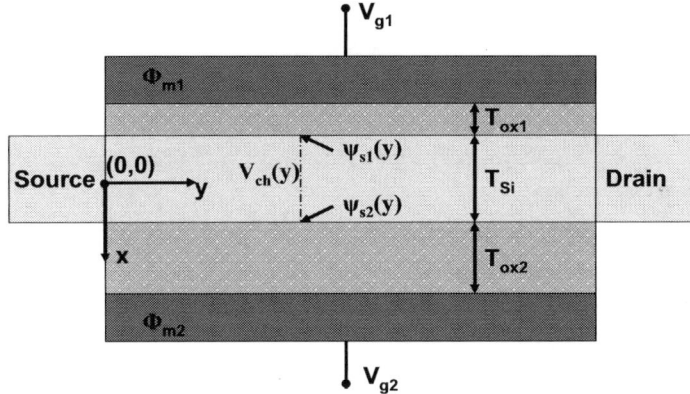

**Fig. 3.27.** Schematic of the independent/asymmetric DG-FET under study showing the asymmetry in dielectric thickness and workfunction.

The electronic potential in the body is obtained by solving the 1-D Poisson's equation together with Gauss's law at the front and back surfaces as the boundary conditions. The 1-D Poisson's equation for the lightly doped body can be written as:

$$\frac{\partial^2 \psi(x,y)}{\partial x^2} = \frac{qn_i}{\varepsilon_{Si}} \cdot e^{\frac{q(\psi(x,y)-V_{ch}(y))}{kT}} \tag{3.54}$$

Gauss's law at the front and back surface imply that

$$V_{g1} - V_{fb1} - \psi_{s1}(y) = -\frac{\varepsilon_{Si}}{C_{ox1}} \cdot \frac{\partial \psi(x,y)}{\partial x}\bigg|_{x=-T_{Si}/2} \tag{3.55}$$

$$V_{g2} - V_{fb2} - \psi_{s2}(y) = \frac{\varepsilon_{Si}}{C_{ox2}} \cdot \frac{\partial \psi(x,y)}{\partial x}\bigg|_{x=+T_{Si}/2}$$

where $\psi_{s1}(y)$ and $\psi_{s2}(y)$ are the front and back surface potentials respectively. The solution to Eq. (3.54) depends on the existence of the zero electric field plane, $\frac{\partial \psi(x,y)}{\partial x}\bigg|_{x=x_0} = 0$.

If the zero electric plane ($x_0$) exists, i.e. $-\infty < x_0 < \infty$, the potential in the body can be written as [22]

$$\psi(x,y) = V_{ch}(y) - \frac{2kT}{q} \cdot \ln\left(\frac{T_{Si}}{2\beta}\sqrt{\frac{q^2 n_i}{2\varepsilon_{Si}kT}}\sin\left(\frac{2\beta x}{T_{Si}} + \alpha\right)\right) \tag{3.56}$$

where $\alpha$ and $\beta$ are a function of the applied bias. The boundary condition for this case can be obtained by substituting Eq. (3.56) in Eq. (3.55).

$$V_{g1} - V_{fb1} - V_{ch}(y) = r_0 + \frac{2kT}{q}\ln\left(\frac{2\beta}{\sin(\alpha-\beta)}\right) + r_1\beta\cot(\alpha-\beta) \tag{3.57}$$

$$V_{g2} - V_{fb2} - V_{ch}(y) = r_0 + \frac{2kT}{q}\ln\left(\frac{2\beta}{\sin(\alpha+\beta)}\right) - r_2\beta\cot(\alpha+\beta)$$

where $r_o = \frac{kT}{q}\ln\left(\frac{2\varepsilon_{Si}kT}{q^2 n_i T_{Si}^2}\right)$ and $r_{1(2)} = \frac{4kT\varepsilon_{Si}}{qC_{ox1(2)}T_{Si}}$. The unknown parameters $\alpha$ and $\beta$ in Eq. (3.56) are obtained by solving the system of equations in Eq. (3.57). It is interesting to note that for a symmetric DG-FET, when the zero-field plane lies in the middle of the silicon film ($x_0 = 0$), $\beta = \pi/2$ and the Eq. (3.57) is identical to the surface potential equation for a lightly doped DG-FET, Eq. (3.19).

If the device is heavily asymmetrical due to the workfunction difference between the two gates and different bias applied to the two gates, the zero electric field plane may not exist at all. In this case, the potential profile in the body is given by:

$$\psi(x,y) = V_{ch}(y) - \frac{2kT}{q} \cdot \ln\left(\frac{T_{Si}}{2\beta}\sqrt{\frac{q^2 n_i}{2\varepsilon_{Si} kT}} \sinh\left(\frac{2\beta x}{T_{Si}} + \alpha\right)\right) \quad (3.58)$$

The unknowns $\alpha$ and $\beta$ are determined by solving the new boundary conditions which are obtained by using the definition of $\psi(x,y)$, Eq. 3.58, in Eq. 3.55:

$$V_{g1} - V_{fb1} - V_{ch}(y) = r_0 + \frac{2kT}{q}\ln\left(\frac{2\beta}{\sinh(\alpha-\beta)}\right) + r_1\beta\coth(\alpha-\beta) \quad (3.59)$$

$$V_{g2} - V_{fb2} - V_{ch}(y) = r_0 + \frac{2kT}{q}\ln\left(\frac{2\beta}{\sinh(\alpha+\beta)}\right) - r_2\beta\coth(\alpha+\beta)$$

Eqs. (3.56)-(3.59) together yield the potentials at the front and back surface of an asymmetric DG-FET under all bias conditions. Figure 3.28 shows the surface potentials for the case of a DG-FET with an $N^+$ front gate and a $P^+$ back gate. The back gate is biased at $V_{g2} = 0V$ and the voltage of front gate is varied.

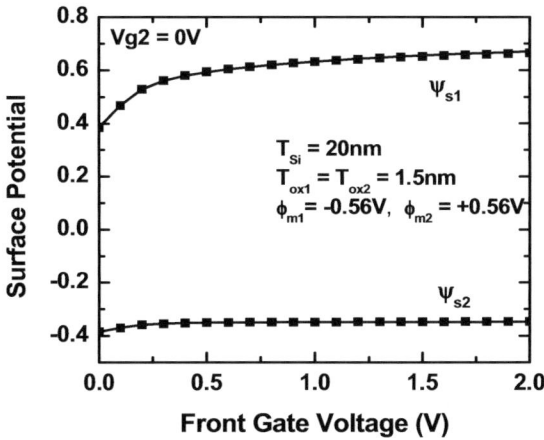

**Fig. 3.28.** Front and back surface potential equation in an asymmetrical, independent-gate DG-FET from the model (lines) and TCAD (symbols).

The surface potential at the source and drain ends are obtained by solving Eqs. (3.56)-(3.59) at $V_{ch}(0) = 0$ and $V_{ch}(L) = V_{ds}$ respectively. The I-V and C-V model can be developed using the surface potentials at source and drain ends.

### 3.5.2 BSIM-IMG Features

Figure 3.29(a) shows the drain current calculated from the model as a function of front gate voltage for different back biasing. BSIM-IMG has excellent accuracy and the threshold voltage variation with back-gate bias is accurately predicted by the model. Figure 3.29(b) shows good agreement between the C-V model of BSIM-IMG and TCAD simulations. The inherent symmetry in the model is clearly indicated by the equality of capacitances $C_{g1d} = C_{g1s}$ at $V_{ds} = 0V$. The verification shown against TCAD in Figure 3.29 did not use any fitting parameters confirming the rooted physics and scalability of the model.

Numerous physical effects are added to the core model to produce an accurate independent/asymmetric multi-gate FET. The similarity in the frameworks allows BSIM-IMG to re-use several physical effects from BSIM-CMG with moderate changes. Table 3.2 lists the physical effects modeled in BSIM-IMG.

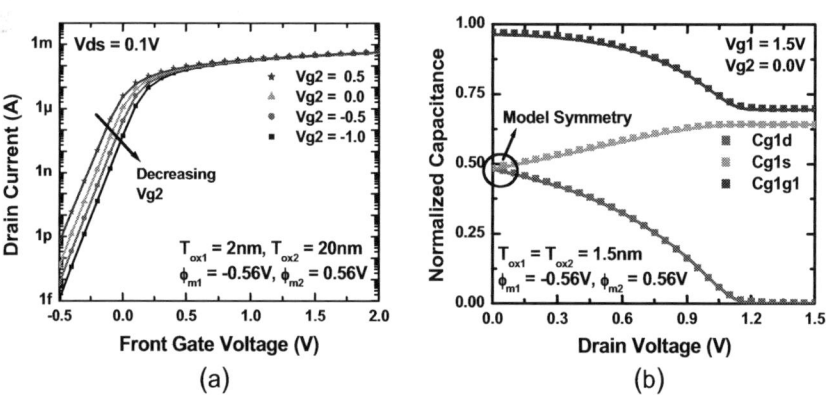

**Fig. 3.29.** Verification of the core BSIM-IMG model prediction (lines) against TCAD simulation (symbols) for the (a) I-V model and (b) C-V model for two different DG-FET structures.

**Table 3.2.** List of physical effects modeled in BSIM-IMG.

1) Quantum Mechanical Effects
2) Short-channel Effects
   a) $V_{th}$ roll-off
   b) DIBL
   c) Sub-threshold slope
   d) Channel length modulation
3) Series resistance
4) Mobility degradation
5) Velocity Saturation
6) Velocity Overshoot/Source-End Velocity Limit
7) Gate Induced Drain (Source) Leakage (GIDL, GISL)
8) Impact Ionization
9) S/D Junction leakage
10) Gate tunneling
11) Parasitic capacitance

## 3.6 Summary

The BSIM-CMG and BSIM-IMG together constitute a versatile multi-gate MOSFET compact model. It is ready to serve multi-gate CMOS technologists and circuit designers to facilitate the transition from the conventional planar MOSFET structure to multi-gate FET structures such as the FinFETs.

## References

1. Y. Taur, D. A. Buchanan, Wei Chen, D. J. Frank, K. E. Ismail, Shih-Hsien Lo, G. A. Sai-Halasz, R. G. Viswanathan, H.-J. C. Wann, S. J. Wind, Hon-Sum Wong: CMOS Scaling into the Nanometer Regime. Proceedings of IEEE **85** 486 (1997)
2. D.J. Frank, R. H. Dennard, E. Nowak, P.M. Solomon, Y. Taur and H.-S.P. Wong: Device Scaling Limits of Si MOSFETs and Their Application Dependencies. Proceedings of IEEE **89**, 259 (2001)
3. D. Hisamoto, Wen-Chin Lee, J. Kedzierski, H. Takeuchi, K. Asano, C. Kuo, E. Anderson, Tsu-Jae King, J. Bokor, Chenming Hu: FinFET – A Self-Aligned Double Gate MOSFET Scalable to 20nm. IEEE Trans. Electron Devices **47**, 2320 (2000)

4   H.-S.P. Wong, D.J. Frank, P.M. Solomon: Device Design Considerations for Double-Gate, Ground Plane and Single-Gate Ultra-Thin SOI MOSFETs at the 25nm Channel Length Generation. *IEDM Tech. Digest*, 407 (1998)
5   Y. Taur: Analytic Solutions of Charge and Capacitance in Symmetric and Asymmetric Double-Gate MOSFETs. IEEE Trans. Electron Devices **48**, 2861 (2001)
6   G.D.J. Smit, A.J. Scholten, N. Serra, R.M T. Pijper, R. van Langevelde, A. Mercha, G. Gildenblat, D.B.M. Klassen: PSP-based compact FinFET model describing dc and RF measurements. *IEDM Tech. Digest*, 175 (2006)
7.  J.-M. Sallese, F. Krummenacher, F. Pregaldiny, C. Lallement, A. Roy, C. Enz: A design-oriented charge-based current model for symmetric DG MOSFET and its correlation with the EKV formalism. Solid-State Electronics **49**, 485 (2005)
8   Mohan V. Dunga, Chung-Hsun Lin, Xumei. Xi, Darsen D. Lu, Ali M. Niknejad, Chenming Hu: Modeling Advanced FET Technology in a Compact Model. IEEE Trans. Electron Devices **53**, 1971 (2006)
9   Mohan V. Dunga, Chung-Hsun Lin, Darsen D. Lu, Weize Xiong, C. R. Cleavelin, P. Patruno, Jiunn-Ren Hwang, Fu-Liang Yang, Ali M. Niknejad, Chenming Hu: BSIM-MG: A Versatile Multi-Gate FET Model for Mixed-Signal Design. *Proceedings of the VLSI Technology Symposium*, 80 (2007)
10  J. R. Brews: A charge sheet model of the MOSFET. Solid-State Electronics **21**, 345 (1978)
11  D. Ward, R. Dutton: A charge-oriented model for MOS transistor capacitances. IEEE J. Solid State Circuits **13**, 703 (1978)
12  Y. King, H. Fujioka, S. Kamohara, K. Chen, C. Hu: DC electrical oxide thickness model for quantization of the inversion layer in MOSFETs. Semicond. Sci. Technol. **13**, 963 (1998)
13  B. Majkusiak, T. Janik, J. Walczak: Semiconductor thickness effects in the double-gate SOI MOSFET. IEEE Trans. Electron Devices **45**, 1127 (1998)
14  L. Ge, J. Fossum: Modeling of Quantization and Volume Inversion in Thin Si-Film Double-Gate MOSFETs. IEEE Trans. Electron Devices, **49**, 287 (2002)
15  V. Trivedi, J. Fossum: Quantum-Mechanical Effects on the Threshold Voltage of Undoped Double-Gate MOSFETs. IEEE Electron Devices Lett. **26**, 579 (2005)
16  Y. Li, S.-M. Yu: A unified quantum correction model for nanoscale single- and double-gate MOSFETs under inversion conditions. Nanotechnology **15**, 1009 (2004)
17  D. Vasileska, Z. Ren: SCHRED-2.1 Manual (2000)
18  K. Suzuki, T. Tanaka, Y. Tasaka, H. Horie, Y. Arimoto: Scaling theory for double-gate SOI MOSFETs. IEEE Trans. Electron Devices **40**, 2326 (1993)
19  X. Liang, Y. Taur: A 2-D Analytical Solution for SCEs in DG MOSFETs. IEEE Trans. Electron Devices **51**, 1385 (2004)
20  R. H. Yan, A. Ourmazd, and K. F. Lee: Scaling the Si MOSFET;from bulk to SO1 to bulk. IEEE Trans. Electron Devices **39**, 1704 (1992)

21. G. Pei, J. Kedzierski, P. Oldiges, M. Ieong, E. Kan: FinFET Design Considerations Based on 3-D Simulation and Analytical Modeling. IEEE Trans. Electron Devices **48**, 1441 (2002)
22. H. Lu, Y. Taur: An Analytic Potential Model for Symmetric and Asymmetric DG MOSFETs. IEEE Trans. Electron Devices **53**, 1161 (2006)

# 4 Physics of the Multigate MOS System

Bogdan Majkusiak

## 4.1 Device electrostatics

The schematic structure in Fig. 4.1 illustrates the geometry of the multigate MOS system in its simplest form. The semiconductor body in the middle of the structure is the transport region, through which electrons travel from the source to the drain. The semiconductor region is surrounded by the gate stack.

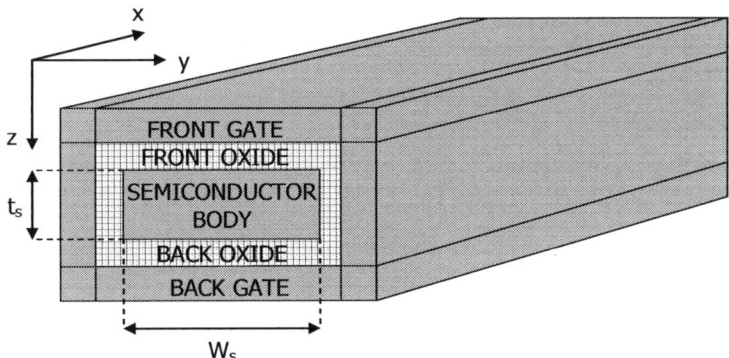

**Fig. 4.1.** Geometry of the multigate MOS system.

The gates can be connected together or can be electrically isolated. If there are no side gates, the structure is a double-gate device. If the back oxide is very thick, the structure can be regarded as a single-gate silicon-on-insulator device. If there are side gates and if the back oxide is very thick, the structure is the triple-gate device. The cross section of the multigate system can be non-rectangular, for instance cylindrical or triangular, but in this chapter, a rectangular channel cross section will be considered since it is easier to use to illustrate the physics of the device.

Studying the electrostatics of the device involves the determination of the electrostatic potential $\phi$ and the charge density $\rho$ in the structure under given bias conditions. The charge density comprises contributions from electrons, holes, positively and negatively ionized impurities with concentrations $n$, $p$, $N_d^+$, and $N_a^-$, respectively:

$$\rho(r) = q\left[p - n + N_d^+ - N_a^-\right] \qquad (4.1)$$

and is tied to the electrostatic potential through Poisson's equation:

$$\nabla^2 \phi = -\frac{\rho}{\varepsilon} \qquad (4.2)$$

where $\varepsilon$ is the electrical permittivity.

In the quantum-mechanical approach, an electron in a given state is described by a wave function $\varphi(r)$ that depends on the position $r$, since electrons are not strictly localized in the geometrical space. The spectrum of the wave functions $\varphi(r)$ and the allowed energy states can be found by solving Schrödinger's equation using the effective mass approximation:

$$-\frac{\hbar^2}{2} \nabla\left(m_e^{-1} \nabla \varphi\right) + E_C(r)\varphi = E\varphi \qquad (4.3)$$

The boundary conditions for this equation are set by imposing continuity for $\varphi$ and for the product $m_e^{-1}\nabla\varphi$ at the boundaries of the different regions of the device.

In (4.3), $m_e^{-1}$ is the reciprocal electron effective mass tensor. The bottom of the silicon conduction band can be represented by six valleys located on the wave vector axes corresponding to the motion along [100]-equivalent directions in the crystal. For a (100)-oriented crystal surface and a (100) oriented the channel cross-section, these axes coincide with the directions $x$, $y$, $z$ of the device, as shown in Fig. 4.2, and the reciprocal effective mass tensor is diagonal. The dynamics of electrons near the valley edges can be approximated by parabolic $E(k)$ relationships with a longitudinal effective mass $m_l = 0.916m_0$ and a transverse effective mass $m_t = 0.19m_0$ as parameters. Therefore, there are two spectra of allowed energies for the motion of electrons in any given direction of the silicon crystal, resulting from solving (4.3) with either $m_e = m_l$ or $m_e = m_t$.

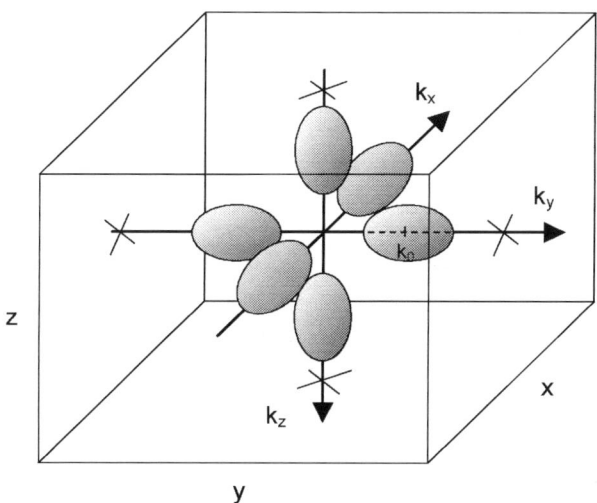

**Fig. 4.2.** Constant energy ellipsoids of the conduction band for the (100) oriented silicon surface against a background of the geometrical space.

The concentration of electrons in a given place $r$ is obtained calculating the sum of the product of the occupation probability of a given electron state $f(E)$ by the state degeneracy factor $g_n$, and the density of probability $|\varphi(r)|^2$ of finding the electron at location $r$:

$$n(r) = \sum_n g_n f(E) |\varphi_n(r)|^2 \tag{4.4}$$

over all allowed states. The set of the states can be discrete or quasi-continuous depending on the geometry of the system.

Since there is no dimensional confinement for motion in the direction of the transport along the channel, $x$, the wave vector $k_x$ and the related kinetic energy component are a quasi-continuous function of $x$. The allowed states in the wave vector space create conduction modes at the intersection of the allowed $k_y$ and $k_z$ planes, as illustrated in Fig. 4.3. In this case, the summation over the energy levels connected with the motion in the $x$ direction must be replaced by integration over the quasi-continuously changing $E_x$ spectrum with the use of the one-dimensional density-of-state function $\sigma_{1D}$, starting from the bottom of each subband $E = E_n$ corresponding to the $n$-th wave function $\varphi_n(y,z)$:

$$n(r) = \sum_n |\varphi_n(y,z)|^2 \int_{E_n}^{\infty} \sigma_{1D}(E) f(E) dE \qquad (4.5)$$

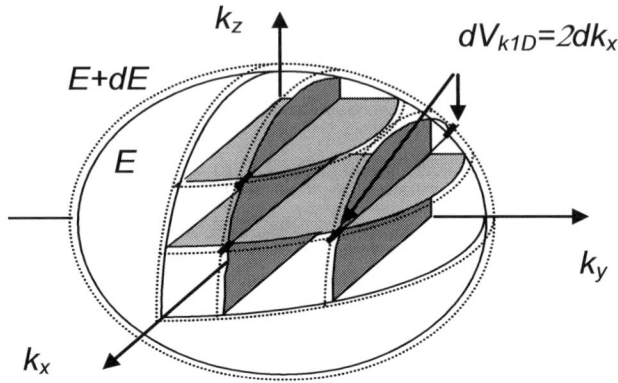

**Fig. 4.3.** Change of the one-dimensional wave vector space volume with an increase of the total energy $E$ under conservation of $k_y$ and $k_z$.

Having a one-dimensional density of states (DOS) $\sigma_{1D}$ means that the number of allowed states in the unit is confined to a one-dimensional geometrical space. The unit increment of the energy connected with motion is also confined in this space. Any increase of the total energy $E$ with conserved $k_y$ and $k_z$ wave vectors corresponds to the change of the one-dimensional wave vector space volume $dV_{k1D} = 2dk_x$ at two sides of each ellipsoid, as shown in Fig. 4.3. Taking into account the density of states $2/2\pi$ of the phase space $xk_x$ (the factor 2 results from the spin degeneracy and $2\pi$ is the volume of the elementary cell in this space), one obtains:

$$\sigma_{1D}(E) = \frac{dN_{1D}}{dE}\bigg|_{k_y,k_z} = \frac{2}{2\pi} \frac{dV_{k1D}}{dE} = \sum_n \frac{\sqrt{2m_x}}{\pi\hbar} \frac{1}{\sqrt{E-E_n}} \theta_n(E-E_n) \qquad (4.6)$$

where $m_x = m_l$ or $m_t$ depending on which ellipsoid in the silicon conduction band is being considered, and $\theta$ is the step function.

Under thermal non-equilibrium, the occupation probability function $f(E)$ changes with position. However, the current flow in the directions perpendicular to the gate surfaces ($y$ and $z$) is practically negligible, and for small perturbations of the thermal equilibrium in the transport direction $x$ the occupation probability function can be approximated by the quasi-equilibrium Fermi-Dirac distribution function:

$$f(E) = f(E,x) = \frac{1}{1+\exp\left[\dfrac{E-E_{Fn}(x)}{k_B T}\right]} \quad (4.7)$$

with the electron quasi-Fermi level $E_{Fn}$ as $x$-position dependent parameter.

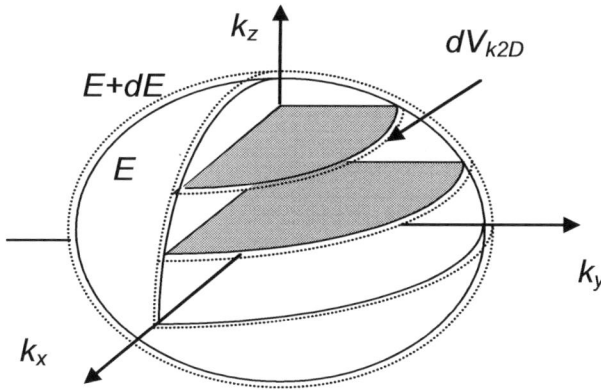

**Fig. 4.4.** Change of the two-dimensional wave vector space volume with an increase of the total energy $E$ under conservation of $k_z$.

If the width $W_s$ of the semiconductor body is large enough for the spectrum of allowed states connected with the motion in the $y$ direction to be nearly continuous, free motion planes are created in the wave vector space in conjuction with with the $k_x$ space, as shown in Fig. 4.4. In such a case, the electrostatics problem can be simplified and one can skip the need for solving Schrödinger's equation in the $y$-direction as well. Then, expression (4.4) must be modified and the summation over the energy connected with the motion in the $x$ and $y$ directions can be replaced by the following integration over the quasi-continuous energy spectrum:

$$n(\mathbf{r}) = \sum_n |\varphi_n(z)|^2 \int_{E_n} \sigma_{2D}(E) f(E,x) dE \quad (4.8)$$

where $\sigma_{2D}$ is the two-dimensional density-of-state function, which describes the number of allowed states in the two-dimensional unit of the geometrical space and the unit increment of the total energy under conservation of the $k_z$ component:

$$\sigma_{2D}(E) = \frac{dN_{2D}}{dE}\bigg|_{k_z} = \frac{2}{(2\pi)^2}\frac{dV_{k2D}}{dE} = \sum_n \frac{m_{d2D}}{\pi\hbar^2}\theta_n(E-E_n) \qquad (4.9)$$

In (4.9), $m_{d2D} = (m_x m_y)^{1/2}$ is the two-dimensional density-of-state mass for a given energy ellipsoid. For the (100)-equivalent silicon layer surface orientations $m_{d2D}$ is equal to $m_t$ for electrons in the ellipsoids centered on the $k_z$ axis and to $(m_l m_t)^{1/2}$ for electrons in the ellipsoids centered on the $k_x$ or $k_y$ axes. Summation over the allowed $E_z$ energy levels must take into account the two spectra ($j =1,2$) obtained from Schrödinger's equation for two possible values of $m_z$, including the appropriate degeneracy factor $g_v = 2$ for $m_z = m_l$ and $g_v = 4$ for $m_z = m_t$. Since the two-dimensional density of states is a constant quantity, the integration in (4.8) with the quasi-equilibrium Fermi-Dirac function (4.7) can be performed analytically and the electron concentration distribution in the silicon can be expressed as:

$$n(r) = \sum_{j=1,2}\sum_i \frac{g_{vj} m_{d2Dj}}{\pi\hbar^2} k_B T \ln\left[1 + \exp\left(\frac{E_{Fn}-E_{ij}}{k_B T}\right)\right] |\varphi_{ij}(z)|^2 \qquad (4.10)$$

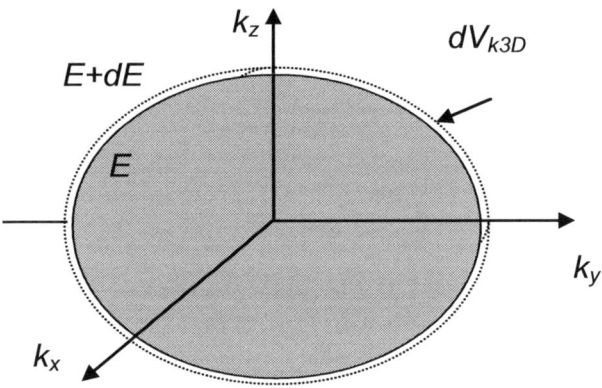

**Fig. 4.5.** Change of the three-dimensional wave vector space volume with an increase of the total energy $E$.

Furthermore, if the thickness of the semiconductor body is large enough, such that the spectrum of allowed wave vectors and energies is quasi-continuous in all three directions, as illustrated in Fig. 4.5, one can avoid the need for solving the Schrödinger equation in the $z$ direction as well, which yields the classical expression for the 3D the electron concentration:

$$n(r) = \int_{E_C(r)}^{\infty} \sigma_{3D}(E) f(E, x) dE \qquad (4.11)$$

where the three dimensional density-of-states function is expressed by:

$$\sigma_{3D}(E) = \frac{dN_{3D}}{dE} = \frac{2}{(2\pi)^3} \frac{dV_{k3D}}{dE} = \frac{1}{2\pi^2} \left(\frac{2m_{d3D}}{\hbar^2}\right)^{3/2} \sqrt{E - E_C} \qquad (4.12)$$

where $m_{d3D}$ is the three-dimensional density-of-states mass, which is the equivalent spherical band mass giving the same state population as six ellipsoidal bands:

$$m_{d3D}(E) = \left(\sum_{j=1}^{6} \sqrt{m_{xj} m_{yj} m_{zj}}\right)^{2/3} = 6^{2/3} \left(m_l m_t^2\right)^{1/3} \qquad (4.13)$$

Equation (4.11) is usually written in the form:

$$n = N_C \mathfrak{I}_{1/2}\left(-\frac{E_C - E_{Fn}}{k_B T}\right) \qquad (4.14)$$

where $N_C$ is the effective density of states for the conduction band:

$$N_C = \frac{1}{4}\left(\frac{2m_{d3D} k_B T}{\pi \hbar^2}\right)^{3/2} \qquad (4.15)$$

and $\mathfrak{I}_{1/2}(\eta)$ represents the Fermi integral of the ½ order:

$$\mathfrak{I}_{1/2}(\eta) = \frac{2}{\sqrt{\pi}} \int_0^{\infty} \frac{\xi^{1/2}}{1 + \exp(\xi - \eta)} d\xi \qquad (4.16)$$

In case of a non-degenerated conduction band, i.e., when $\eta = -(E_C - E_F)/k_B T < 2$, the Fermi-Dirac distribution can be approximated by the Maxwell-Boltzmann distribution:

$$n = N_C \exp\left(-\frac{E_C - E_{Fn}}{k_B T}\right) \qquad (4.17)$$

Approaches based on the solution of the Schrödinger equation require applying appropriate boundary conditions. The problem is not trivial since the quantum system is open for energy levels above the bottom of conduction band in the bulk of the gate electrodes and the electron wave functions can propagate within them infinitely. This renders the normalization of the wave functions impossible. The most common approximation used to solve this problem is based on the assumption that the wave functions vanish at a chosen plane: at the gate-insulator interface or at the insulator-semiconductor interface. The latter choice is equivalent to the assumption that the potential barrier 'seen' by electrons from the semiconductor side of the surface is infinite. Cutting the wave functions at a chosen plane results in a neglect of the probability represented by part of the wave functions propagating beyond this plane but also affects the resulting spectrum of allowed energies in the semiconductor region.

Poisson's equation, which is the equation at the core of our electrostatics problem, can also undergo some simplifications. It cannot, however, be simplified by separation into a set of one-dimensional equations, because the electrical charge density $\rho(r)$ is not ascribed to any distinguished direction.

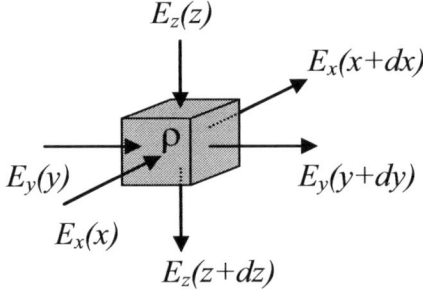

Fig. 4.6. Electric charge as a source of changes of the electric field.

Poisson's equation can be written as:

$$\frac{\partial E_x}{\partial x} + \frac{\partial E_y}{\partial y} + \frac{\partial E_z}{\partial z} = -\frac{\rho(x,y,z)}{\varepsilon} \tag{4.18}$$

Fig. 4.6 shows that the electric charge density is the source of electric field variations. Some simplification of the problem is possible if the electric field is constant or nearly constant in some directions. The first situation takes place in single- and double-gate devices if the channel width $W_s$ large enough for any edge effects to be negligible. In that case, the potential is practically constant along the $y$ direction and the electric field component $E_y$ is zero. Another approximation that is frequently used to describe long-channel transistors and that is known as the gradual channel approximation. It relies on the assumption that the variation of the lateral electric field $dE_y/dy$ along the channel is much smaller than the variation of the transverse electric field $dE_x/dx$.

In conclusion to this section, it is worthwhile noting that solving the electrostatics problem in the multigate MOS system can be simplified by eliminating the need for solving Schrödinger's equation in one or two directions. Such a "classical" simplification has some implications that can be either insignificant or, on the contrary, quite important for the accuracy of the simulation, depending on the geometry and the dimensions of the structure. This will be investigated in the next chapter in the basis on numerical simulations of the double-gate MOS system.

## 4.2 Double gate MOS system

### 4.2.1 Modeling assumptions

Figure 4.7 presents the energy band diagram and the allowed electron energy spectrum $E_{ij}$ in a symmetrical double-gate device, while Fig. 4.8 shows the electron concentration distribution in the semiconductor at a gate voltage $V_G = 1$ V. The (100)-oriented silicon layer has a thickness $t_s = 10$ nm and an acceptor atom concentration $N_A = 10^{17}$ cm$^{-3}$. The gate stack consist of a SiO$_2$ layer having a thickness $t_{ox} = 1$ nm and a metal electrode with midgap work function $\phi_M = 4.6$ eV.

The energy levels $E_{i1}$ corresponding to the longitudinal effective mass (solid lines in Fig. 4.7) are located closer to one another than levels $E_{i2}$, which correspond to the transverse mass (dashed lines). The electron Fermi level $E_{Fn}$ is assumed to overlap with the majority carrier Fermi level $E_F$ which ties the semiconductor region to the external voltage source. The energy bands relax to equilibrium outside the gate region. At the gate voltage under consideration, the charge density in the whole semiconductor region is dominated by the negative charge of electrons and the ionized acceptors. As a result, the potential energy exhibits a maximum in the middle of the semiconductor layer and decreases at the semiconductor surfaces, creating two quantum subwells.

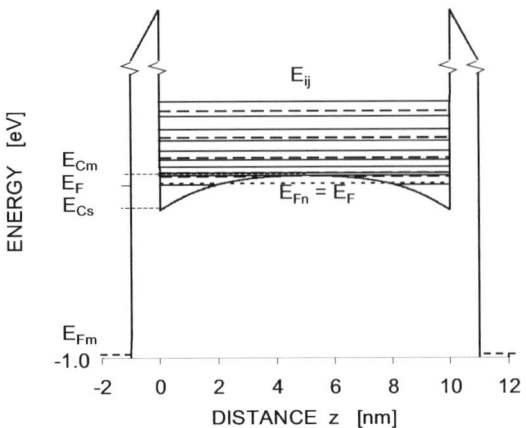

**Fig. 4.7.** Band diagram of a symmetrical double-gate (DG) MOS system at $V_G=1$V.

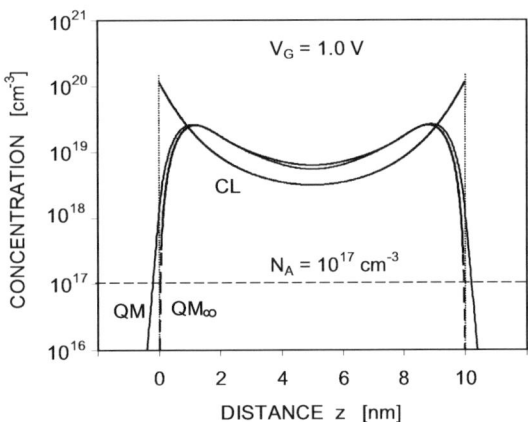

**Fig. 4.8.** Electron concentration distribution in the silicon body for different approaches: QM – quantum-mechanical, QM$_\infty$ – quantum-mechanical with no penetration of the barrier region, CL – classical.

In order to emphasize the importance of the modeling assumptions that are being used, Fig. 4.8 shows the electron concentration distributions in the semiconductor which are obtained according to three approaches: QM – quantum-mechanical (4.10), QM$_\infty$ – quantum-mechanical with no penetration of the barrier region by the electron wave functions, which is equivalent to infinitely high oxide barrier, and CL – classical, based on expression (4.14) which neglects energy quantization effects. The difference between the distributions becomes qualitative near the semiconductor interfaces. The distribution obtained from the classical approach exhibits maxima located exactly at the interfaces, correspondingly to the lowest potential energy, while the distributions resulting from the quantum-mechanical approaches exhibit peaks at a certain distance from the surfaces and decrease to a zero electron concentration ($QM_\infty$) or a near-zero concentration (QM) at the interfaces.

The simplified quantum-mechanical approach that neglects the penetration of the electron wave functions in the gate oxide results in higher positions of the allowed energy levels and larger distances between them. The distance between energy levels affect the rate of transitions in scattering processes, which in turn determines the effective mobility of carriers. Fig. 4.9 shows the lowest longitudinal mass levels as a function of the total surface density of electrons in the semiconductor.

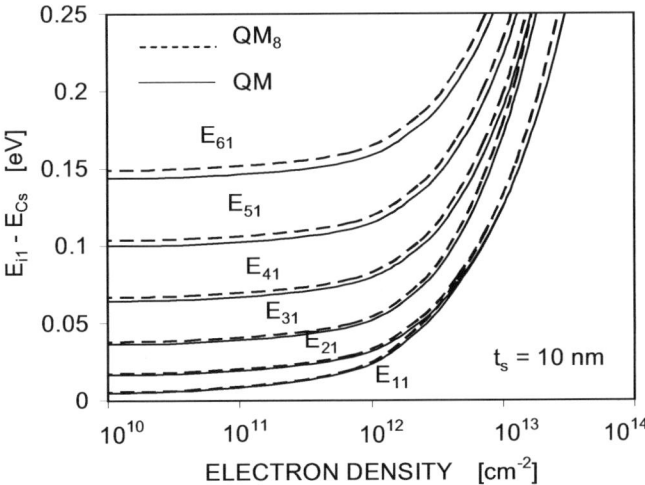

**Fig. 4.9.** Lowest longitudinal mass levels $E_{il}$ as a function of the electron surface density for two approaches: QM – quantum-mechanical, and $QM_\infty$ – quantum-mechanical with no penetration of the gate oxide.

The change of the electron concentration distribution due to the assumption that the electrons do not no penetrate the gate oxide also results in a larger average distance of electrons from the semiconductor surface, which for the case of the symmetrical double gate MOS structure can be defined as:

$$z_{av} = \int_0^{t_s/2} zn(z)dz \bigg/ \int_0^{t_s/2} n(z)dz \qquad (4.19)$$

As it can be seen in Fig. 4.10, the classical approach results in a significantly less average distance $z_{av}$ due to the location of the peak of the electron concentration at the semiconductor surface.

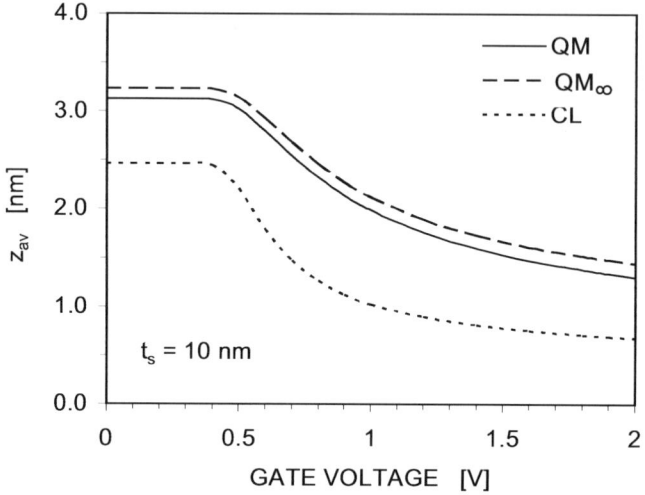

**Fig. 4.10.** Average distance of the electrons from the semiconductor surface for different modeling approaches.

An important physical approximation used in modeling multigate MOS structures of very small dimensions is the effective mass approximation. The energy band structure of a semiconductor crystal with dimensions that are only on the order of a few multiples of the lattice constant is expected to be dependent on the actual dimensions of the crystal. This problem has been tackled by many researchers and results based of first-principle (*ab initio*) calculations of the energy structure of low-dimensional silicon crystals can be found in the literature.[1-4] The rather optimistic conclusion from [4] is that the effective mass theory based on the bulk silicon effective masses gives good description of the electronic states in silicon quantum wires with dimensions down to at least 1 nm. Other calculations [5-6] suggest that the effective mass in silicon layers confined between $SiO_2$ regions increases significantly when the silicon thickness is decreased below approximately 6 nm. Furthermore, the silicon energy gap increases and becomes direct in very thin crystals. Only experimental results can tell us which model is closest to reality.

## 4.2.2 Gate voltage effect

Fig. 4.11 presents the dependence of the electron energy levels $E_{ij}$ on gate voltage, with the Fermi level $E_{Fn} = E_F$ being taken as reference energy. The minimum energy of the conduction band edge at the interfaces $E_{Cs}$ and in the middle of the semiconductor film $E_{Cm}$ is shown as well.

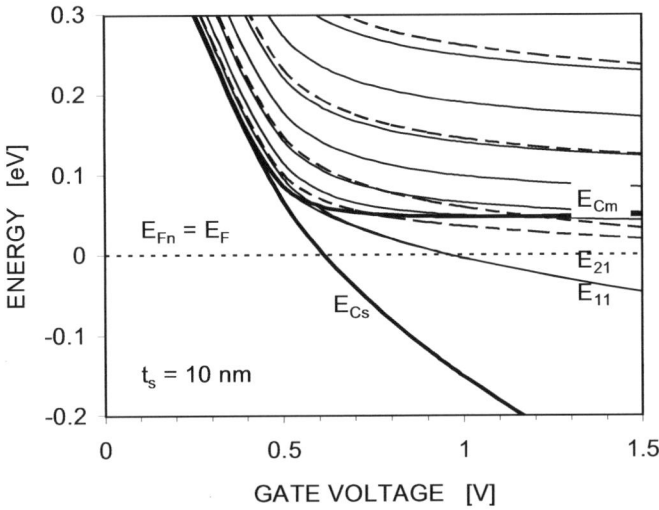

**Fig. 4.11.** Electrons energy levels $E_{ij}$ and conduction band edge at the interfaces, $E_{Cs}$, and in the middle of the semiconductor film, $E_{Cm}$, vs. gate bias.

Fig. 4.12 shows the distribution of electron concentration in the semiconductor region at different gate voltages. As can be seen, at low enough gate voltages, the electron concentration is lower than the acceptor concentration in the entire semiconductor film, and the surface subwells have a parabolic shape with a depth $E_{Cm}-E_{Cs} = qN_A(t_s/2)^2/2\varepsilon_s$. For $t_s = 10$ nm and $N_A = 10^{17}$ cm$^{-3}$ this band bending is small and the semiconductor quantum well is practically rectangular, as shown in Fig. 4.13a. In this voltage range, the shape of the semiconductor potential well and the spectrum of the energy levels referred to the bottom of this well do not change when the gate voltage is increased. When the gate voltage increases, the energy of all the levels in the semiconductor decrease with respect to the electron Fermi level $E_{Fn}$ and their occupation probability increases, resulting in an exponential increase of the electron concentration in subthreshold operation.

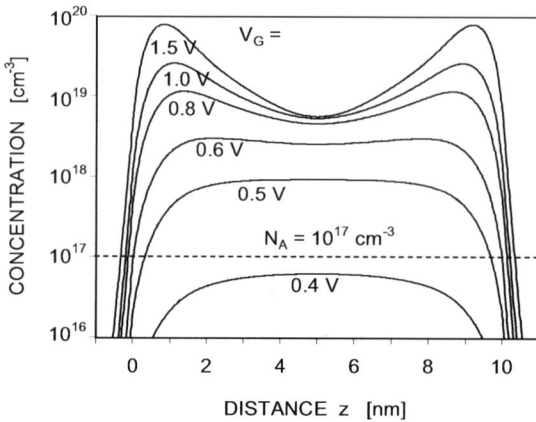

**Fig. 4.12.** Effect of gate voltage on electron concentration distribution in the silicon body of the symmetrical DG MOS structure.

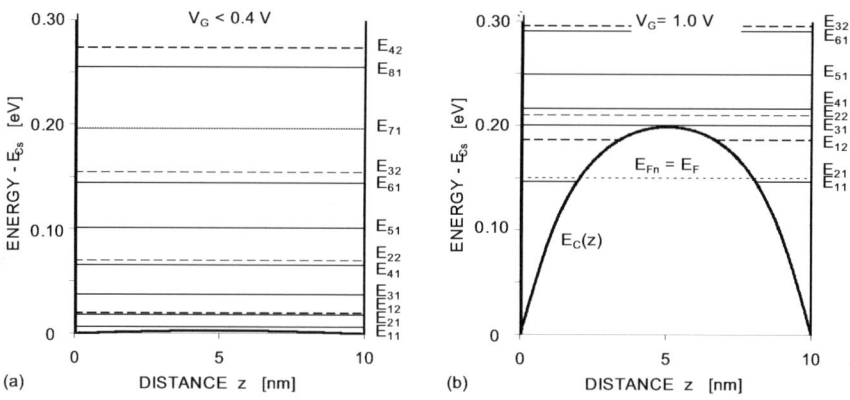

**Fig. 4.13.** Distribution of the conduction band edge and the electron energy levels $E_{ij}$ at two gate voltages (a): $V_G<0.4\text{V}$; (b) $V_G=1\text{V}$.

When the gate voltage is further increased, the electron concentration exceeds the acceptor concentration and the negative charge due to inversion electrons starts deepening the surface subwells, while the potential in the middle of the semiconductor $E_{Cm}$ saturates. The lowest energy levels enter the surface potential wells and gradually pair up, as e.g., $E_{11}$ and $E_{21}$, describing the electron states in more and more insulated surface subwells. At $V_G = 1$ V, the levels $E_{11}$ and $E_{21}$ practically overlap, as shown in Fig. 4.13b. The occupation probability of the levels confined in the wells is the highest, resulting in two maxima of electron concentration, as can be seen in Fig. 4.12 at high gate voltage.

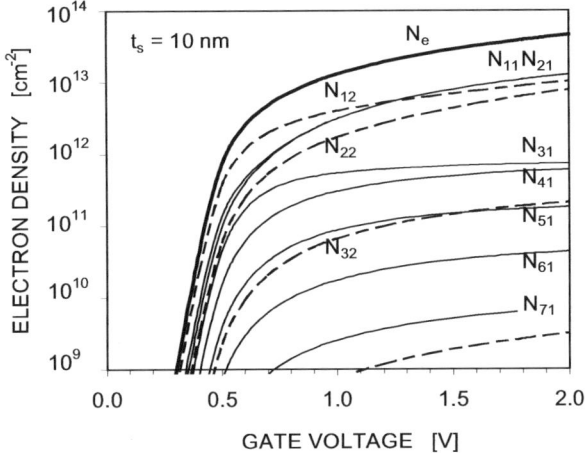

**Fig. 4.14.** Gate voltage dependence of the electron densities in the subbands $E_{ij}$ and of the total electron density $N_e$.

Fig. 4.14 shows the gate voltage dependence of the contributions of different subbands $N_{ij}$ to the total electron surface density $N_e$. $N_{ij}$ is determined by the position of a given level $E_{ij}$ relative to the electron Fermi level $E_{Fn}$, the valley degeneration factor $g_v$, and the two-dimensional density-of-state mass $m_{d2D}$:

$$N_e = \sum_{j=1,2}\sum_i N_{ij} = \sum_{j=1,2}\sum_i \frac{g_{vj}m_{d2Dj}}{\pi\hbar^2} k_B T \ln\left[1+\exp\left(\frac{E_{Fn}-E_{ij}}{k_B T}\right)\right] \quad (4.20)$$

Although the level with the lowest longitudinal mass level, $E_{11}$, is occupied with the greatest probability, the first transverse mass level $E_{12}$ can hold more electrons due to the higher $g_v m_{d2D}$ product, provided the energy distance between $E_{12}$ and $E_{11}$ is not too large. This is the case for low gate voltages. Since the energy distances between the levels increase with increasing the gate voltage and the lowest levels enter the surface subwell, most of the electron density, $N_e$, is concentrated in the degenerate levels $E_{11} = E_{12}$, at higher gate voltages.

## 4.2.3 Semiconductor thickness effect

The thickness of the semiconductor layer is one of the most important parameters of the double gate structure provided it is less than the sum of widths of the space charge regions induced by gate electrodes at the two semiconductor surfaces (*i.e.*, provided the device is fully depleted). In the opposite case (*i.e.*, in a partially depleted device) the double gate MOS structure can be regarded as two separate bulk MOS structures connected in parallel by a common semiconductor substrate. The maximum width of the semiconductor space-charge region in a single-gate MOS structure $x_{dmax}$ is obtained at the onset of strong inversion. It iss equal to about 100 nm for $N_A = 10^{17}$ cm$^{-3}$, for instance. It is worth noting that $x_{dmax}$ can be further increased under non-equilibrium, transient deep-depletion conditions. In general, one can say that the semiconductor region of a symmetrical DG structure is either fully depleted (FD) if $t_s < 2x_{dmax}$, or partially depleted (PD) if $t_s > 2x_{dmax}$.

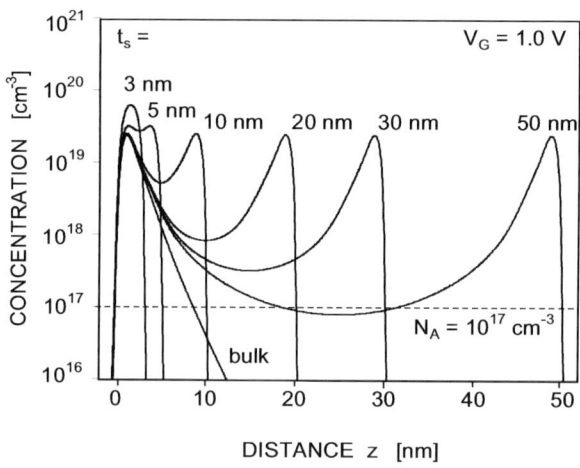

**Fig. 4.15.** Electron concentration profile at constant gate voltage ($V_G$=1V) for different values of semiconductor film thickness, $t_s$.

Thinning the semiconductor layer results in different physical effects, particularly resulting from the quantum-mechanical confinement of electrons.[7-10] Fig. 4.15 shows the distribution of the electron concentration in the symmetrical DG MOS structure at a gate voltage of 1V, with the semiconductor thickness $t_s$ as parameter.

Fig. 4.16 presents the dependence of the electron energy levels on the semiconductor thickness while Fig. 4.17 compares the contribution of different subbands $N_{ij}$ to the total electron surface density $N_e$.

If the semiconductor layer is thicker than $2x_{dmax}$, the electron concentration distribution is the same as in two bulk MOS structures. The lowest energy levels create pairs confined in the surface subwells. At any given gate voltage, the greatest contribution to the electron concentration is brought by the lowest levels corresponding to the longitudinal mass.

As it can be seen in Fig.4.15, when the semiconductor layer is thinner than approximately 50 nm, the semiconductor region is in volume inversion [11] since that the electron concentration in the entire thickness of the semiconductor region is larger than the acceptor concentration.

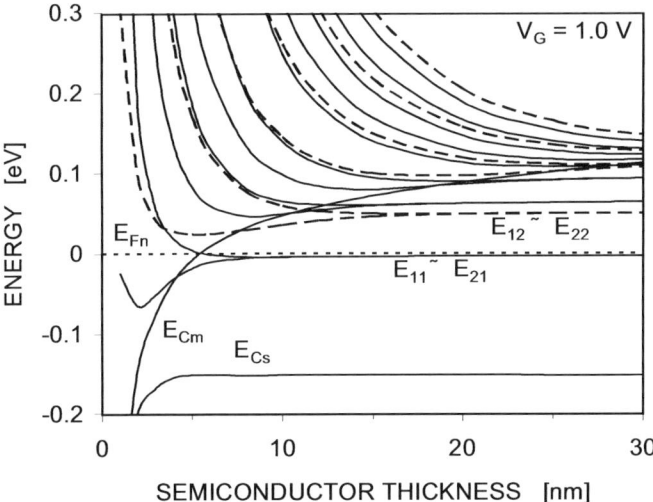

**Fig. 4.16.** The semiconductor thickness dependence of the energy levels $E_{ij}$ and the conduction band minima $E_{Cs}$ and $E_{Cm}$.

Further reduction of the semiconductor thickness results in decreasing the surface well depth $E_{Cm} - E_{Cs}$, as shown in Fig. 4.16. As a result, the electron concentration distribution becomes more uniform. When the semiconductor film is very thin, -3 nm or less- the electron concentration distribution can exhibit a single maximum located in the middle of the layer. In semiconductor films with a thickness ranging from about approximately 5nm to 10nm, the main contribution to the total surface carrier density comes from electrons in the lowest transverse mass level $E_{12}$, owing to the changes of distance between the energy levels. For thickness values lower than 5 nm, the lowest energy level with longitudinal mass, $E_{11}$, recovers its role as main the contributor to the total population of electrons, but the total electron surface density decreases due to the strong rise of all energy levels in a very narrow quantum well.

The predominance of a given level in the total population of electrons has an important influence on the conduction properties of the semiconductor channel since the effective mass of electrons in the transport direction $x$ for the longitudinal mass levels $E_{i1}$ is smaller than for the transverse mass levels $E_{i2}$. The averaged effective mobility of electrons will thus change with the semiconductor thickness, independently on its effect on scattering rates, due to the changes of participation of different subbands in the total electron population. The gate voltage dependence of the total semiconductor charge density with the semiconductor thickness as parameter is shown in Fig. 4.18 for a semiconductor body film doped with $N_A = 10^{17}$ cm$^{-3}$, and in Fig. 4.19 for an undoped device.

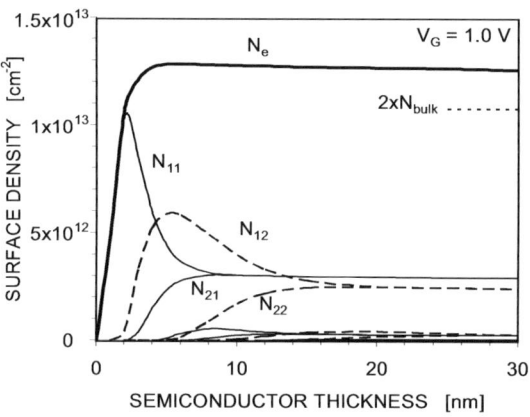

**Fig. 4.17.** Dependence of the contribution of the different subbands $E_{ij}$ to the total electron surface density $N_e$ on the semiconductor film thickness.

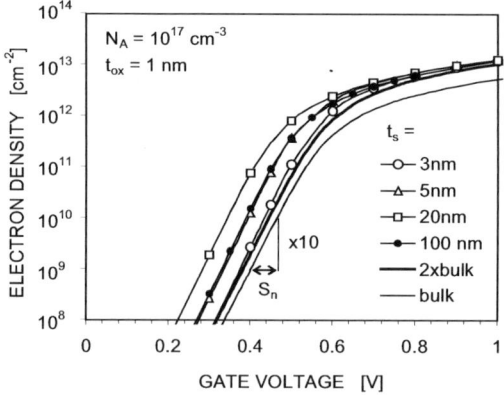

**Fig. 4.18.** Electron surface density *vs.* gate voltage for different values of the semiconductor film thickness. The doping concentration is $N_A = 10^{17}$ cm$^{-3}$.

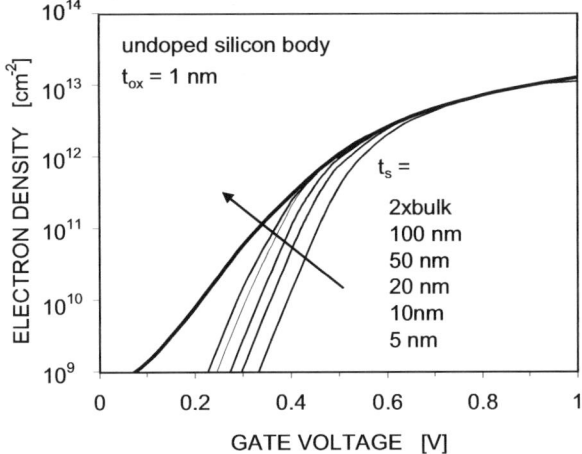

**Fig. 4.19.** Electron surface density *vs.* gate voltage for different values of the semiconductor film thickness. The semiconductor film is undoped.

The use of a double gate MOS system insures good gate control of the electron concentration in the semiconductor region even if the semiconductor film is undoped. This is a very beneficial feature since the scaling rules for bulk MOS transistors at the nanometer scale require extremely high substrate doping levels, which is detrimental to carrier mobility. In addition, the threshold voltage of very small transistors suffers from random doping fluctuation effects due to the discrete nature of doping atom distribution.[12]

The total electron density in the channel region of a double-gate structure is larger than twice the density in the equivalent bulk MOS structure, unless the semiconductor thickness is in the range of only a few nanometers, in which case the strong quantization of the electron energy pushes the energy levels upwards, resulting in a decrease of the electron concentration.

Fig. 4.18 illustrates another important property of the double-gate MOS system working in the full depletion regime: one can observe a reduction of the electron *concentration* subthreshold swing, $S_n$, of the due to the elimination of the depletion layer capacitance. In analogy to the more familiar *current* subthreshold current swing $S$, $S_n$ is defined as the increase in gate voltage that is required to increase the electron surface density tenfold, in subthreshold operation.

The theoretical dependence of the threshold voltage of the double-gate MOS structure on the semiconductor thickness has been investigated in Ref. [10]. In that paper, the threshold voltage is defined as the intersection of the linearly extrapolated $N_e(V_G)$ curve with the gate voltage axis. Fig. 4.20 shows the semiconductor thickness dependence of the threshold voltage defined as the gate voltage corresponding to the maximum value of $d^2N_e/dV_G^2$, which corresponds to the maximum of $dg_m/dV_G$ if the electron mobility is constant.[13-14] When the semiconductor film is fully depleted, the threshold voltage decreases with decreasing $t_s$ as a result of a decrease of the acceptor ion charge (depletion charge). This effect is not observed if the semiconductor film is undoped. When the semiconductor is thin enough, on the other hand, the threshold voltage increases with decreasing the semiconductor thickness due to the increase of electron quantization energy. The latter effect is independent of the dopant concentration.

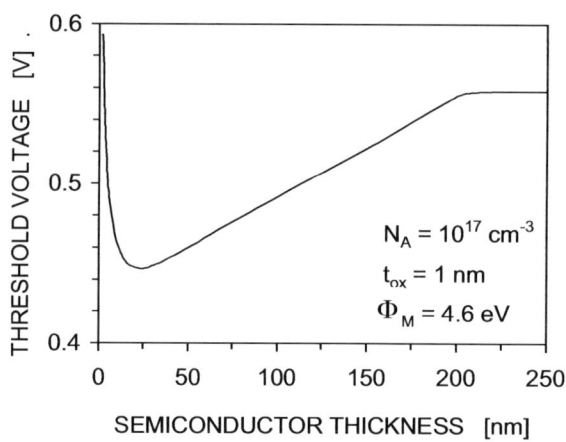

**Fig. 4.20.** Threshold voltage of the doped DG MOS structure *vs.* semiconductor thickness.

### 4.2.4 Asymmetry effects

Asymmetry in a double gate MOS structure can result from the use of different composition oxide/gate stacks, usually due to the use of different gate electrode materials, for instance $n^+$-type and $p^+$-type polysilicon.[15] Asymmetrical distributions of carriers in the surface regions can also be induced by using different bias conditions on the gates, provided they can be polarized individually (Multiple Independent-Gate FET, MIGFET).[16-18] The asymmetry can also result from compound composition of the channel, as in the velocity modulation transistor [19] having the channel consisting of two semiconductor layers with strongly different transport properties.

An asymmetrical architecture can be a consequence of the complexity of the double gate MOS fabrication process - this especially concerns planar DG MOS structures, in which the front and the back gate stacks are produced by different techniques - or can be designed intentionally. For instance, the ground plane SOI transistor is an asymmetrical transistor with a highly conductive layer under the buried oxide [20-22], which plays the role of a back gate connected to a fixed potential. The conductive plane under the buried oxide screens the semiconductor channel from the electric field lines originating in the drain, reducing drain-induced barrier lowering effects (DIBL).[23] The thinner the insulator between the ground plane and the semiconductor film (*i.e.*: the thinner buried oxide in the SOI structure), the more efficient the screening effect. [24]

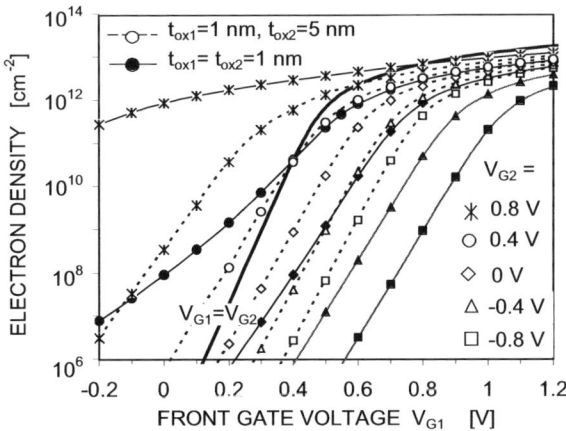

**Fig. 4.21.** Electron density in the channel as a function of gate voltage in a double-gate device. The second gate voltage, $V_{G2}$, is used as a parameter. $t_s = 10$ nm, $N_A = 10^{17}$ cm$^{-3}$, $\Phi_M = 4.6$ eV.

Modulating the work function of the back gate material or the potential of the back gate can help tuning the threshold voltage of the DG MOS transistor.[25-27] This is possible in fully depleted DG MOS structures because of the electrical coupling between the back gate and the front channel. The thinner the semiconductor layer, the stronger the influence of $V_{G2}$ on the threshold voltage. In order to illustrate this effect, Fig. 4.21 shows the simulated density of electrons at the surface of DG SOI device as a function of front gate voltage, with the back gate voltage as parameter.

Tuning the threshold voltage using a back gate bias is a very useful tool, but there is a penalty associated with it: there is a significant degradation of the subthreshold swing, compared to the case where the gates are connected together ($V_{G1} = V_{G2}$), which is indicated by the thick solid line in Fig. 4.21. If the back gate voltage is too negative, an accumulation layer can be induced at the back surface of the semiconductor body, and the subthreshold slope is degraded due to the voltage loss needed to deplete the accumulation region using the front gate. In the opposite case, *i.e.*, when the back gate voltage is positive, the back interface of the semiconductor can be driven in strong inversion. In that case, it is difficult to switch off the channel by decreasing the front gate voltage, which results in subthreshold slope degradation as well.

The detrimental effect of the fixed potential of the back gate on the subthreshold slope can be reduced by the use of asymmetrical gate stacks. It has been experimentally confirmed that the subthreshold slope can be improved if a thicker back oxide is used [28], as shown in Fig. 4.21. The coupling of the front gate electrode with the channel at the back surface of the semiconductor region is more efficient if the back gate oxide thickness is increased. And as a result, electrons in the inversion layer at the back surface of the semiconductor can be more effectively removed from this region by an increase of front gate bias when the back gate is positively biased. Fig. 4.22 compares the effect of the front gate potential on the electron concentration distribution in a semiconductor region for the cases of symmetrical gate stacks: $t_{ox1} = t_{ox2} = 1$ nm and asymmetrical gate stacks: $t_{ox1} = 1$ nm, $t_{ox2} = 5$ nm.

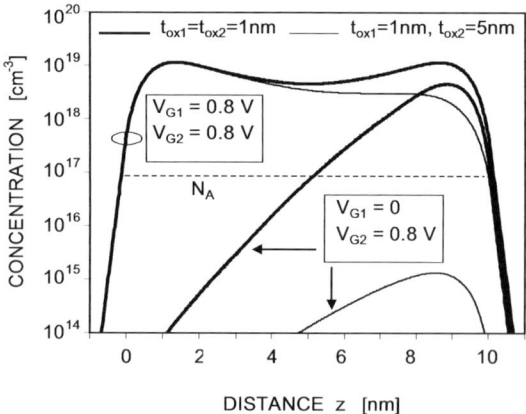

**Fig. 4.22.** Electron concentration distribution in the DG MOS structure for different bias conditions. $t_s = 10$ nm, $N_A = 10^{17}$ cm$^{-3}$.

The lack of symmetry in the DG SOI structure can result not only in the loss of the volume inversion and subthreshold slope degradation, but also in an increase of the transverse electric field, which has a detrimental effect on mobility of carriers in the channel. Fig. 4.23 shows the simulated dependence of the transverse effective field in a DG MOS structure as a function of the electron surface density, $N_e$, for different bias conditions. As can be observed, the symmetrical structure and symmetrical bias conditions in the DG MOS structure correspond to the lowest transverse effective field for a given total density of electrons in the channel. The effective field is defined as the local transverse electric field averaged over the electron population in the semiconductor film:

$$E_{eff} = \int_0^{t_s} |E_z(z)| n(z) dz \Big/ \int_0^{t_s} n(z) dz \qquad (4.21)$$

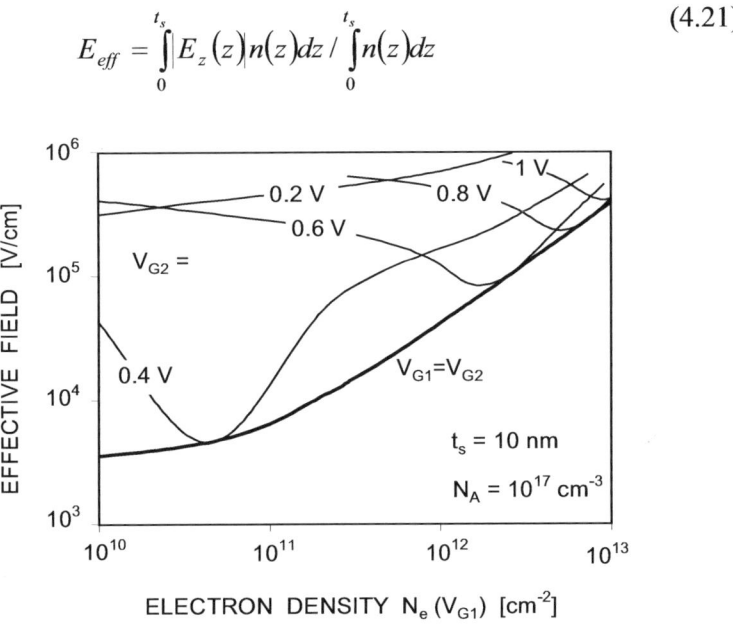

**Fig. 4.23.** Effective transverse electric field in the semiconductor body as a function the electron surface density induced by the front gate voltage, for a fixed back gate voltage.

As a conclusion to this section on asymmetry effects, it can be said that the design parameters and the back-gate bias of the asymmetrical DG MOS transistors should be optimized to obtain a good trade-off between the beneficial effect of the structure on DIBL and the detrimental effects on the subthreshold slope and the carrier mobility, as it is in the case with ground-plane SOI transistors.[29]

## 4.2.5 Oxide thickness effect

The insulator layer in the MOS system electrically separates the semiconductor region from the gate electrode, while simultaneously serving as a medium transferring the field effect of the gate electrode in the semiconductor region. The thinner the insulator layer, the higher the electron concentration induced at the semiconductor surface. Additionally, thinning the insulator layer shields the channel carriers from the fringing field lines originated in the drain, thereby reducing short-channel effects. However, reduction of the insulator layer thickness below 2 nm results in an increase of gate tunnel leakage current. The boundary on minimum gate insulator thickness in the DG MOS system is set by three conditions: (i) if the insulator is too thin, the total tunnel current in the circuit becomes too large and significantly increases the standby power consumption, (ii) if the insulator is too thin, the current of electrons tunneling from the transistor channel in the gate becomes comparable to the drain current in magnitude, (iii) if the insulator is too thin, coupling of the wave function between the semiconductor and the gate regions becomes too large, resulting in a loss of electron localization. The first of these limitations is defined by requirements at the circuit level, the second one jeopardizes the correct operation of the transistor, while the last one constitutes a fundamental quantum-mechanical limit for the minimum insulator thickness in the MOS system.

The quantum-mechanical limit for the minimum oxide thickness in the MOS system has been addressed in Ref. [30] using the effective mass approximation. The calculation is based on a simple representation of the double-gate vacuum/Si/SiO$_2$/Si/SiO$_2$/Si/vacuum system formed by a central silicon quantum well sandwiched between two gate-stack quantum wells. The electron effective mass in the vacuum, oxide and silicon regions is assumed to be $m_v = m_0$ in the vacuum, $m_b = 0.5m_0$ in the oxide, and $m_w = 0.19m_0$ in the silicon, respectively.

Fig. 4.24 presents the electron energy levels in the system as a function the gate oxide barrier thickness. The dimensions of the quantum wells are 10 nm (gate wells) and 5 nm (central well). Such small dimensions allow for the direct observation of the distance between the discrete energy levels. In addition, the central well levels $E_s$ can be distinguished from the gate levels $E_g$. When the gate oxide thickness is smaller than approximately 0.5 nm, the central well levels become a function of the oxide thickness, and the splitting of the gate levels is observed.

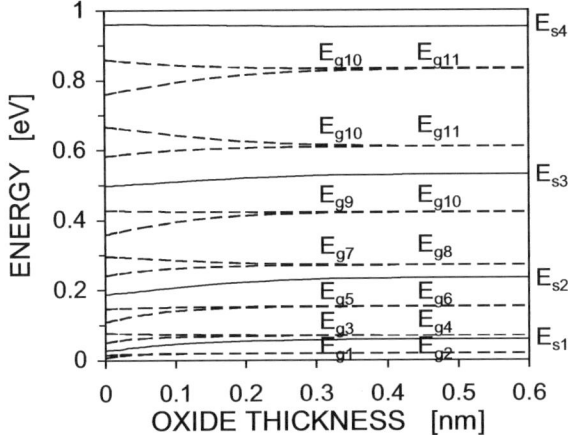

**Fig. 4.24.** Dependence of the energy levels on the gate oxide thickness in a rectangular double-gate quantum well system.

Fig. 4.25 shows the probability that the electrons at different energy levels be located within the boundaries of the central silicon well as a function of the gate oxide thickness. If the gate oxide barriers are thicker than approximately 0.5 nm, the electrons from the $E_s$ energy levels (solid lines) are mostly located within the central well and the electrons from the gate levels $E_g$ (dashed lines) are very likely to be located outside the central well, *i.e.*, in the gate wells. When the oxide thickness is reduced below 0.5 nm, however, the wavefunctions of the electrons located in the $E_s$ subbands leak out in the gate and wavefunctions located in the $E_g$ subbands leak in the central well. Using this simplified model and simulation based on the effective mass approximation one can determine the limit for the SiO$_2$ thickness in the MOS system due to the uncertainty of the electron location to be about 0.4-0.5 nm.

Fig. 4.25 clearly illustrates the physical problem of the uncertainty on the electron location. The limit of the oxide thickness can be evaluated in a simple manner, using the exponential term of the simplified WKB formula for the tunneling probability:

$$P \approx \exp(-2\kappa_{ox} t_{ox}) \gg 1 \qquad (4.22)$$

where $\kappa_{ox}$ is the imaginary wave vector in the oxide barrier. According to this expression, the tunneling probability through the 0.5nm-thick SiO$_2$ layer with an energy barrier of 3eV is equal to $2 \times 10^{-3}$.

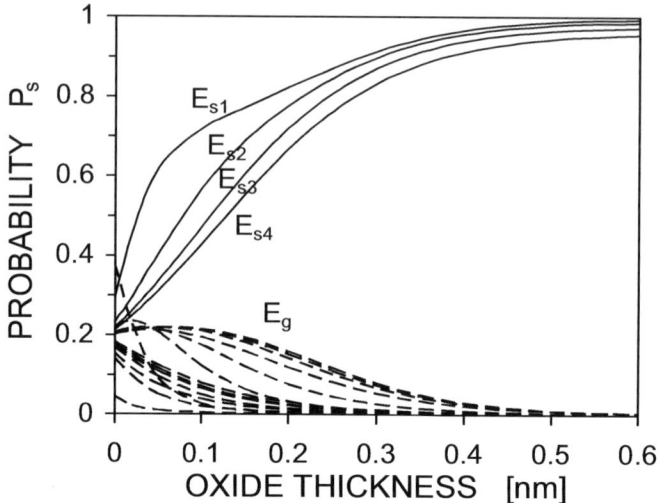

**Fig. 4.25.** Semiconductor thickness dependence of the probability located within the central quantum well of the DG MOS system.

### 4.2.6 Electron tunnel current

The current of electrons tunneling from the semiconductor region to the gate electrodes is given by summation of contributions from the different energy subbands:

$$J_e = \sum_{j=1,2} \sum_i q \frac{N_{ij}}{\tau_{ij}} \qquad (4.23)$$

where $\tau_{ij}$ is the lifetime (or: the escape time) of electrons from the $ij$-th subband. It is assumed that the lifetimes for all states corresponding to the same energy of the motion in the direction perpendicular to the potential barrier are identical.

There are different approaches to determine the lifetime of electrons.[31] The simplest quantum-mechanical approach is based on the Heisenberg uncertainly principle:

$$t_{ij} = \frac{\hbar}{\Gamma_{ij}} \qquad (4.24)$$

According to the quantum transmitting boundary method [32] and the transverse resonance method [33], $\Gamma/2$ is the imaginary part of the complex energy eigenvalue obtained from solution of the Schrödinger equation. In case of the double barrier system $\Gamma$ can be determined as the half width of the resonant tunneling probability peak [34], as illustrated in Fig. 4.26, in which the probability of tunneling through the double barrier system is shown. The resonant peaks correspond to the quasi-bound levels of the central quantum well. Ref. [35] proposes to calculate $\Gamma$ as the half width of the peak of the derivative of the phase of the complex reflection coefficient. In turn, Ref. [36] calculates $\Gamma$ using the perfectly matched layer method, in the analogy to the formalism used in electromagnetics.

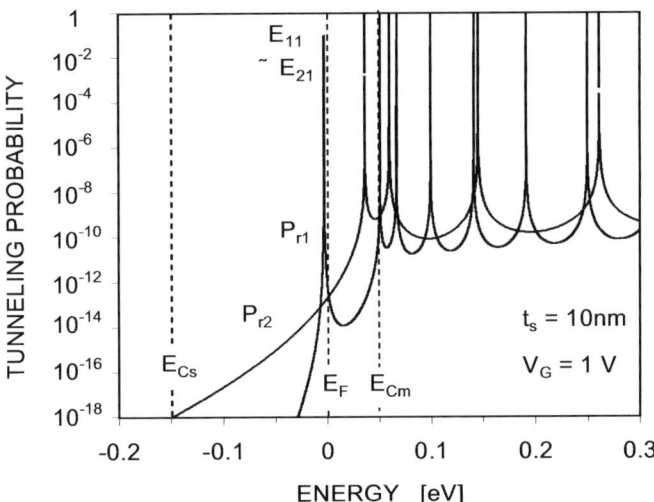

**Fig. 4.26.** Resonant tunneling probability through a symmetrical double-barrier system. $V_G = 1$V, $t_s = 10$ nm, $t_{ox} = 1$ nm, $N_A = 10^{17}$cm$^{-3}$.

All these methods are difficult to use in practice because of they involve a complicated calculation procedure. It is much easier to use the quasi-classical approach [37] which expresses the reciprocal of the lifetime as the product of the frequency of impacts of electrons against the barrier and the probability of tunneling from the quantum well $P(E_x)$. Since the escape of electrons from the semiconductor region of the double-gate structure may happen due to tunneling through the front or through the back oxide with a probability $P_1$ and $P_2$, respectively, the lifetime for the quasi-bound state in the double gate structure can be expressed according to the quasi-classical approach as:

$$\tau_{ij}^{-1} = f_{imp}(P_1 + P_2) = \frac{1}{t_T}(P_1 + P_2) \qquad (4.25)$$

where the impact frequency is expressed as a reciprocal of the travel time between the subsequent impact events, i.e., the time of travel around the quantum well:

$$t_t = 2 \int_{L_{well}} \frac{dz}{v_g(z)} = 2 \int_{L_{well}} \frac{\sqrt{m_{zj}/2}}{\sqrt{E_{ij} - E_c(z)}} dz \qquad (4.26)$$

Theoretical comparisons of the lifetimes calculated with the use of the quantum-mechanical and quasi-classical approaches are in good quantitative agreement.[38-39] In order to illustrate this agreement, Fig. 4.27 shows the lifetime of electrons in different energy subbands of the semiconductor well. The well thickness is 40 nm and the gate voltage is 1V. The lifetime is calculated using both the resonant tunneling peak method and the quasi-classical approach. The lifetime of electrons on the transverse mass levels $E_{i2}$ is shorter than that of electrons on the longitudinal mass levels $E_{i1}$ due to the lower effective mass. One can observe the excellent agreement between the two calculation methods.

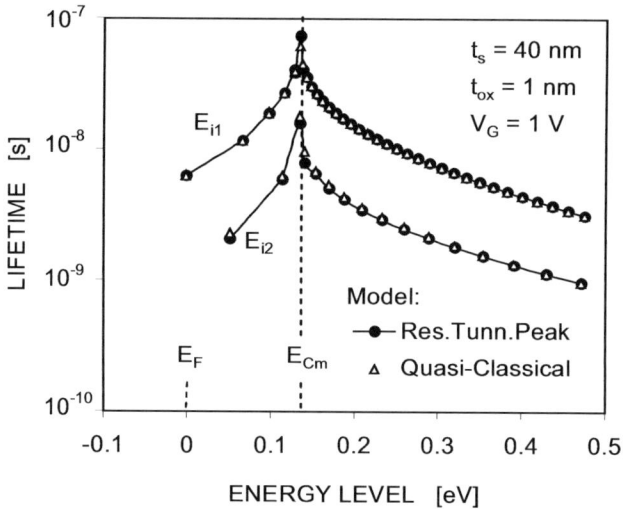

**Fig. 4.27.** Lifetime of electrons in the quasi-bound state according to the resonant tunneling peak approach and the quasi-classical approach.

For the quasi-bound states above the top of the central barrier $E_{ij} > E_{Cm}$, an increase of the energy results in a decrease of the lifetime due to the decrease of the travel time (higher kinetic energy) and the increase of the tunneling probability (lower barrier). In the opposite case, i.e., for $E_{ij} < E_{Cm}$ corresponding to confinement of electrons in the surface subwells, an increase of the energy results in an increase of the lifetime due to the increase of the travel time.

Fig. 4.28 shows the gate voltage dependence of the current resulting from the tunneling of electrons from the semiconductor region into the gate electrodes for different values of the semiconductor film thickness. The tunnel current from the quasi-bound states in a classical bulk MOS structure is also presented for comparison. A curve representing twice the gate current in the bulk device is also shown, for the sake of fair comparison with the double-gate structure. At high gate voltages, when electrons mainly are localized in surface subwells, the tunneling current in the double gate structure is equal to the (doubled) current in the bulk device. For a given gate voltage, a decrease of the semiconductor thickness brings about an increase of the gate tunnel current. This is especially visible for low voltages where the semiconductor film is depleted, and for semiconductor layers thinner than 20 nm. These results do not fully agree with measurement of the gate tunnel current made on FinFETs [40], highlighting the need for further research in that area.

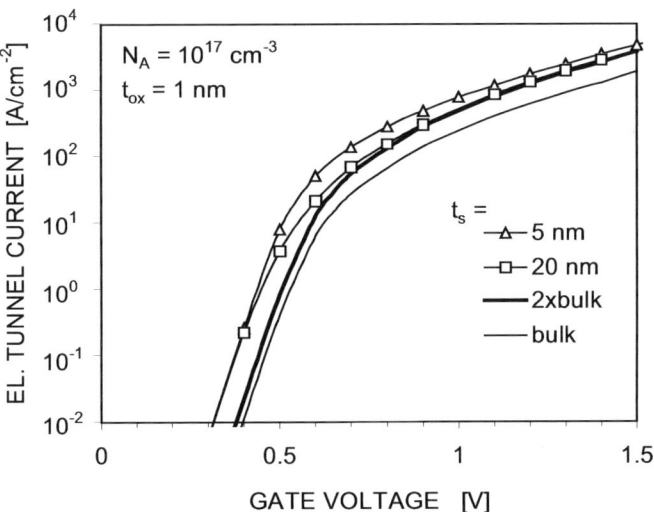

**Fig. 4.28.** Electron tunnel current *vs.* gate voltage with the semiconductor thickness as parameter.

Fig. 4.29 shows the electron tunnel current as a function of the electron surface density, $N_e$. Note that in the *2×bulk* curve the electron density is multiplied by two as well (*2×Ne*). The curves shown in Fig. 4.29 indicate that the double gate structure is more advantageous than the bulk device, since the tunnel current is lower for any given value of the electron surface density, unless the semiconductor layer is ultrathin. This conclusion is in agreement with experiment results found in the literature.[41]

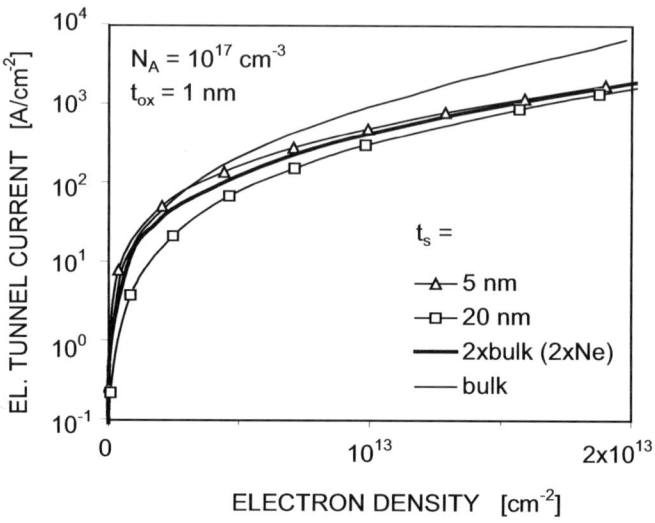

**Fig. 4.29.** Electron tunnel current *vs.* the electron density with the semiconductor thickness as a parameter.

## 4.3 Two-dimensional confinement

If both the thickness $t_s$ and the width $W_s$ of the semiconductor region are very small and the energy of the electrons in these directions is significantly quantized, the device becomes a quantum wire and the electrons in the channel form a one-dimensional electron gas (1DEG). The Schrödinger and Poisson equations can be solved self-consistently in two dimensions to calculate the electron concentration profile. The result of such calculations are qualitatively similar to those obtained for one-dimensional confinement.[42-45] This similarity can be exemplified by the two-dimensional distribution of electrons calculated in a silicon nanowire with a cylindrical section.[46-47] At high gate voltage, the electron distribution along the diameter of the channel section exhibits two maxima located in a certain distance from the surfaces, just like in the case of a symmetrical double gate structure. Like the double-gate MOS structure, the quantum wire exhibits volume inversion provided the dimensions of the wire section are small enough. The minimum energy of the subbands increases when the wire section is reduced, due to two-dimensional confinement effects, and the threshold voltage is larger that what is predicted by classical theory. Furthermore, the threshold voltage increases when the cross section of the channel is reduced.[48]

In case of the rectangular geometry of the semiconductor channel, the two-dimensional distributions of the electron concentration exhibit peaks near the corners of the cross-section of the semiconductor region due to the superimposed field effect of two sides of the gate electrode. These peaks are the intrinsic conductive part of the active structure and introduce specific features on the transistor current-voltage characteristics – sometimes, there appears a the second peak on the $dg_m/dV_G$ characteristics, which indicates different threshold voltages at the corners of the device and at the top/sidewall interfaces (corner effect). This effect also degrades the subthreshold slope. The corner effect vanishes out if a low channel doping concentration is used and/or if the corners are rounded.[49] In that case the electron concentration distribution in the corners is similar to that at the top and sidewall interfaces.[50]

Another specific effect resulting from quantum-mechanical one-dimensional confinement of electrons in low-dimensional multigate MOS channels is formation of one-dimensional energy subbands. The formation of subbands results in a phenomenon called the "inter-subband scattering" in which the presence of electrons in a given subband affects the mobility of electrons in other subbands. Inter-subband scattering manifests itself in the form of oscillations in the $I_D(V_G)$ characteristics of nanowire transistors a low temperature.[51] If the section of the device is small enough, the oscillations can be even be observed at room temperature.[52]

## References

1. X. Zhao, C. Wei, L. Yang, M. Y. Chou: Quantum confinement and electrical properties of silicon nanowires. Physical Review Letters **92**, 236805 (2004)
2. Y. J. Ko, M. Shin, S. Lee, K. W. Park: Effects of atomic defects on coherent electron transmission in silicon nanowires: Full band calculations. Journal of Applied Physics, **89**, 374 (2001)
3. J. Wang, A. Rahman, A. Ghosh, G. Klimeck, M. Lundstrom: Performance evaluation of ballistic silicon nanowire transistors with atomic-basis dispersion relations. Applied Physics Letters **86**, 093113 (2005)
4. S. Horiguchi: Validity of effective mass theory for energy levels in Si quantum wires. Physica B **227**, 336 (1996)
5. K. Nehari, N. Cavassilas, J. L. Autran, M. Bescond, D .Munteanu, M. Lannoo: Influence of band structure on electron ballistic transport in silicon nanowire MOSFET's: an atomistic study. Solid-State Electronics **50**, 716 (2006)

6. P. V. Sushko, A. L. Shluger: Electronic structure of insulator-confined ultrathin Si channels. Microelectronic Engineering **84**, 2043 (2007)
7. T. Quisse: Self-consistent quantum-mechanical calculations in ultrathin silicon-on-insulator structures. Journal of Applied Physics **75**, 5989 (1994)
8. B. Majkusiak, T. Janik, J. Walczak: Semiconductor thickness effects in the double-gate SOI MOSFET. IEEE Transactions on Electron Devices **45**, 1127 (1998)
9. T. Ernst. S. Cristoloveanu, G. Ghibaudo, T. Ouisse, S. Horiguchi, Y. Ono, Y. Takahashi, K. Murase: Ultimately thin double-gate SOI MOSFETs. IEEE Transactions on Electron Devices **50**, 830 (2003)
10. Y. Omura, S. Horiguchi, M. Tabe, K. Kishi: Quantum-mechanical effects on the threshold voltage of ultrathin SOI nMOsFET's. IEEE Electron Device Letters **14**, 569 (1993)
11. F. Balestra, S. Cristoloveanu, M. Benachir, J. Brini, T. Elewa: Double-gate silicon-on-insulator transistor with volume inversion: A new device with greatly enhanced performance. IEEE Electron Device Letters **ED-8**, 410 (1987)
12. H.-S. Wong, Y. Taur: Three-dimensional "atomistic" simulation of discrete random dopant distribution effects in sub-0.1 µm MOSFET's", *Technical Digests of IEDM*, 705 (1993)
13. H.-S. Wong, M. H. White, J. Krutsick, R. V. Booth: Modeling of transconductance degradation and threshold voltage in thin oxide MOSFET's. Solid-State Electronics **30**, 953 (1987)
14. P. Francis, A. Terao, D. Flandre, F. Van de Wiele: Modeling of ultrathin double-gate nMOS/SOI transistors. Solid State Electronics **41**, 715 (1994)
15. L. Mathew, M. Sadd, M.E. White, A. Vandooren, S. Dakshina-Murthy, J. Cobb, T. Stephens, R. Mora, D. Pham, J. Conner, T. White, Z. Shi, A.V-Y. Thean, A. Barr, et al.: FinFET with isolated $n^+$ and $p^+$ gate regions strapped with metal and polysilicon. *Proceedings of IEEE Intl. SOI Conference*, 109 (2003)
16. S. Zhang, X. Lin, R.Huang, R. Han, P.K. Ko, Y.Y. Wang, M. Chan: A self-aligned double-gate MOS transistor technology with individually addressable gates. *Proceedings of IEEE Intl. SOI Conference*, 207 (2002)
17. M. Masahara, Y. Liu, T. Sekigawa, S. Hosokawa, K. Ishii, T. Matsukawa, H. Tanoue, K. Sakamoto, H. Yamauchi, S. Kanemaru, H. Koike, E. Suzuki: Demonstration of threshold voltage control techniques for vertical-type 4-terminal double-gate MOSFETs (4T-DGFET). *Proceedings of ESSDERC*, 73 (2004)
18. X. Liu, M. Masahara, K. Ishii, T. Tsutsumi, T. Sekigawa, H. Takashima, H. Yamauchi, E. Suzuki: Flexible threshold voltage FinFETs with independent double gates and an ideal rectangular cross-section Si-Fin channel. *Proceedings of IEEE Intl. SOI Conference*, 18.8.1 (2003)
19. F. Gámiz, C. Sampedro, A. Godoy, M. Prunnila, J. Ahopelto: DG SOI devices operated as velocity modulation transistors. *Proceedings of IEEE Intl. SOI Conference*, 198 (2004)

20 R. Koh: Buried layer engineering to reduce the drain-induced barrier lowering of sub-0.05 µm SOI-MOSFET. *Japanese Journal of Applied Physics* **38**, 2294 (1999)
21 L.J. Huang, K. Chan, P.M. Solomon, E. Jones, C. D'Emic, W.C. Lee, F.R. McFeely, H.-S. Wong,: Preparation of 200 mm silicon substrates with metal ground-plane for double-gate SOI devices. *Proceedings of IEEE Intl. SOI Conference*, 14 (2001)
22 R.-H. Yan, A. Ourmazd, K.F. Lee: Scaling the Si MOSFETs: from bulk to SOI to bulk. IEEE Transactions on Electron Devices **39**, 1704 (1992)
23 T. Ernst, S. Cristoloveanu: The ground-plane concept for the reduction of short-channel effects in fully-depleted SOI devices. *SOI Technology and Devices IX. Electrochem. Soc.*, Pennington, 1999, p. 329.
24 H.-S. Wong, D.J. Frank, P.M. Solomon: Device design considerations for double-gate, ground-plane, and single-gated ultra-thin SOI MOSFETs at the 25 nm channel length generation. *Technical Digest of IEDM*, 407 (1998)
25 J.G. Fossum, Y. Chong: Simulation-based assessment of 50 nm double-gate SOI CMOS performance. *Proceedings of IEEE Intl. SOI Conference*, 107 (1998)
26 J. Pretet et al.: Scaling issues for advanced SOI devices: gate oxide tunneling, thin buried oxide, and ultra-thin films. *Silicon Nitride and Silicon Dioxide Thin Insulating Films VII. Electrochem. Soc. Proc.* **2003-02**, 476 (2003)
27 J. Pretet, F. Gauge, A. Vandooren, L. Mathew, B. Y. Nguyen, J. Jomaah, S. Cristoloveanu: Substrate effects in SOI FinFETs. *Proceedings of Electrochemical Society* **2003-05**, 231 (2003)
28 M. Masahara, R. Surdeanu, L. Witters, G. Doornbos, V. H. Nguyen, G. Van den bosch, C. Vrancken, K. Devriendt, F. Neuilly, E. Kunnen, E. Suzuki, M. Jurczak, S. Biesemans: Independent double-gate FinFETs with asymmetric gate stacks. Microelectronic Engineering **84**, 2097 (2100)
29 J. Lolivier, J. Widiez, M. Vinet, T. Poiroux, F. Dauge, B. Previtali, M. Mouis, J. Jommah, F. Balestra, S. Deleonibus: Experimental comparison between double gate, ground plane, and single gate SOI CMOSFETs. *Proceedings of ESSDERC*, 77 (2004)
30 B. Majkusiak, J. Walczak: Theoretical limit for the $SiO_2$ thickness in silicon MOS devices. *Science and Technology of Semiconductor-on-Insulator Structures and Devices Operating in a Harsh Environment*. Eds. D. Flandre *et al.*, Kluwer Academic Publishers, 309 (2005)
31 R. Clerk, A. Spinelli, G. Ghibaudo, G. Pananakakis: Theory of direct tunneling current in metal-oxide-semiconductor structures. Journal of Applied Physics, **91**, 1400 (2002)
32 C. L. Fernando, W. R. Frensley: An efficient method for the numerical evaluation of resonant states. Journal of Applied Physics **76**, 2881 (1994)
33 S.-H. Lo, D. A. Buchanan, Y. Taur, W. Wang: Quantum-mechanical modeling of electron tunneling current from the inversion layer of ultra-thin-oxide nMOSFET's. IEEE Electron Device Letters **18**, 290 (1997)

34 G. Gildenblat, B. Gelmont, and S. Vatannia: Resonant behavior, symmetry, and singularity of the transfer matrix in asymmetric tunneling structures. Journal of Applied Physics **77**, 6327 (1995)
35 E. Cassan: On the reduction of direct tunneling leakage through ultrathin gate oxides by a one-dimensional Schrödinger-Poisson solver. Journal of Applied Physics **87**, 7931 (2000)
36 M. Karner, A. Gehring, H. Kosina: Efficient calculation of lifetime based direct tunneling through stacked dielectrics, Journal of Computational Electronics **5**, 161 (2006)
37 F. Rana, S. Tiwari, D. A. Buchanan: Self-consistent modeling of accumulation layer and tunneling currents through very thin oxides. Applied Physics Letters **69**, 1104 (1996)
38 A. Dalla Serra, A. Abramo, P. Palestri, L. Selmi, F. Widdershoven: Closed- and open-boundary models for gate-current calculation in n-MOSFETs. IEEE Transactions on Electron Devices **48**, 1811 (2001)
39 R. Clerk, A. Spinelli, G. Ghibaudo, G. Pananakakis: Theory of direct tunneling current in metal-oxide-semiconductor structures. Journal of Applied Physics **91**, 1400 (2002)
40 T. Rudenko, V. Kilchytska, N. Collaert, M. Jurczak, A. Nazarov, D. Flandre: Experimental evidence for reduction of gate tunneling current in FinFET structures and its dependence on the fin width. *Proceedings of ESSDERC*, 375 (2006)
41 L. Chang, K. J. Yang, Y.-C. Yeo, I. Polishchuk, T.-J. King, C. Hu: Direct-tunneling gate leakage current in double-gate and ultrathin body MOSFETs. IEEE Transactions on Electron Devices **49**, 2288 (2002)
42 J. Wang, E. Polizzi, M. Lundstrom: A three-dimensional quantum simulation of silicon nanowire transistors. Journal of Applied Physics **96**, 2192 (2004)
43 E. Gnani, A. Marchi, S. Reggiani, M. Rudan, G. Baccarani: Quantum-mechanical analysis of the electrostatics in silicon-nanowire and carbon-nanotube FETs. Solid-State Electronics **50**, 709 (2006)
44 A. Godoy, A. Ruiz-Gallardo, C. Sampedro, F. Gámiz: Self-consistent solution of the 2D Schrödinger-Poisson equations in multiple-gate SOI MOSFETs. *Proceedings of EUROSOI Workshop*, 17 (2006)
45 M. Bescond, K. Nehari, J. L. Autran, N. Cavassilas, D. Munteanu, M.Lannoo: 3D quantum modeling and simulation of multiple-gate nanowire MOSFETs. *Technical Digests of IEDM*, 617 (2004)
46 A. Marchi, E. Ghani, S.Reggiani, M. Rudan, G. Baccarani: Investigating the performance limits of silicon-nanowire and carbon-nanotube FETs. Solid-State Electronics **50**, 78 (2006)
47 L. Wang, D. Wang, P. M. Asbeck: A numerical Schrödinger-Poisson solver for radially symmetric nanowire core-shell structures. Solid-State Electronics **50**, 1732 (2006)
48 J.-P. Colinge, J. C. Alderman, C. R. Cleavelin: Quantum-mechanical effects in trigate SOI MOSFETs. IEEE Transactions on Electron Devices **53**, 1131 (2006)

49 W. Xiong, J. W. Park, J. P. Colinge: Corner effect in multiple-gate SOI MOSFETs. *Proceeding of IEEE International SOI Conference*, 111 (2003)
50 A. Godoy, A. Ruiz-Gallardo, C. Sampedro, F. Gámiz: Quantum-mechanical effects in multi-gate MOSFETs. Journal of Computational Electronics **6**, 145 (2007)
51 J.-P .Colinge, A. J. Quinn, L. Floyd, G. Redmond, J. C. Alderman, W. Xiong, C. R. Cleavelin, T. Schulz, K. Schruefer, G. Knoblinger, P. Patruno: Low-temperature electron mobility in trigate SOI MOSFETs. IEEE Electron Device Letters. **27**, 120 (2006)
52 J.-P. Colinge, W. Xiong, C. R. Cleavelin, T. Schulz, K. Schrűfer, K. Matthews, P. Patruno: Room-temperature low-dimensional effects in Pi-gate SOI MOFETs. IEEE Electron Device Letters. **27**, 775 (2006)

# 5 Mobility in Multigate MOSFETs

Francisco Gámiz and Andrés Godoy

## 5.1 Introduction

The continuing scaling of CMOS (Complementary Metal-Oxide-Semiconductor) technology requires significant innovations in different fields, from short channel effect suppression to carrier transport enhancement [1-5]:

i) Firstly, multi-gate devices exhibit a scaling advantage due to better gate control of the channel, as has been widely discussed in previous chapters of this book. However, this in itself is not enough.

ii) The second key to the further improvement of CMOS technology is the enhancement of carrier mobility in the device channel.[1] In recent years, much research activity has been focused on this task, with the use of specific doping profiles, the growth of lightly doped epitaxial layers on highly doped substrates [6], and even the use of silicon-related materials instead of silicon. In relation to the latter proposal, a significant step was taken with the introduction of strained silicon to build the MOSFET (Metal-Oxide-Semiconductor Field Effect Transistor) channel.[7]

In this chapter we analyze the behavior of electron mobility in different multigate structures comprising double-gate transistors, FinFETs, and silicon nanowires. The effect of technological parameters on carrier mobility is broadly analyzed, and its behaviour physically explained. Our main goal is to show how mobility in multiple gate devices compares to that in single-gate devices and to study different approaches for improving the mobility in these devices, such as different crystallographic orientations, strained Si channels, etc.

We considered two kinds of multiple gate devices: quantum-well based devices, and quantum-wire based devices. Section 5.2 is devoted to the first group of devices, where carriers are confined in only one dimension but can move freely in the other two dimensions. Section 5.3 focuses on multigate silicon nanowires, where carrier confinement is produced in two dimensions. In this case, carriers are drifted in the other dimension.

The different nature of these two types of device led us to develop different approaches to study the electrostatics and transport in each device.We have considered only n-channel devices. The behaviour of hole mobility in multigate devices is of course of great importance.[8-9] However hole mobility has been theoretically much less studied than electron mobility because of the complexity of the valence band, and large research efforts are still required.

## 5.2 Double-Gate MOSFETs and FinFETs

Several versions of multi-gate device are discussed extensively in the literature.[2-4] There are two varieties: planar and vertical structures [2] (Figure 5.1). The former group contains ground plane and back-gate devices, which are derivatives of the SOI device.[3] The vertical structures contain the FinFET [9-11], Omega FET [12], and the Tri-Gate.[13]

A Planar Dual-Gate-Silicon-On-Insulator (DGSOI) structure consists, basically, of a silicon slab sandwiched between two oxide layers. A metal or a polysilicon film contacts each oxide (Figure 5.1 (left)). Each of these films acts as a gate electrode (front and back gate), which can generate an inversion region near the Si-SiO$_2$ interfaces if an appropriate bias is applied. Thus, we would have two MOSFETs sharing the substrate, source and drain. The outstanding feature of these structures lies in the concept of volume inversion, introduced by Balestra *et al.* [14]: if the Si film is thicker than the sum of the depletion regions induced by the two gates, no interaction is produced between the two inversion layers. The operation of this device is similar to that of two conventional MOSFETs connected in parallel (Figure 5.2 (left)).

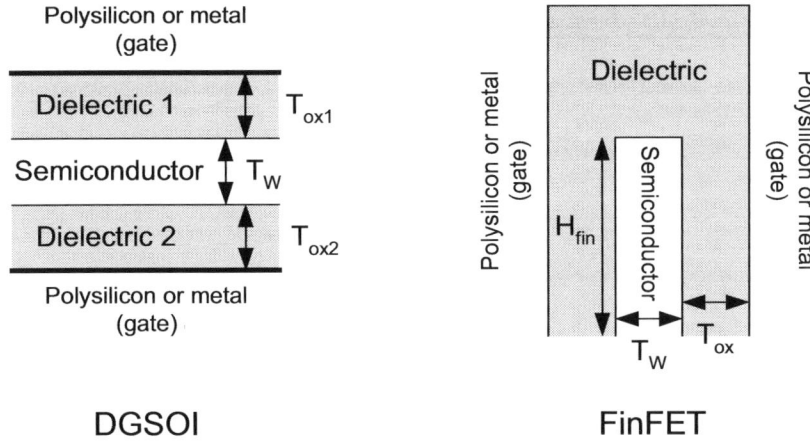

**Fig. 5.1.** Schematic representation of a Double-Gate transistor (left) and a FINFET (right).

However, if the Si thickness is reduced, the whole silicon film is depleted and an important interaction occurs between the two potential wells. In such conditions the inversion layer is formed not only at the top and bottom of the silicon slab (*i.e.*, near the two silicon-oxide interfaces) but throughout the entire silicon film thickness. It is then said that the device operates in '*volume inversion*', *i.e.*, carriers are no longer confined at the Si/SiO$_2$ interface, but distributed throughout the entire silicon volume (Figure 5.2 (right)).

In a FinFET, if the height of the fin is greater than its thickness and the dielectric thickness on the upper side is greater than that on the lateral sides, the FinFET can be analyzed as a vertical Double-gate MOSFET (Figure 5.1 (right)). Otherwise, carriers are quantized in two dimensions and therefore the approach developed in Section 5.3 would have to be considered.

Several authors have claimed that volume inversion presents a significant number of advantages, such as: i) enhancement of the number of minority carriers; ii) increase in carrier mobility and velocity due to the reduced influence of scattering associated with oxide and interface charges and surface roughness; iii) as a consequence of the latter, an increase in drain current and transconductance; iv) a decrease in low-frequency noise, and v) a large reduction in hot-carrier effects.[14] In addition, like other dual-gated devices, DGMOSFETs are claimed to be more immune to short channel effects (SCE) than bulk silicon MOSFETs or even than single gate fully depleted SOI MOSFETs. This is due to the fact that the two gate electrodes jointly control the carriers, thus screening the drain field from the channel.[15] This latter feature would permit a much greater scaling down of these devices than was ever imagined in conventional MOSFETs.

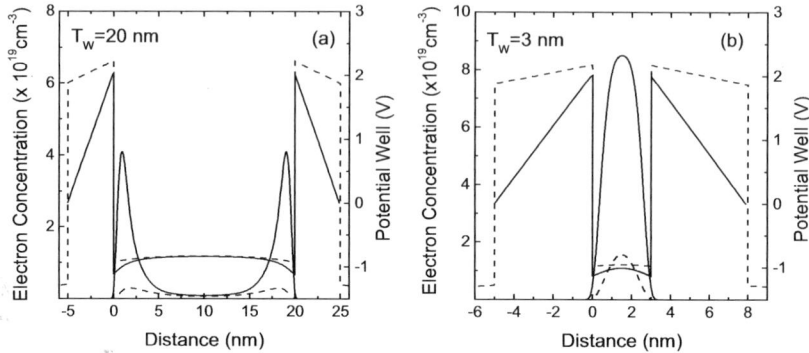

**Fig. 5.2.** Electron distribution and potential well for two DGSOI devices with different silicon-layer thicknesses and two values of the inversion charge concentration. Dashed lines correspond to $N_{inv}=1\times10^{12}\text{cm}^{-2}$, and solid lines correspond to $N_{inv}=8\times10^{12}\text{cm}^{-2}$.

Some steps have been taken in the theoretical study of these devices:

- i) Ouisse has self-consistently solved Poisson's and Schrödinger's equations in ultrathin silicon-on-insulator structures and has studied the interaction between the front and back inversion layers as a function of the silicon film thickness, electron concentration and temperature.[16]
- ii) Majkusiak et al. have studied the dependence of the carrier distribution on the silicon thickness in DGSOI devices.[17-18]
- iii) Taur analytically studied the electrostatic of double gate MOSFETs.[19]
- iv) Frank et al. [15] have performed a Monte Carlo simulation of a 30 nm gate length DGMOSFET with a channel thickness of $T_w$=5nm. The main finding of this study is that these ultrashort devices offer excellent properties for use in digital logic.
- v) Shoji et al. [20-21] have also studied the electronic structure of these devices. In addition, using a relaxation time approximation method, they have calculated the phonon-

limited electron mobility in double gate devices. In particular, they have shown that in DGMOSFETs, as silicon thickness is reduced, phonon-limited mobility gradually increases to a maximum around $T_w$=10nm, decreases in the $T_w$=5nm-10nm range to values below the value this parameter presents in conventional bulk MOSFETs, rises rapidly to another maximum in the vicinity of $T_w$=3nm and finally falls. They have also shown that the value of these maxima is highly dependent on the total electron concentration. According to these results, the optimum thickness of the silicon layer in DGMOSFETs is around $T_w$=10nm (from the point of view of phonon-limited mobility).

vi) Esseni *et al.* [22-25] have studied both experimentally and theoretically DGSOI devices. These authors experimentally observed the improvement of the electron mobility in DGSOI devices for silicon thicknesses of 10nm and below compared to single gate operation.

vii) Ravaioli *et al.* [26-28] have developed full-band Monte Carlo simulations to analyze the performance of scaled n-channel double-gate (DG) MOSFETs. They observed that the sheet charge in DG devices does not decrease as much as expected in bulk devices when quantum-mechanical effects are included. The average carrier velocity in the channel was also somewhat reduced by quantum effects, as a second-order effect. According to their simulations in a DG structure, quantum effects tend to concentrate the charge density in the center of the channel, where transverse fields are lower. Because of this, interface scattering appears to be less pronounced when quantum effects were included.

viii) Dollfus *et al.* [29-30] have also simulated the behaviour of the mobility in different DGSOI structures by using a Monte Carlo simulation scheme.

ix) Sverdlov *et al.* [31-32] have theoretically studied the effects of the strain and the crystallographic orientation of the silicon layer on different DGSOI structures.

x) Fossum *et al.* [33-35] have also studied and modeled the behavior of the electron mobility in DGSOI devices.

xi) Z. Ren *et al.* [36] have used a non-equilibrium Green's function (NEGF) approach to examine the double gate (DG) MOSFET and similar structures. NEGF simulations are

xii) used to examine: i) choice of body thickness, ii) effect of body thickness variations, iii) the required junction abruptness, iv) sensitivity of the device to gate-S/D (source/drain) over/underlap, and v) the impact of metal-semiconductor contact resistance.

xii) Our group at the University of Granada has also carried out a large research activity around these devices.[37-40] This chapter summarizes our main results.

Phonon scattering is not the only scattering mechanism present in multigate devices. Although other scattering mechanisms (namely, those associated with the Coulomb interaction with oxide and interface charges and with the roughness of the silicon-oxide interfaces) are likely to be weakened by a volume inversion operation [14-26], their contribution has to be taken into account. The weakness of these scattering mechanisms is justified, at least *a priori*, by the spread of the electrons throughout the whole silicon region. Nevertheless, we must not forget that in order to achieve volume inversion, both channels must interact strongly, and this only happens in the medium-high transverse electric field range when the silicon slab between the two oxides is thin enough (below 20 nm as pointed out by [16-20]). In these thin devices, although electrons are certainly spread along the whole silicon layer, they may not be far enough from the interfaces and may therefore be significantly affected by surface scattering mechanisms; much more so, in fact, than in bulk MOSFETs, since they are now interacting with two interfaces. This means that scattering mechanisms may play a very important role in the electron mobility in ultra-thin DGMOSFETs, contrary to what was previously believed. This imposes a serious limitation on the minimum silicon thicknesses which can be used in these devices, in addition to the limitations already presented by other physical and technological issues, as detailed elsewhere.[15,20]

We have used a one-electron Monte Carlo method to study the stationary electron transport properties in DGSOI and FinFET inversion layers, focusing our attention on evaluation of the stationary drift velocity and the low-field mobility at room temperature and lower. Electron quantization in the inversion layer was taken into account in an appropriate manner, self-consistently solving Poisson's and Schrödinger's equations assuming a simple non-parabolic band model for the silicon. Once the electron distribution in the silicon layer was determined, the Boltzmann transport equation was solved by the Monte Carlo method, simultaneously taking into account phonon, surface-roughness and Coulomb scattering. To do this, it was necessary to improve on existing scattering models. The presence of two close silicon-oxide interfaces in a DGMOSFET makes it significantly different from its standard bulk counterparts. Therefore, we had to develop a new model capable of taking into account the effect of the roughness of both interfaces on the total scattering rate. This model has been applied to single-gate SOI MOSFETs.[37] The calculation of mobility curves in DGSOI devices in which Coulomb scattering is accurately included is also performed. The following sections give the mobility results together with a thorough explanation. We will also consider behavior at low temperatures and under asymmetrical operation. Other effects, such as the decrease in silicon thickness will be studied and finally we will consider arbitrary crystallographic orientations, strained silicon channels and the use of metal-gate/high-k stacks.

### 5.2.1 Phonon-limited mobility

To study electron mobility behavior in ultrathin DGSOI inversion layers, we used a one-electron Monte Carlo simulator.[38-40]

Phonon, surface-roughness and Coulomb scattering were taken into account. We used bulk electron-phonon scattering models, considering acoustic deformation potential scattering and intervalley scattering (between both equivalent and non-equivalent valleys). The coupling constants for intervalley phonons and the acoustic deformation potential are the same as in bulk silicon inversion layers.[41-42] The phonon-scattering rates for inversion layers were deduced by using Price's formulation.[43] The use of bulk phonons is questionable, as the presence of $Si/SiO_2$ interfaces undoubtedly alters the dispersion of the phonons, their nature and their coupling with the electrons. Previous studies [44] accounting for these effects in idealized conditions show that phonon-limited mobility is reduced by up to 20% [41] due to the presence of the $Si/SiO_2$ interfaces. However, if such idealized conditions are relaxed, an even lower reduction is expected. For these reasons and due to the difficulty of dealing with the effects of the interfaces on the phonon-scattering rate [41,44], we have assumed that the bulk phonons are not influenced by the layered structure. In any case, the presence of the two $Si/SiO_2$ interfaces becomes more important as the silicon layer thickness is reduced. This effect is analyzed later in this chapter. Finally, the effect of $SiO_2$ polar-phonon remote scattering is initially ignored, although in the later sections we will analyze its effect when high-k dielectrics are used.

In our simulations the electron energy was limited to 0.5 eV, since for higher electron energies the results obtained from the simulation are not likely to be very accurate unless a detailed band structure is used. As the silicon band gap was set to 1.12 eV at room temperature (thus setting the energy threshold for the impact ionization process), impact ionization was not included.

By using this simulator, we were able to calculate the electron mobility in DGSOI inversion layers for different silicon film thicknesses ($T_w$). Our attention was focused on evaluating the stationary drift velocity and the low-field mobility. A comprehensive description of this simulator can be found elsewhere.[37-38,45-47]

Figure 5.3 shows the electron mobility curves versus the transverse effective field for different silicon layer thicknesses, calculated at room temperature and taking only phonon scattering into account. For comparison, the electron mobility in a bulk silicon inversion layer at room temperature is also shown. It can be observed in this figure that there is more than one trend in electron mobility as the silicon slab thickness is reduced and that, in addition, this behavior is highly dependent on the value of the electric field.

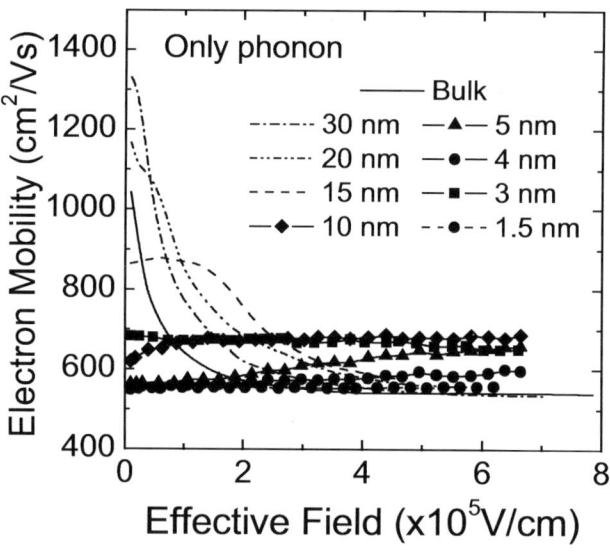

**Fig. 5.3.** Phonon-limited mobility in a DGSOI versus the transverse effective field for different thicknesses of the silicon slab.

To illustrate this more clearly, Figure 5.4 shows the electron mobility versus silicon layer thickness for two different electric field values.

Basically, our Monte Carlo results reproduce, qualitatively, the results obtained by Shoji et al. [20-21] for DGSOI inversion layers using the relaxation time approximation.

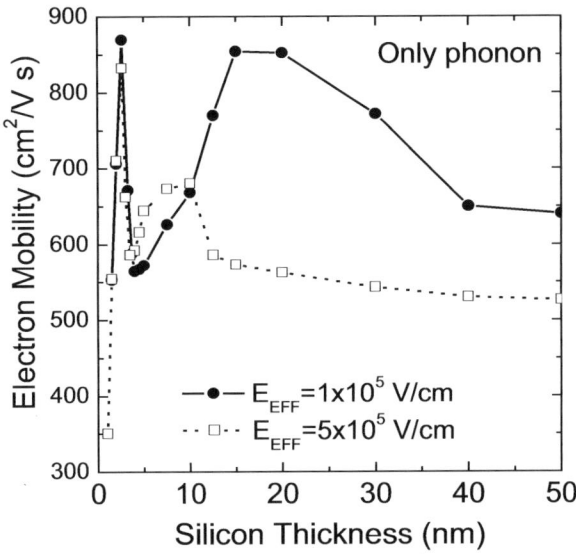

**Fig. 5.4.** Evolution of phonon limited mobility with the silicon thickness at room temperature for a DGSOI transistor.

As silicon thickness is reduced, the phonon-limited mobility increases gradually to a maximum around $T_w$=10nm (for $E_{EFF}$=5x10$^5$V/cm), decreases in the $T_w$=5nm-10nm range to values below those of this parameter in conventional bulk MOSFETs, rises rapidly to another maximum in the vicinity of $T_w$=3nm and finally falls. For higher electric fields, the maximum mobility is smaller and is shifted to lower Si thicknesses. Figure 5.4 reveals the existence of three regions with different behavior in the DGSOI phonon limited-mobility:

i) The first region corresponds to thick silicon slabs. In DGSOI inversion layers the two channels are sufficiently separated and no interaction appears between them. This situation corresponds to two conventional inversion layers in parallel, separated by a large potential barrier. The behavior of electrons in each of these inversion layers is the same as that observed in a bulk silicon inversion layer. As the silicon thickness is reduced, the interaction of the two inversion layers causes the electrons to occupy the entire silicon volume. This is the beginning of the second region, which strongly depends on the value of the transverse effective field, since for high electric fields a potential barrier, which obstructs the mutual influence of the two channels, is formed in the middle of the silicon slab.

ii) In the second region, the electron mobility in DGSOI inversion layers is up to 20% larger than the mobility in SGSOI inversion layers. The limits of this region and the values of the mobility depend on the electric field considered. This is the region in which volume inversion occurs. In this region of silicon thickness, both subband energy levels and wavefunctions vary significantly as a consequence of the two channels interacting. It is for this reason that the form factor which multiplies the phonon scattering rate decreases [40] compared to its value in conventional bulk MOSFETs in the same transverse effective field. (see Figure 5.5). This happens down to a certain value of silicon thickness. For lower thicknesses, although the electrons are distributed throughout the entire silicon layer, their confinement is greater (due to the geometrical confinement), and therefore, the form factor and the phonon scattering rate increase, as shown in Figure 5.5. This marks the beginning of the third region.

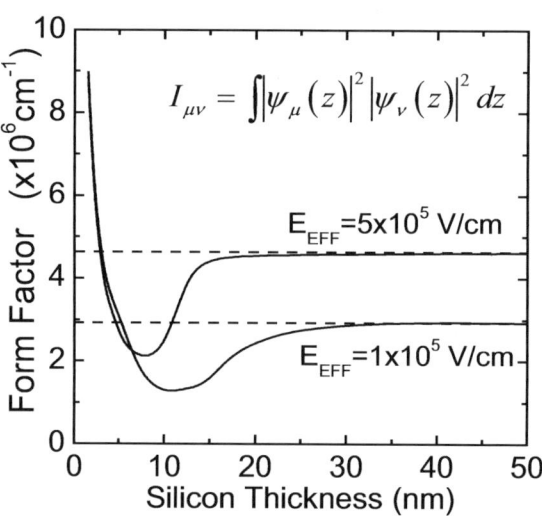

**Fig. 5.5.** Form factor for the ground subband of a DGSOI MOSFET as a function of the silicon thickness for two values of the transverse effective field. Dashed lines correspond to a conventional bulk MOSFET.

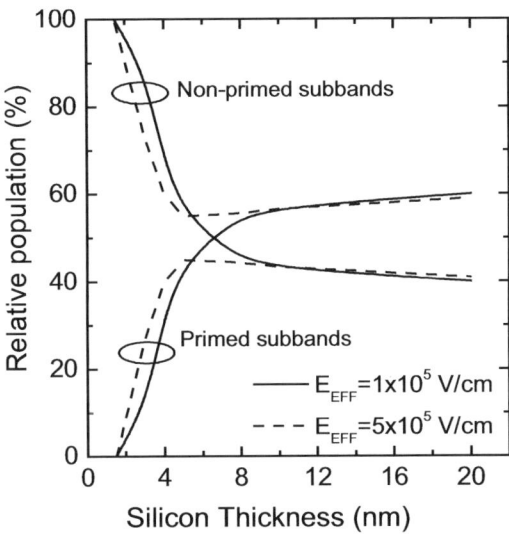

**Fig. 5.6.** Relative population in non-primed subbands and primed subbands in a Si-(100) DGSOI inversion layer for two values of the transverse effective field.

iii) In the third and last region ($T_w<4$nm), the mobility for DGSOI falls abruptly. In this zone, mobility is limited by the thickness of the silicon slab; that is to say, limitations on mobility are imposed by the geometrical confinement of the carriers. The limits of this region do not depend on the transverse electric field. In fact, electron mobility in this region is hardly modified by the transverse electric field. As can be appreciated in the mobility curves of Figure 5.4, electron mobility increases abruptly in the range 4nm→3nm. This is a little surprising since, as stated above, the form factor and the phonon scattering rate were expected to increase. However, the sharp increase in electron mobility can be understood by taking into account the evolution of the phonon scattering rate and the relative population of electrons in primed and non-primed subbands as shown in Figure 5.6. In Figure 5.5, it is shown that from $T_w=4$nm to $T_w=3$nm, the form factor (and therefore the phonon scattering rate) increases by about 20% (as predicted in the discussion above). However, Figure 5.6 shows that the relative population of the non-primed ladder (where electrons have a lower conduction effective mass) also increases by more than 30%, reaching almost 90% of the total. Thus we have two diverging trends, and in the case $T_w$=4nm→3nm, the reduction in the conduction effective mass dominates the increase in the phonon scattering rate. As a consequence, mobility increases. When a smaller silicon slab is considered, the increase in the phonon scattering rate is greater than the reduction in the conduction effective mass, and therefore electron mobility falls abruptly.

### 5.2.2 Confinement of acoustic phonons

Up to now, phonon confinement has been neglected when modeling the interaction of electrons with phonons; the bulk phonon model was believed to provide a good enough approximation for the calculation of electron transport properties. However, this assumption may be questioned when the silicon layer in the devices being considered is only a few atomic layers thick and evidence of confined phonons has been obtained experimentally.[48] If bulk phonons are considered, the electron mobility displays a complex behavior as a function of silicon layer thickness, as shown in Figure 5.4. To check whether the results obtained with the bulk model are still valid when phonon confinement is taken into account, we introduced [49-50] a confined phonon model for ultrathin SOI devices, following the theoretical work on III-V based devices.[51-52] We used this model for the study of DGSOI devices, computing the electron-phonon scattering rates and the electron mobility for several device structures in order to show the influence of phonon confinement and its dependence on the device geometry. The confined phonon model assumes simplified boundary conditions (either rigid or free) at the external surfaces of the two silicon dioxide layers: we have analyzed the difference between the computed mobility for the two cases, and discuss those cases where the results are (almost) independent of the boundary conditions imposed. A detailed explanation on the derivation of the model for confined acoustic phonons can be found elsewhere.[53] We considered the three-layer $SiO_2$/Si/$SiO_2$ structure shown in Figure 5.1. In each layer, acoustic phonons are modeled as elastic waves in an isotropic medium satisfying the wave equation:

$$\frac{\partial^2 u}{\partial t^2} = s_t^2 \nabla^2 u + \left(s_l^2 - s_t^2\right)\nabla(\nabla u) \quad (5.1)$$

where $u$ is the displacement vector, $s_l$ and $s_t$ the longitudinal and transversal sound speeds, respectively. To solve Equation 5.1 we have considered two types of boundary conditions: i) in the case of rigid boundary conditions the external surface of the $SiO_2$ layers are considered rigid and fixed, so the displacement vector, $u$, must vanish; ii) in the case of free boundary conditions, the external surfaces are free to vibrate and unconstrained, so the stress tensor components must vanish.[54] Once phonon states are calculated, the interaction Hamiltonian can be computed. Its matrix elements give us the transition probability between two states, according to the Fermi golden rule.[42] The electron scattering rates are then obtained summing over phonon absorption or emission, over phonon branches, and over the final electron state. The scattering rates for electrons in the first subband are shown in Figure 5.7, normalized by the bulk model scattering rate. Both oxide layers were considered to be 1nm thick. A significant increase in scattering rates compared to the non-confined case is observed. The scattering rates are computed assuming an elastic approximation, that is, the electron energy does not change in a scattering event with a phonon. Indeed, the phonon energy is typically negligible compared to the electron energy (at room temperature) and it can be observed that this assumption does not significantly affect the scattering rate.[50]

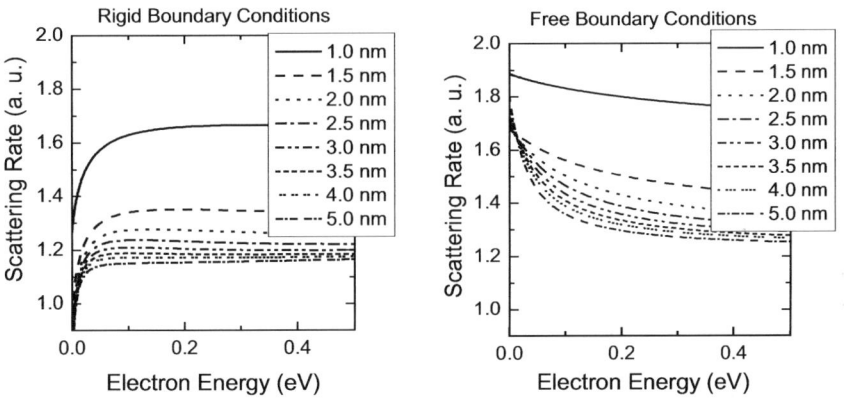

**Fig. 5.7.** Confined phonon-electron scattering rate, normalized to the bulk scattering rate, for the first electron subband, with different $T_w$. Rigid boundary conditions are considered in the left-hand plot, while free BC are assumed for the right-hand one.

Figure 5.7 shows that the confined phonon model with both free and rigid boundary conditions implies a significant increase in the scattering rate for thin layers compared to the bulk model; this effect is highly dependent on $T_w$ and is greater for a very thin layer. The difference between free and rigid boundary conditions is quite small at large electron energies and increases as the energy decreases, the scattering rates being higher in the case of free boundary conditions.

The electron mobility is then shown in Figure 5.8. Mobility is reduced when phonon confinement is taken into account for both sets of boundary conditions considered.

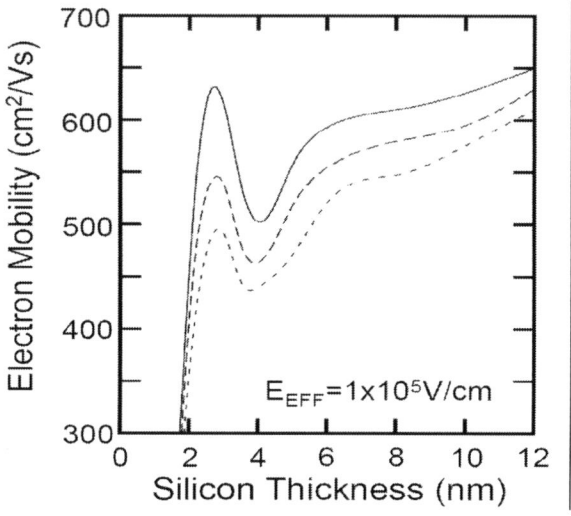

**Fig. 5.8.** Electron mobility in a DGSOI device as a function of the silicon layer thickness, calculated with different boundary conditions.

Up to now, we have considered a fixed value of the silicon dioxide layers. To analyze the effect of oxide thickness, we now fix the silicon layer thickness $T_w = 4$ nm and allow the thickness of the silicon dioxide layers to change. The scattering rates computed for symmetrical DGSOI devices with $T_w = 4$ nm and $T_{ox} = 1$ or 2 nm are shown in Figure 5.9 (left). The corresponding electron mobility is shown in Figure 5.9 (right). While for $T_{ox} = 1$ the difference between free boundary conditions and rigid boundary conditions is quite large, an oxide layer thickness of 2 nm greatly reduces the difference between the two cases. As seen for thick oxide layers, it is predictable that the boundary conditions used become almost irrelevant because the more extensive the external surfaces, the less important the boundary conditions. However, it is quite remarkable that for such thin oxide layers, the computed mobility curves are so close to each other.

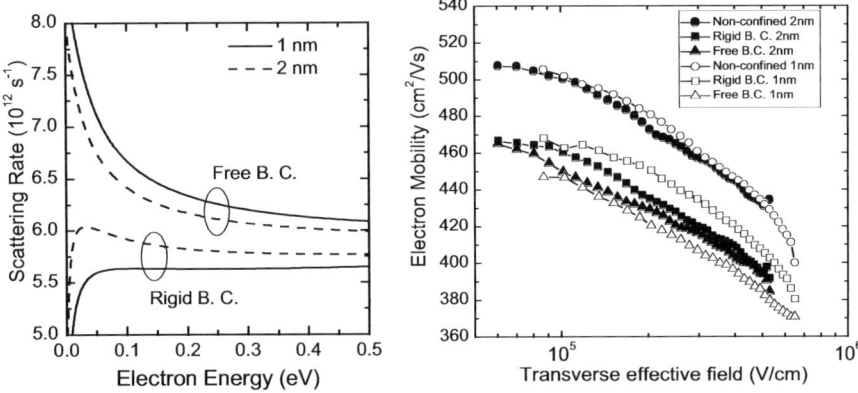

**Fig. 5.9. Left:** Scattering rates for the first electron subband for symmetrical double gate devices with $T_w = 4$ nm and different values of $T_{ox}$. **Right:** Electron mobility for symmetrical double gate devices with $T_w = 4$ nm and different values of $T_{ox}$.

### 5.2.3 Interface roughness scattering

Thus far, we have seen that the volume inversion operation increases electron mobility by up to 20%. Nevertheless, phonon scattering is not the only scattering mechanism in a silicon inversion layer. In [40] it was shown that in a DGSOI structure and in certain conditions, electrons are spread throughout the silicon volume. However, it was also shown that in order for this to happen, the silicon slab has to be so narrow that electrons are very close to the interfaces and must therefore be affected by surface scattering mechanisms, namely surface roughness and Coulomb scattering, due to the charges trapped at the Si-SiO$_2$ interface. The issue now is to determine how these scattering mechanisms affect mobility and whether volume inversion plays a role in the process by modifying the contribution of the surface scattering mechanisms to the total scattering rate.

Different authors have studied the effect of surface roughness scattering on electron transport properties in extremely thin silicon-on-insulator inversion layers.[37, 55] The main conclusion of such studies is that if the silicon layer is thin enough (less than 15 nm) the presence of a second Si-SiO$_2$ interface (bottom interface) plays a very important role. In fact, it has been shown that the buried interface has a double effect since, on the one hand, it modifies the surface roughness scattering rate due to the front-gate interface and on the other, itself provides a non-negligible scattering rate. Thus, it is absolutely necessary to take into account the presence of the second Si-SiO$_2$ interface when studying the electron transport properties of very thin SOI devices.

In order to reach these conclusions, we had to modify the normal surface-roughness scattering model used for bulk silicon-inversion layers. In Refs. [37, 56], we showed that if the silicon layer is very thin, the normal surface roughness scattering model used to simulate bulk silicon inversion layers fails. In particular, in Ref. [37] it was observed that the bulk model overestimates the effect of surface-roughness scattering due to the gate interface as a consequence of the minimal thickness of the silicon layer. Therefore, it was necessary to modify this model in order to simulate correctly the effect of surface-roughness scattering in very thin SOI inversion layers. A detailed description of these models can be found elsewhere.[37, 56]

We calculated electron mobility curves for different silicon layer thicknesses and different surface roughness heights at room temperature. Figure 5.10 shows the electron mobility curves as a function of the transverse effective field. Different values of the silicon thickness, $T_w$, were considered. In addition, for each $T_w$ value, different sets of surface roughness parameters were assumed: (solid line): No surface roughness scattering, *i.e.*, the two interfaces were assumed ideal, and therefore the only scattering mechanism is phonon scattering; (solid squares): $\Delta_1=\Delta_2=0.25$nm; $L_1=L_2=1.5$nm; (solid circles): $\Delta_1=\Delta_2=0.5$nm; $L_1=L_2=1.5$nm.

Note that the value sets chosen for surface roughness scattering parameters belong to the interval shown in the work of Goodnick *et al.*[57] as the usual values for (100) Si-SiO$_2$ interfaces. The first fact to note in this figure is that the mobility curves fall significantly as a consequence of the contribution of surface roughness mobility, mainly at high transverse effective fields. In addition, the importance of surface roughness scattering increases as the silicon layer thickness decreases. For the thickest samples ($T_w$=20nm and $T_w$=10nm), surface roughness scattering only affects electron mobility at high transverse effective fields, even for the roughest sample. However, for the thinnest samples, the effect of surface-roughness scattering is noticeable even at very low transverse effective fields. Therefore, as the silicon layer decreases, the importance of the surface-roughness effect grows. This is due to the fact that for the thinner samples, the confinement of the electrons near the Si-SiO$_2$ interfaces is due to the thinness of the silicon layer.

By applying the Mathiessen rule we isolated the contribution of surface-roughness to electron mobility:

$$\frac{1}{\mu_{SR}} = \frac{1}{\mu_{total}} - \frac{1}{\mu_{phonon}} \qquad (5.2)$$

where $\mu_{SR}$ is the mobility due to surface roughness scattering alone, $\mu_{phonon}$ is the mobility due to phonon scattering alone, and $\mu_{total}$ is the mobility due to both surface-roughness and phonon scattering.

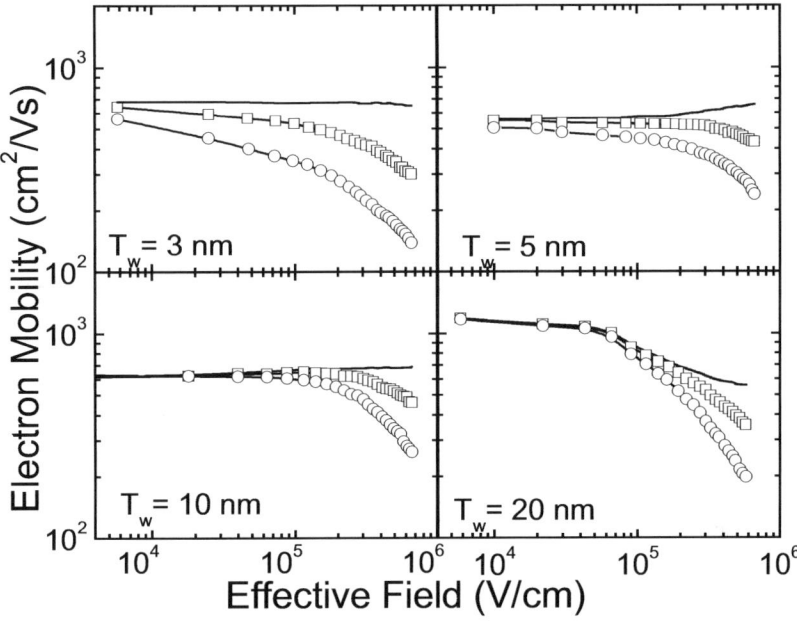

**Fig. 5.10.** Electron mobility curves versus the transverse effective field for different values of silicon thickness, $T_w$. For each $T_w$ value, different sets of surface roughness parameters were assumed: (solid line): No surface roughness scattering, i.e., the two interfaces were assumed ideal, and therefore the only scattering mechanism is phonon scattering; (open squares): $\Delta_1=\Delta_2=0.25$nm, $L_1=L_2=1.5$nm; (open circles): $\Delta_1=\Delta_2=0.25$nm; $L_1=L_2=1.5$nm.

Figure 5.11 shows that at low transverse effective fields, the surface-roughness mobility decreases as the silicon layer decreases. However, at high inversion charge concentrations, a different behavior is observed. As the silicon layer is reduced, $\mu_{SR}$ increases until a maximum value in the range 10nm-5nm (depending on the $E_{EFF}$ value) is reached. This is the region where the interaction between the two channels causes the electrons to spread along the silicon layer, thus reducing the surface roughness scattering rate. For lower $T_w$ values, the surface roughness mobility decreases abruptly, due to the high geometrical confinement of the electrons.

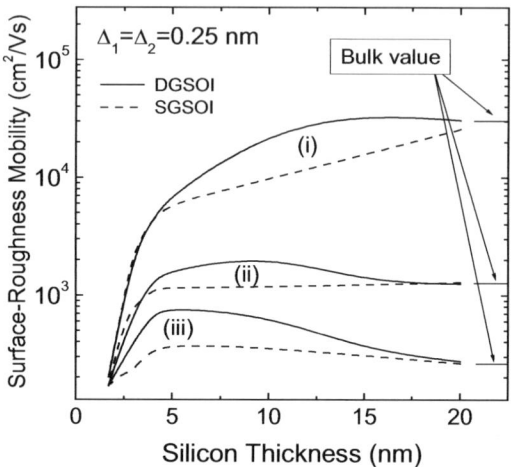

**Fig. 5.11.** Surface roughness limited mobility versus the silicon layer thickness for different values of the transverse effective field. Solid line: DGSOI; dashed line: SGSOI. (i):$E_{EFF}=10^5$V/cm; (ii):$E_{EFF}=5\times10^5$V/cm; (iii):$E_{EFF}=10^6$V/cm.

Finally, we compared the surface roughness mobility in DGSOI and Single Gate SOI (SGSOI) devices. Figure 5.11 also compares $\mu_{SR}$ for the two devices as a function of the silicon layer thickness for different values of the transverse effective field. In this figure, the following three regions can be observed:

i) There is a region for large $T_w$ values where no interaction between the two channels is produced and the mobility curves for DGSOI and SGSOI coincide. These mobility values are very similar to the values obtained for bulk silicon inversion layers.

ii) In the second region (intermediate values of $T_w$), the electron mobility of DGSOI inversion layers is greater than that of SGSOI inversion layers. This region, between 3nm<$T_w$<15nm, is where volume inversion occurs. In this region, surface-roughness scattering is reduced as a consequence of the spread of electrons throughout the silicon volume (volume inversion).

iii)   Finally, for $T_w<3$nm, the electron mobility curves of the two structures coincide again. In this region, controlled by geometric effects, mobility decreases abruptly.

### 5.2.4 Coulomb scattering

Although the channels of FinFETs and Double-Gate devices are usually undoped, the charges trapped at Si-SiO$_2$ interfaces can still play an important role, making it necessary to take their effects into account.[58] In the Coulomb scattering models for bulk silicon inversion layers, it has been shown that Coulomb scattering is highly dependent on factors such as: a) the distribution of electrons in the inversion layers, b) the geometrical distribution of external charged centers, c) the screening of charged centers by mobile carriers, d) the charged-center correlation, and e) image charges.[39,41,59] On the other hand, we know from the self-consistent solution of Poisson and Schrödinger equations in ultrathin SOI devices.[46,58], that the mutual influence of the two Si–SiO$_2$ interfaces means that the electron distribution in ultrathin DGSOI and FinFET devices differs greatly from that found in bulk inversion layers. This would, *a priori*, lead to a different screening effect, a different relative position between carriers and charged centers and, as a consequence, a very different Coulomb scattering effect on the electron mobility. We improved on a previous Coulomb scattering model to make it applicable to SOI inversion layers. The model developed is valid for single-gate silicon-on-insulator devices (SGSOI) and also for double-gate silicon-on-insulator (DGSOI) devices and FinFETs.[58]

Figure 5.12 shows the mobility curves for different silicon slab thicknesses versus the transverse effective field. The same interface charge concentration, $N_{it}$, has been assumed at both interfaces. ($N_{it}=5\times10^{10}$cm$^{-2}$). When Coulomb scattering is considered, this, rather than phonon scattering, is the main scattering mechanism at low inversion charge concentrations (low transverse effective field). Here again it is necessary to note the effect of surface scattering mechanisms on DGSOI mobility. Even for such a low interface trap concentration, Coulomb scattering significantly reduces electron mobility (mainly at low transverse effective fields) as a consequence of a weak screening of charged centers. As the effective field increases, the Coulomb scattering rate is reduced by the effect of the screening of scattering charge centers by the electrons themselves. As seen, Coulomb scattering has to be taken into account if we want to obtain the accurate electron mobility in these devices.

We have used Matthiessen's rule to isolate the effect of Coulomb scattering, and thus to observe the effect of the silicon slab thickness.

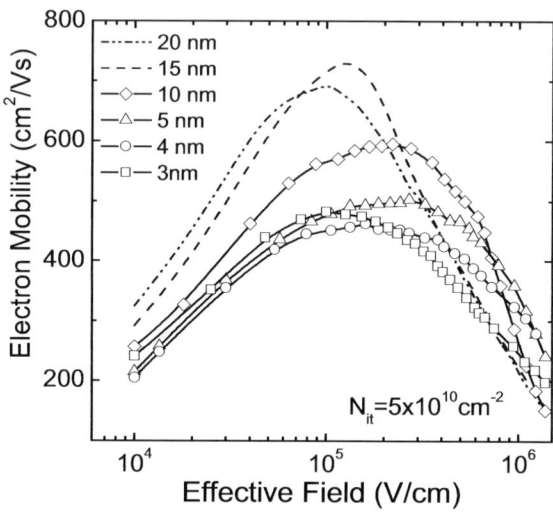

**Fig. 5.12.** Electron mobility curves in a DGSOI versus the transverse effective field for different thicknesses of the silicon slab. Phonon, surface roughness and Coulomb scattering were all taken into account.

Figure 5.13 shows the Coulomb limited mobility for two values of $T_w$ and two different interface charge concentrations. From this figure, the following facts can be observed:

i) As the electric field increases, the Coulomb-limited mobility increases due to the screening of the charged centers by the carriers in the channel.

ii) For larger $N_{it}$ values, the Coulomb-limited mobility is lower as a consequence of a stronger Coulomb interaction.

iii) The Coulomb-limited mobility increases more quickly for smaller $T_w$. This means that screening is more effective for thinner silicon slabs. Here again, volume inversion plays an important role.

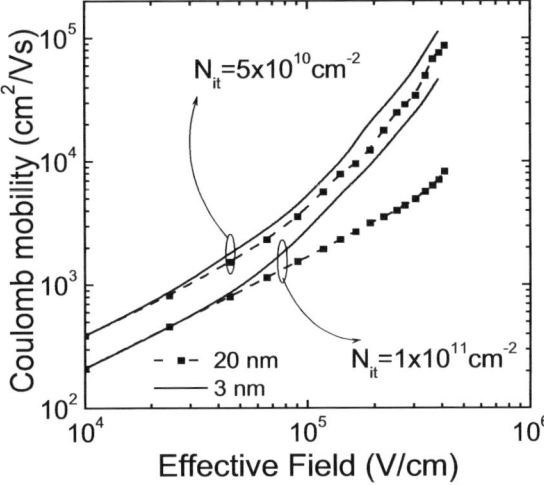

**Fig. 5.13.** Coulomb limited mobility obtained by applying Matthiessen's rule to the mobility curves of Figure 5.12.

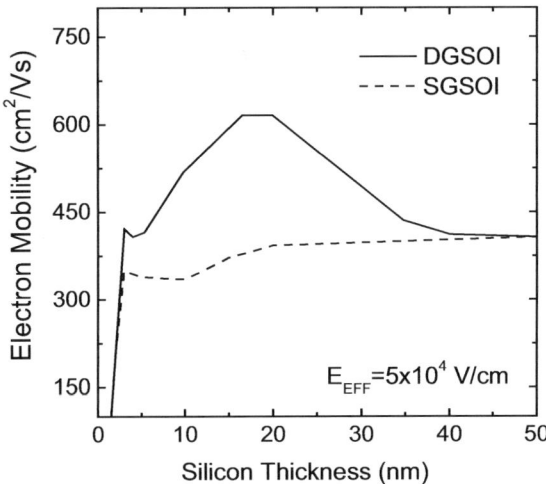

**Fig. 5.14.** Evolution of electron mobility for a DGSOI (solid line) and a SGSOI (dashed line) with the thickness of the silicon layer. All the scattering mechanisms were taken into account.

We also compared the electron mobility in DGSOI and in SGSOI inversion layers, considering the contribution of Coulomb scattering. Figure 5.14 shows the total electron mobility as a function of the silicon layer thickness for both DGSOI (solid line) and SGSOI (dashed line) inversion layers. In this figure, the electric field was assumed to be $E_{EFF}=5 \times 10^4$V/cm (small enough for Coulomb scattering not to be screened by the other scattering mechanisms).

As can be seen, the shapes of the curves are similar to those observed in Figure 5.4, when only phonon scattering was taking into account. Therefore we can still refer to the three regions described above. Nevertheless, in this case the differences between the electron mobility for the two types of structure are greater in the volume inversion region (region ii). This is due to a more efficient screening of the trapped centers by the carriers in the DGSOI structure, a result in agreement with our third comment regarding Figure 5.13.

### 5.2.5 Temperature dependence of mobility

Thus far, we have studied electron mobility at room temperature. We show the existence of a range of thicknesses of a silicon layer (between 5 nm and 20 nm) in which electron mobility increqases by 25% or more, due to volume inversion. Therefore, volume inversion plays an important role in electron mobility in these DGSOI devices when the silicon thickness is reduced to below 20 nm. The distribution of electrons in the silicon layer plays a very important role in the determination of transport properties. It is well known that the distribution of electrons in inversion layers varies greatly with changes in temperature.[60] We speculated as to whether this mobility behavior at room temperature is modified as the temperature decreases. For example, *a priori*, it is well known that as temperature decreases, quantum-size effects become more important, even for bulk silicon inversion layers, where at low temperatures the population of electrons in non-primed subbands is very high.

As shown by Ando *et al.* [60], the fraction of electrons in the non-primed subbands in a bulk silicon inversion layer is 1 at very low temperatures. This means, for example, that the influence of subband-modulation effects becomes increasingly weaker as the temperature is reduced. To shed some light on the matter in question, we carried out an in-depth study of the temperature behavior of electron mobility in DGSOI devices.[61] Figure 5.15 shows electron mobility curves versus the inversion charge sheet for different values of silicon thickness. Different temperatures were considered, namely $T = 25$ K, 77 K, 130 K and 300 K. It can be observed that at all the temperatures considered, the electron mobility shows more than one trend as the silicon thickness decreases. In addition, these trends are highly dependent on the inversion electron density.

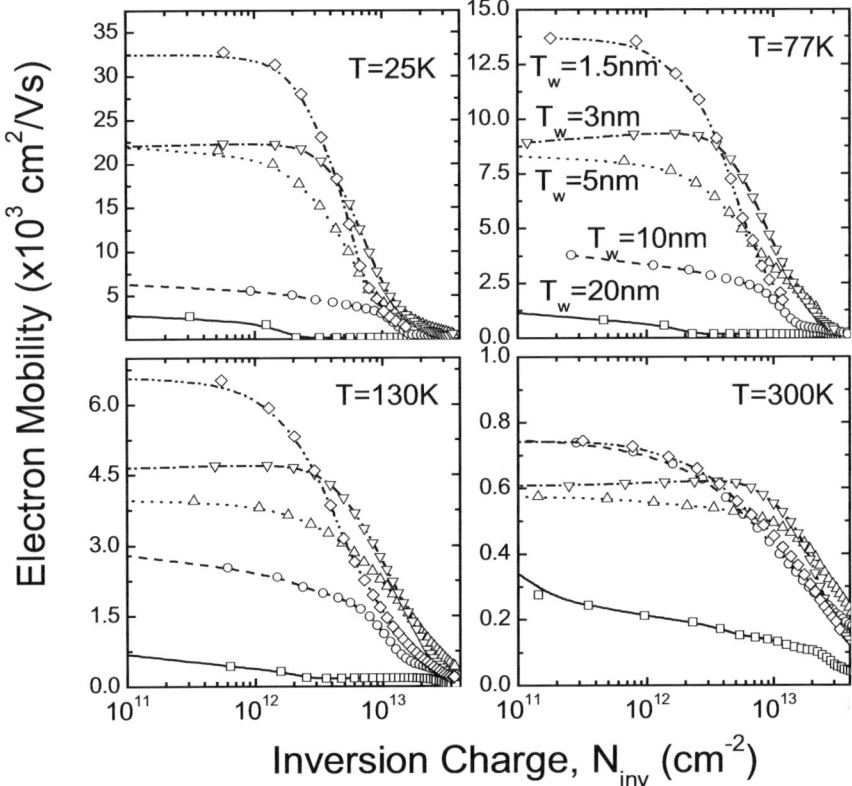

**Fig. 5.15.** Electron mobility curves in a DGSOI inversion layer as a function of $N_{inv}$ for different temperatures.

To make it clearer, Figure 5.16 (left) shows the evolution of mobility with silicon layer thickness for the temperatures considered in Figure 5.15. Note that all the curves take the shape described in Section 5.2.1 for the behavior of the mobility in DGSOI devices at room temperature, *i.e.*, there are three different regions: (i) an initial region for thick silicon layers ($T_w >$ 20–30 nm), where mobility decreases as $T_w$ increases and approaches the bulk value. (ii) As $T_w$ decreases, we show that volume inversion modifies the electron transport properties by reducing the effect of all scattering mechanisms. Accordingly, the electron mobility in DGSOI inversion layers increases by an important factor that depends on the silicon thickness, the transverse effective field and the temperature. (iii) Finally, for very small thicknesses, the limitations to electron transport are due to geometrical effects and therefore the mobility curves fall abruptly at all temperatures.

To obtain a clearer idea of the effect of volume inversion, we have normalized each mobility curve shown in Figure 5.16 (left) by the corresponding value of the mobility in a bulk silicon inversion layer at the same temperature. The results of this are shown in Figure 5.16 (right). As can be observed, the improvement in mobility in a DGSOI inversion layer, $\mu_{DGSOI}$, compared to the mobility in a bulk silicon inversion layer, $\mu_{bulk}$, becomes greater as the temperature is reduced: the maximum improvement, $\mu_{DGSOI}/\mu_{bulk}$, is 1.35 at room temperature for $T_w = 10$ nm, this quotient reaching 2.33 for $T = 25$ K for the same silicon thickness.

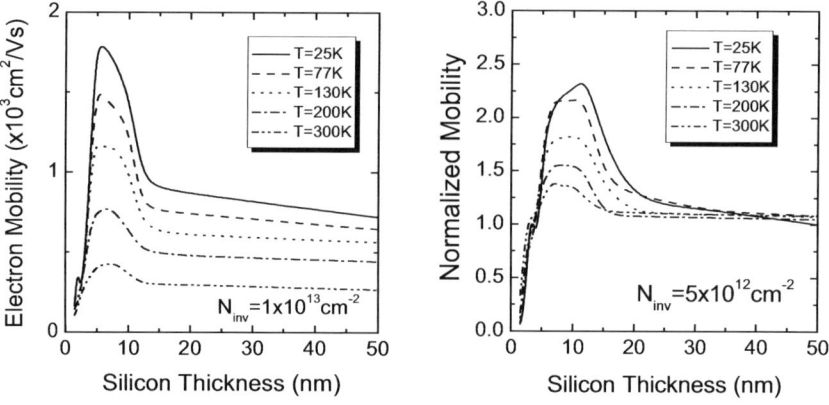

**Fig. 5.16.** Evolution of the electron mobility in a DGSOI inversion layer with the silicon thickness for different temperatures.

## 5.2.6 Symmetrical and asymmetrical operation of DGSOI FETs

Until now, we have considered the symmetrical operation of DGSOI transistors, *i.e.*, both gates identical and with the same bias applied to each of them. There are two main types of DG MOSFET: 1) a symmetrical type with both gates having identical work functions so that the two surface channels turn on at the same gate voltage and 2) an asymmetrical type with different work functions for the gates and only one channel turning on at the threshold voltage.[62-66] The threshold voltage of the symmetrical device is determined by the work function of the gate material and depends to only a negligible extent on the silicon thickness, the silicon doping concentration or the oxide thickness.[63] Figure 5.17 shows the threshold voltage of a symmetrical DGSOI structure using $p^+$-poly gates (dotted line) and $n^+$-poly gates (dashed line) as a function of the silicon thickness, $T_w$. Both oxide thicknesses were considered to be $t_{ox}=1$nm, the silicon doping was taken to be $N_A=10^{15}$cm$^{-3}$ and the doping of the poly gate $N_{D,poly}=N_{A,poly}=1\times10^{20}$cm$^{-3}$.

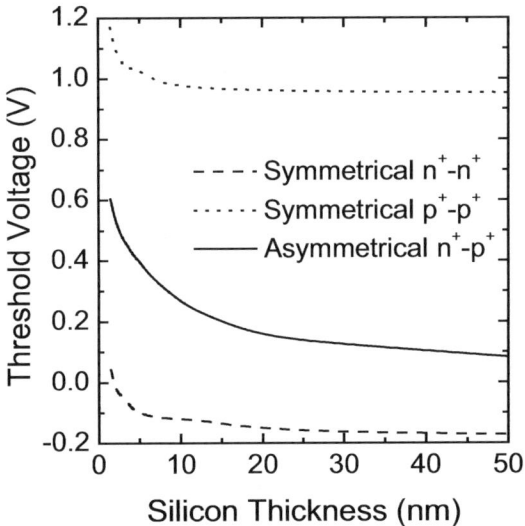

**Fig. 5.17.** Threshold voltage versus silicon thickness for different DGSOI structures.

To calculate these curves, we self-consistently solved the Poisson and Schrödinger equations in the DGSOI structure. The threshold voltage $V_{th}$ was taken as the gate voltage required to induce an inversion charge concentration of $N_{inv}=10^{11}$cm$^{-2}$ in the silicon layer. The threshold voltage values obtained with the symmetrical structure are, depending on the type of doping of the polygate, too high (~1V for $p^+$-poly gates) or too low (~-0.1V for $n^+$-poly gates) and, in any case, inadequate for state-of-the-art technology.

Therefore, it is necessary to look for new gate materials if we want these symmetrical DGSOI devices to have a threshold voltage suitable for use in low-power and high-speed applications. However, Suzuki *et al.* [63] proved that it is still possible to control the threshold voltage of a DGSOI device if an asymmetrical $n^+$-$p^+$ structure is used. Figure 5.17 (solid line) shows the voltage threshold for an asymmetrical $n^+$-$p^+$ DGSOI device as a function of the silicon thickness. As observed, the interaction between the two gates allows control of the threshold voltage in this structure to attain suitable values for state-of-the-art applications. From this viewpoint, it seems, therefore, that an asymmetrical configuration could prove superior to a symmetrical one.

In previous sections, we studied the electron mobility in a symmetrical double-gate silicon-on-insulator (DGSOI) device as a function of the transverse effective field and silicon layer thickness. The contributions of the main scattering mechanisms (phonon scattering, surface roughness scattering due to both Si-$SiO_2$ interfaces and Coulomb interaction with the interface traps of both interfaces) were taken into account and carefully analyzed. We demonstrated the existence of a range of thicknesses of the silicon layer (between 5nm and 20nm) in which the electron mobility in symmetrical DGSOI inversion layers is improved by 25% or more owing to volume inversion. Therefore, volume inversion and the symmetry of the device play an important role in the electron mobility in these symmetrical DGSOI devices when the silicon thickness is reduced below 20nm. Of course, symmetry is lost in asymmetrical DGSOI devices, raising the question of whether this also implies the loss of the volume inversion effect and, as a consequence, the loss of their advantages in terms of mobility. As mentioned above, it seems, from electrostatic studies, that asymmetrical DGSOI devices show a better behavior than their symmetrical DGSOI counterparts. However, the following question then needs to be answered: How does electron mobility in asymmetrical DGSOI devices behave compared to that in symmetrical DGSOI devices?

We undertook an in-depth study of electron mobility behavior in asymmetrical DGSOI devices and compared it with the mobility in their symmetrical counterparts.[67]

From the comparison of electrostatic behavior in the two kinds of devices (symmetrical and asymmetrical), we drew the following conclusions for devices with $T_w>5$nm:

i) As the silicon thickness is reduced, the conduction effective mass of electrons in the asymmetrical case is lower than that in the symmetrical case (Figure 5.18 (left)). This would contribute to an increase in electron mobility in the asymmetrical devices.

ii) The greater confinement of electrons in the asymmetrical case, produced by the deeper potential well in the $n^+$-gate side, produces an increase in the phonon scattering rate (Figure 5.18 (right)). This would contribute to a decrease in mobility in the asymmetrical case.

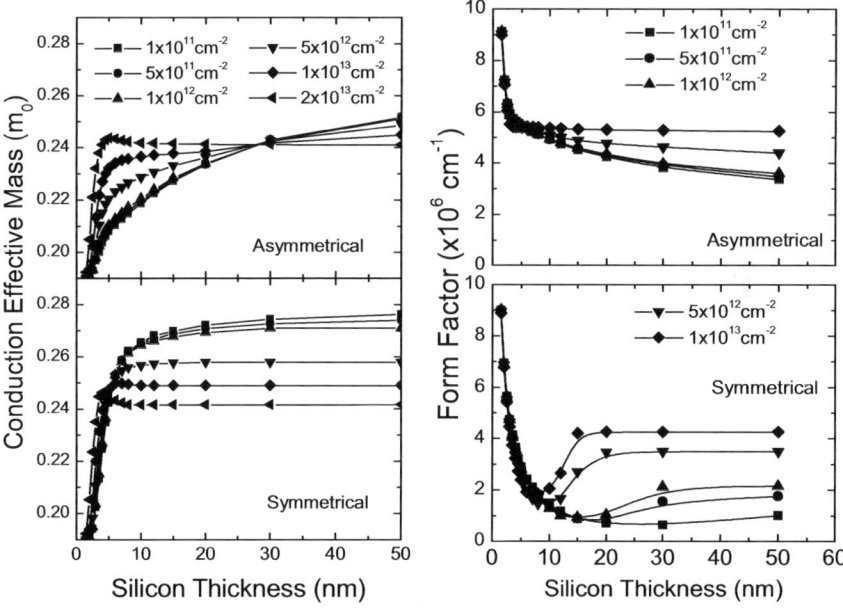

**Fig. 5.18. Left:** Average conduction effective mass for electrons in an asymmetrical $n^+$-$p^+$ (resp. symmetrical $n^+$-$n^+$) DGSOI inversion layer as a function of the silicon thickness for different values of the inversion charge concentration **Right:** Form factor for the ground subband in an asymmetrical $n^+$-$p^+$ (resp. symmetrical $n^+$-$n^+$) DGSOI inversion layer.

Taking these two facts into account, *a priori*, it is not possible to predict the behavior of the mobility until a solution of the Boltzmann transport equation is provided. Figure 5.19 shows the evolution of the electron mobility versus the silicon thickness for two values of the inversion charge concentration (symmetrical-gate devices in solid line and asymmetrical-gate devices in dashed line).

Phonon and surface roughness scattering are simultaneously taken into account. We see that the lack of symmetry, together with volume inversion, and the greater effect of surface roughness scattering, means that the electron mobility in asymmetrical DGSOI devices is considerably below the mobility curves corresponding to symmetrical devices in the whole range of silicon thicknesses. The difference is greatest in the 5-25nm range, where electron mobility in symmetrical DGSOI inversion layers is greatly improved by the volume inversion effect.

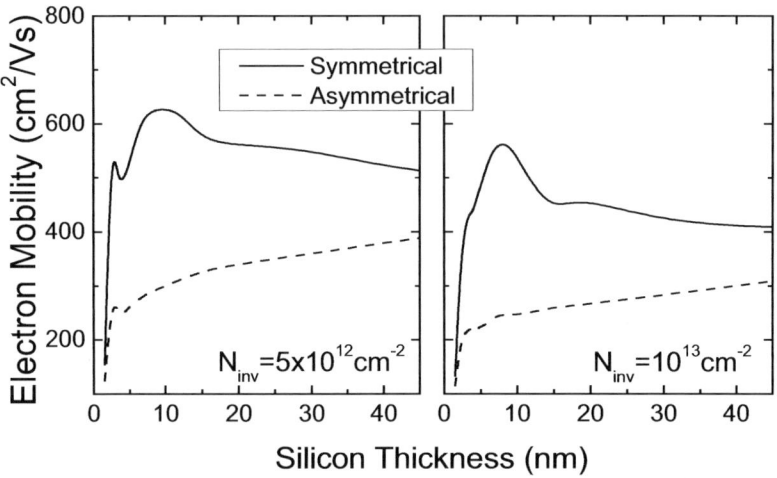

**Fig. 5.19.** Evolution of electron mobility for a symmetrical $n^+$-$n^+$ DGSOI (solid line) and an asymmetrical $n^+$-$p^+$ DGSOI (dashed line) with the thickness of the silicon layer.

### 5.2.7 Crystallographic orientation

Up to now, we have considered the surface orientation of the device to be the usual 100 orientation. However, with the adoption of vertical device structures such as the FinFET, different surface orientations can easily be achieved without using non-standard wafer substrates.[68] In a standard (100) wafer, where the gate and active fin area are aligned either perpendicularly or parallel to the wafer flat, the device channel lies in the (110) plane. However, if the transistor layout is rotated 45° in the plane of the wafer, then the resulting orientation of the device channel is (100). An intermediate rotation yields electron mobilities between those observed in the (100) and (110) orientations, imitating the mobility behavior of a (111) surface.[68] When non-(100) surface orientations are used, the electron and hole mobilities are modified due to asymmetry of the carrier effective masses in the silicon crystal lattice.[69-70] Table 5.1 summarizes the values of the masses for the common surface orientations, *i.e.* (100) (110) and (111). $m_3$ is the effective mass perpendicular to the surface and therefore the quantization mass, while $m_1$ and $m_2$ are the principal masses of the constant energy ellipse in the surface. The degeneracy of each set of ellipses is $n_v$. [70]

Table 5.1. Effective mass in silicon for different surface orientations.

| Surface orientation | $m_1$ | $m_2$ | $m_3$ | $n_v$ |
|---|---|---|---|---|
| (100) | $m_t$ | $m_t$ | $m_l$ | 2 |
|  | $m_t$ | $m_l$ | $m_t$ | 4 |
| (110) | $m_t$ | $(m_t+m_l)/2$ | $2m_tm_l/(m_t+m_l)$ | 4 |
|  | $m_t$ | $m_l$ | $m_t$ | 2 |
| (111) | $m_t$ | $(m_t+2m_l)/3$ | $3m_tm_l/(m_t+2m_l)$ | 6 |

Figure 5.20 shows the constant-energy ellipses associated with motion parallel to the interface for the three common surface orientations (100) (110) and (111).[70]

It is well known that in the case of Si-100, two of the six bulk constant energy ellipsoids will give $m_3=m_l$ (non-primed subbands) while the other four will give $m_3=m_t$ (primed subbands).[60,70] As $m_t<m_l$, non-primed subbands will have a lower energy and lower conduction mass than primed subbands. Similarly, in the case of the (110) orientation, two sets of subbands are also obtained: two bulk energy ellipsoids have an effective mass perpendicular to the interface $m_3=m_t$, while the other four have higher quantization masses. Therefore, these four ellipsoids will provide the lower energy subbands. Finally, in the case of the (111) orientation, the six ellipsoids have the same quantization mass, and therefore only one set of subbands is possible.

Due to the different quantization-masses for different crystallographic orientations, electron distribution, subband energy levels, wavefunctions, form factors and scattering rates depend on the surface orientation.[71] In addition, as observed in Figure 5.20, the effective masses of the constant-energy ellipses associated with motion in the parallel direction can, like in bulk silicon, be somewhat anisotropic. As a consequence, one can expect to see anisotropic conduction, *i.e.* once the quantization direction is fixed, conduction, and therefore, mobility, depend on the direction of the drift electric field, *i.e.*, the direction of the channel.

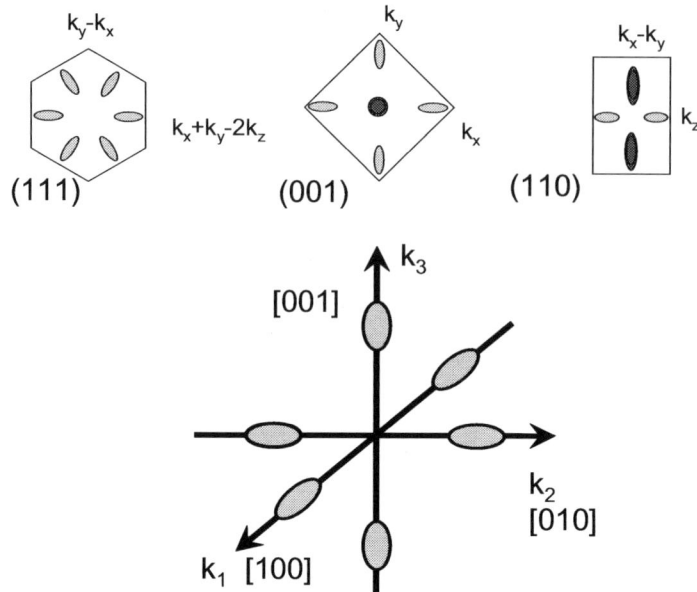

**Fig. 5.20.** Two-dimensional schematic constant-energy ellipses for (001) (110) and (111) Si conduction band.[70] Lower energy subbands are shown in black.

For all our calculations, we assumed a FinFET with an undoped silicon channel, a gate oxide thickness equal to 1nm and different silicon thicknesses and surface orientations. Figure 5.21 shows the electron distribution for the three common orientations in a FinFET with a silicon thickness of $T_W=5$nm and an inversion charge concentration of $N_{inv}=2.5 \times 10^{12}$ cm$^{-2}$. As observed in Figure 5.21, the electron distributions for (111) and (110) orientations are quite similar. However, it is completely different from the electron distribution for (100). This fact can be explained by looking at Table 5.1. The quantization masses, $m_3$, for (110) and (111) orientations are comparable, but very different to that for $m_{3,(100)}$. The higher quantization mass in Si(100) produces lower energy subbands in the two valleys with a longitudinal mass perpendicular to the interface. For thin silicon layers or high inversion charge concentrations (*i.e.* when quantum effects are more important), these two subbands are more populated than the others and the electrons become closer to the Si-SiO$_2$ interfaces.

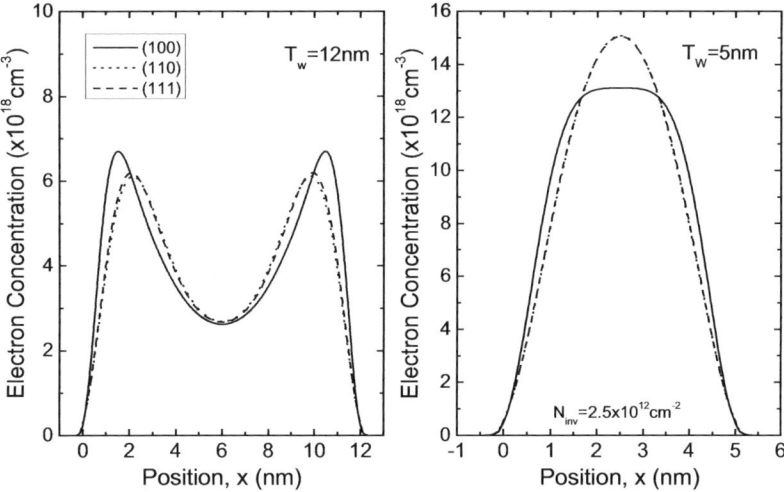

**Fig. 5.21.** Electron distribution in a FinFET for different surface orientations and different silicon thicknesses.

Once we had evaluated the electron distribution in the different structures as a function of surface orientation and silicon thickness, we calculated electron mobility by using the one-electron Monte Carlo simulator. Phonon scattering and surface-roughness scattering were taken into account. As the silicon layers remained undoped, we did not consider Coulomb scattering in the simulation. The following parameters were considered in modeling the surface roughness of the Si-SiO$_2$ interfaces [37, 56]: $L_{sr}$=1.5nm, $\Delta_{sr}$=0.4nm. Figure 5.22 shows the electron mobility curve for different surface orientations and channel orientations for two values of the fin thickness (4nm and 9nm).

Observation of Figure 5.22 is very illuminating and reveals the following facts:

i) The first result we see is the strong dependence of electron mobility on surface orientation.[31-32] Higher mobility values are obtained for the (100) orientation [68], corresponding to lower conduction effective mass, as shown in Table 5.1. The other two orientations have larger conduction effective masses and therefore lower mobility values. As observed experimentally [68], electron mobility can be degraded by a factor of two for the (110)/<1-10> orientation. However, this degradation is much less if a different channel direction (110)/<001> is selected for the same surface orientation (110).

ii) As just shown above, one remarkable result is the strong anisotropy of the mobility in the (110) surface orientation for different channel directions. The mobility value for a <001>-channel orientation is 50% larger than that for a <1-10>-channel direction. Taking into account Figure 5.20 and Table 5.1 for the (110) case, the subbands with lower energy values (higher $m_3$) correspond to the four ellipses with major axes in the <1-10> direction. Therefore, these subbands become more populated and dominate the transport. If we apply a drift electric field in the <001> direction, these subbands have an effective mass in the drift direction equal to $m_{drift}=m_1=0.19m_0$. On the other hand, if we apply a drift electric field in the <1-10> direction, the conduction effective mass is $m_{drift}=m_2=0.553m_0$, almost three times higher than in the <001> direction. This produces the big difference in mobility shown in Figure 5.22 for these two channel directions. In addition, the anisotropy becomes stronger as the silicon thickness decreases.

iii) However, this anisotropy of electron mobility was not observed when we considered other surface orientations. In the case of the (100) surface orientation, the subbands with lower energy corresponded to the two bulk ellipsoids with $m_3=m_l$, which became the two circles shown in Figure 5.20. These two subbands showed the same drift mass regardless of the drift electric field orientation. On the other hand, the other four subbands (corresponding to $m_3=m_t$) present a different effective mass in the x and y directions. Only when the contribution of these four high-energy subbands to conduction becomes significant, do we observe a dependence of mobility on channel orientation. For example, a slightly anisotropic behavior can be seen in Figure 5.22 for $T_w$=9nm: a slight dependence of the mobility curves for <001>- and <011>- channel directions at low inversion charge concentrations is observed for the (100) surface orientation. This effect does not appear in Figure 5.22, where the silicon thickness is $T_w$=4nm.

iv) No anisotropy is observed in the (111) surface orientation. In this case, the anisotropy shown in Figure 5.20 does not lead to an anisotropic conductivity because the ellipses are symmetrically placed.

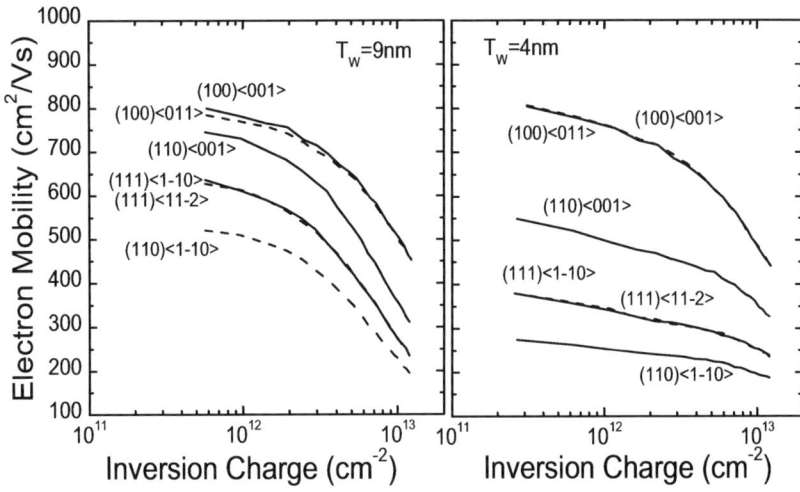

**Fig. 5.22.** Electron mobility as a function of the inversion charge in a FinFET for different surface orientations (hkl) and different channel orientations <hkl> for two values of the silicon thickness $T_w$=9nm (left) and $T_w$=4nm (right).

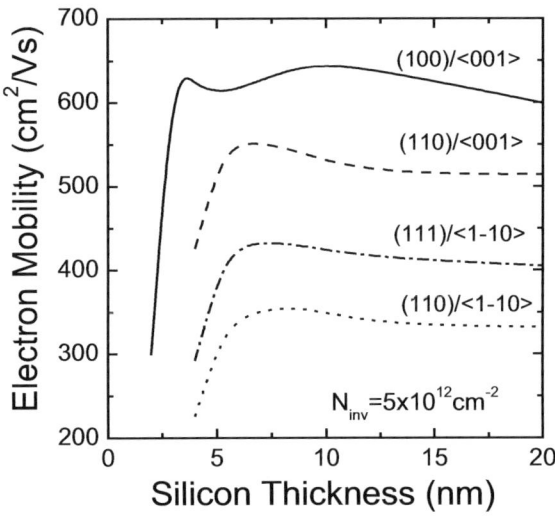

**Fig. 5.23.** Evolution of the electron mobility in a FinFET with the fin thickness for different combinations of surface orientation (hkl) and channel direction <hkl>.

We have also studied the evolution of electron mobility with silicon thickness for the different surface and channel orientations considered. The results are shown in Figure 5.23. As in the case of DGSOI (100) devices [40], electron mobility increases as the silicon thickness decreases until a maximum is reached, then falls abruptly for smaller silicon thicknesses.[40] As can be seen, for all surface orientations and channel directions there is a range of fin thicknesses where electron mobility is higher than in bulk MOSFETs. However, the silicon thickness interval where electron mobility is improved depends on the particular surface/channel combination.

### 5.2.8 High-k dielectrics

As a way to continue the progressive scaling down of MOSFET dimensions, the use of dielectrics with high permittivity (high-k dielectrics) instead of silicon dioxide has been proposed. [72] However, in spite of the reduction in the gate leakage tunnel current, high-k materials also negatively affect the electron mobility in MOSFET channels. In fact, higher dielectric constant is associated with higher electronic mobility degradation due to polar phonons in the dielectric (remote polar phonon scattering).[73-74]

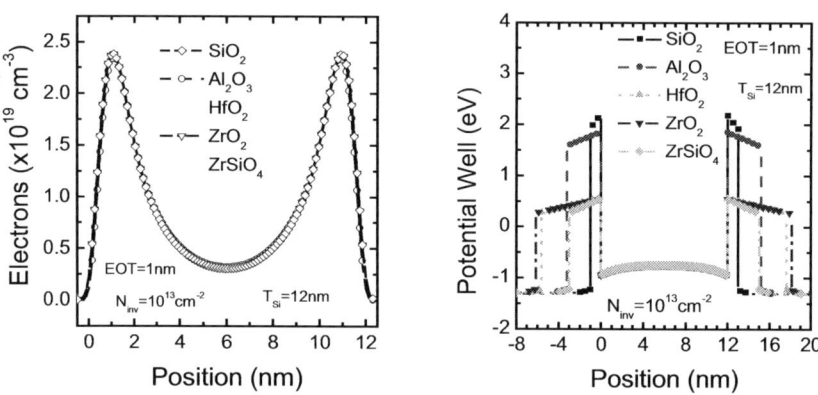

**Fig. 5.24. Left:** Electron distribution in a FINFET with different high-k dielectrics. In all cases, silicon thickness was fixed at $T_W$=12nm. **Right:** Potential well for the FINFETs.

Figure 5.24 shows the potential well and the electron distribution in a FINFET with a silicon thickness of 12nm with different high-k dielectrics as gate insulator. In all cases, the inversion charge concentration was assumed to be $N_{inv}=10^{13}$cm$^{-2}$. In all cases, the insulator thickness was calculated to provide an equivalent oxide thickness (EOT) equal to 1nm. As observed, in Figure 5.24 (left), the use of high-k dielectrics hardly modifies the electron distribution. Only in the $ZrO_2$ and $ZrSiO_4$ cases could a greater penetration of the wavefunctions in the insulator layer be observed, because of the lower barrier seen by the electrons from the channel.

We used a Monte Carlo simulator to study the effect of high-k insulators on the electron mobility in DGSOI devices. Our simulation results are disappointing from the point of view of electron mobility, as shown by the comparison of Figures 5.25 (left) and 5.25 (right). In both cases only phonon and surface roughness were considered. As can be seen, the effect of remote polar scattering strongly degrades the electron mobility when insulators with high permittivity are used. The comparison of Figure 5.25(left) and 5.25 (right) reveals that the important degradation of electron mobility shown in Figure 5.25 (right) is produced by remote polar phonon scattering.

In summary, volume inversion does not help to improve the mobility degradation produced by remote polar phonon scattering when high-k dielectrics are used.

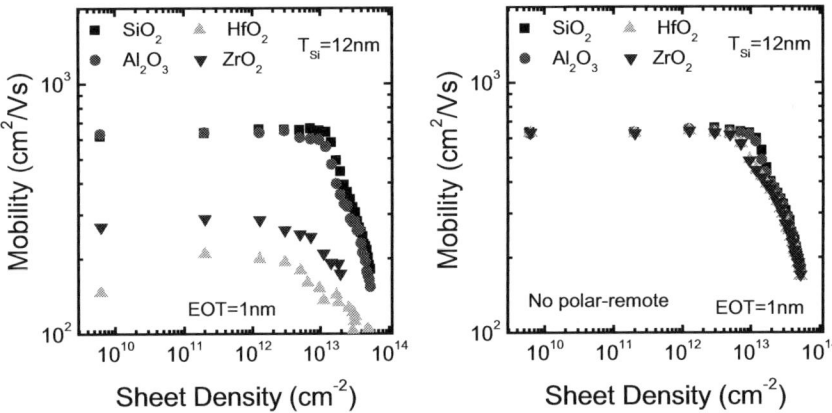

**Fig. 5.25. Left:** Electron mobility in a symmetrical DGSOI device with different insulator, taking into account the effect of remote phonon scattering. **Right:** Electron mobility in a symmetrical DGSOI device with different insulator, not taking into account the effect of remote phonon scattering.

## 5.2.9 Strained DGSOI devices

In recent years, much research activity has been focused on the enhancement of carrier mobility in the channel. In this sense, a significant step was taken with the introduction of strained silicon to build the MOSFET channel. Both theoretical and experimental studies have shown spectacular electron mobility enhancements when silicon is grown pseudomorphically on relaxed $Si_{1-x}Ge_x$.[7,75-76] The strain causes the six-fold degenerate valleys of the silicon conduction band minimum to split into two groups: two lowered valleys with the longitudinal effective mass axis perpendicular to the interface and four raised valleys with the longitudinal mass axis parallel to the interface (Figure 5.26).[77-78]

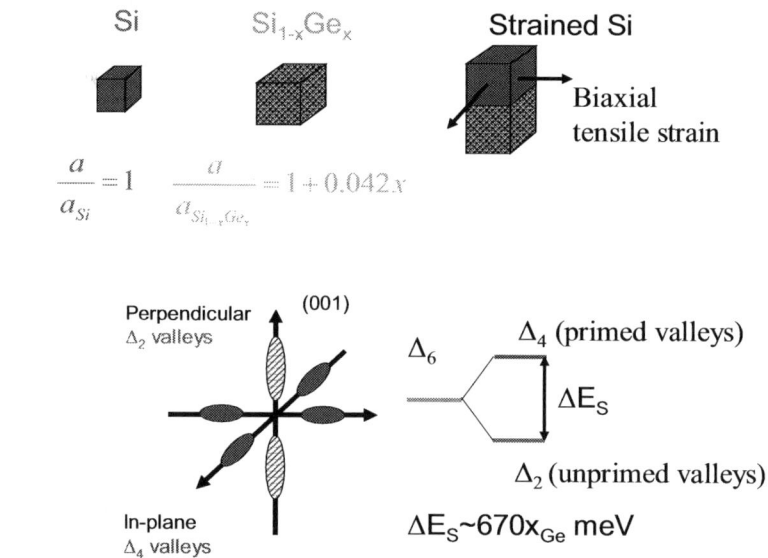

**Fig. 5.26.** Schematic representation of biaxial tensile strain produced when a silicon layer is pseudomorphically grown on a (100) SiGe layer.

This leads to a redistribution of the carriers between the different valleys. In the lowered valleys, which are more densely populated in the strained case, electrons show a smaller conduction effective mass (transverse mass) in transport parallel to the interface. In addition, the splitting between the valleys is enough to suppress the intervalley transitions of electrons from lower valleys to upper valleys, thus reducing the inter-valley phonon scattering rate compared with that of unstrained silicon. The combination of a lower effective mass and reduced intervalley scattering gives rise to higher electron mobility. Moreover, the lower intervalley-scattering rates make energy-relaxation times higher, causing spectacular electron velocity overshoot, as shown elsewhere.[79] In spite of these important advantages, bulk strained $Si/Si_{1-x}Ge_x$ CMOS technology still suffers from some of the limitations of standard silicon CMOS technology for sub-0.1μm applications.

In contrast to strained-Si/SiGe technology, SOI technology is fully compatible with existing standard silicon fabrication facilities and, *a priori*, CMOS circuit designs could be translated to ultrathin SOI technology without much difficulty.[2-3,80]

Recently these two options have been merged and ultra-thin body SOI devices with strained Silicon have been proposed.[76] In particular, the possibility of obtaining strained-Silicon on insulator without the need for a SiGe layer has been demonstrated, thus avoiding the drawbacks caused by the presence of this alloy and allowing thinner semiconductor layers to suppress short-channel effects.[81-83]

Self-consistent calculations have been performed for strained DGSOI for different values of the silicon thickness and different grades of strain. Following the experimental work of the IBM group [81] and our previous results with single gate strained-Si/SiGe–on-Insulator structures [82-83], in this work we only considered ultrathin strained Si directly on insulator DGSOI MOSFETs. In these devices, the strained-Si layer is directly sandwiched between two oxide layers and no SiGe layer is needed. The structure is fabricated by the layer transfer technique [81 and references therein]: a thin layer of strained Si is epitaxially grown on a buffer layer of relaxed $Si_{1-x}Ge_x$. The amount of strain in the Si layer will depend on the Ge mole fraction of the buffer layer. Once the strain layer is formed on the SiGe buffer, it is separated from the SiGe layer and bonded to the handle substrate. In the following, we will use this parameter, Ge mole fraction of the buffer layer, to indicate the amount of strain in the Si layer. In all cases, we assumed two 1nm-thick silicon-dioxide layers and midgap metal gates.

Figure 5.27 shows the electron density and the potential well in a strained 100-DGSOI MOSFET.

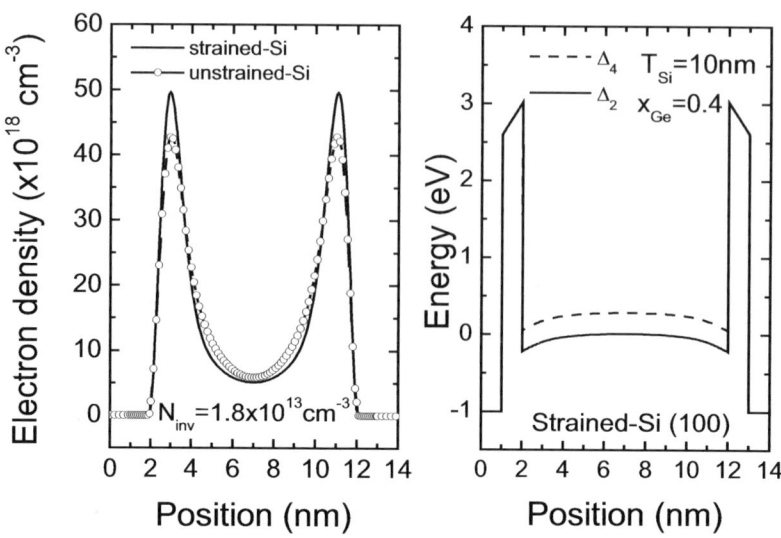

**Fig. 5.27. Left:** Electron density in a strained DGSOI MOSFET (solid line) and an unstrained DGSOI MOSFET (dashed line) for the same value of the inversion charge concentration. **Right:** Potential well in a strained DGSOI MOSFET. Silicon thickness was assumed to be 10 nm. Solid line shows the conduction band edge for the two valleys with longitudinal effective mass perpendicular to the interface ($\Delta_2$) while dashed line represents the conduction band edge for the four valleys with transverse effective mass perpendicular to the interface.

Figure 5.28 shows the electron distribution in a strained DGSOI MOSFET for a silicon thickness of $T_w=3$nm and a strain corresponding to $x_{Ge}=0.4$. For the sake of comparison, the electron distribution for the unstrained case is also shown.

A first result observed in Figure 5.27 (left) and 5.28 is that strain tends to reduce the volume inversion effect since the electron concentration in the center of the channel decreases as the strain increases.[84] Figure 5.29 shows the evolution of the energy level of the ground subband of each subband ladder (unprimed, $\Delta_2$, subbands and primed $\Delta_4$ subbands. It can be observed that the strain produces a greater separation between unprimed and primed levels. The effect caused by the strain is similar to that produced in unstrained DGSOI devices as the silicon thickness decreases.[40] The greater separation between unprimed and primed valleys produces a redistribution of the carriers among the different subbands and as a consequence, the unprimed valley tend to be much more populated, as shown in Figure 5.30.

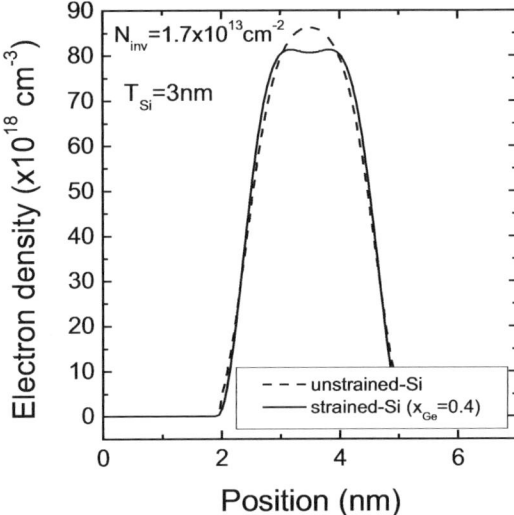

**Fig. 5.28.** Electron density in a strained DGSOI MOSFET (solid line) and an unstrained DGSOI MOSFET (dashed line) for the same value of the inversion charge concentration. Silicon thickness was assumed to be 3nm.

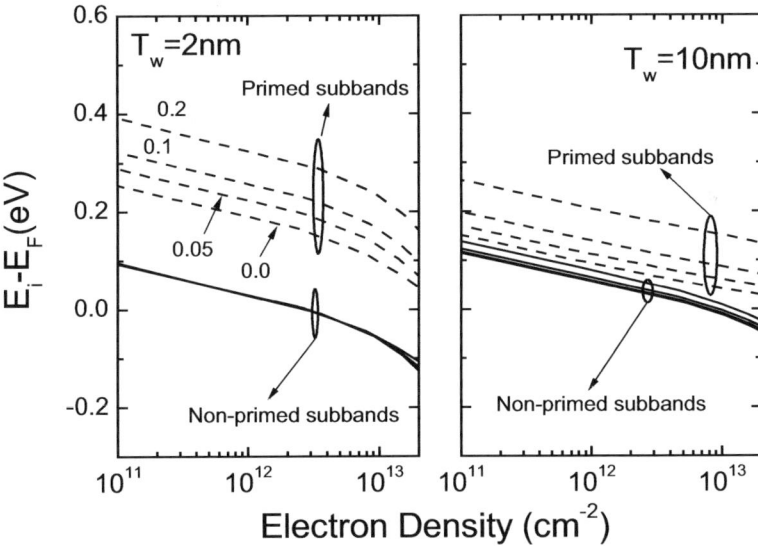

**Fig. 5.29.** Difference between the energy level of ground subband for unprimed and primed valleys and the Fermi level as a function of the electron density. Two values of the strained-Si thickness have been considered: $T_w=2$ nm (left) and $T_w=10$ nm (right). Different Ge mole fractions (between parentheses) have been considered.

This effect is greater for the thinner silicon layers, which feature a larger population of the unprimed subbands even in the absence of strain. This redistribution of carriers among the different subbands has an important effect on the electron transport properties because the conduction effective mass of the electrons in unprimed subbands is much lower than the conduction effective mass in primed subbands and, as a consequence, the average conduction effective mass decreases as the strain increases.

Figure 5.31 shows the average conduction effective mass as a function of the strain (Ge mole fraction). The reduction of the conduction effective mass implies a decrease in the inertia of the carriers and therefore an increase in electron mobility. However, note that this effect saturates for Ge mole fractions greater than 0.2~0.3, which means that for larger increases of the strain, the average conduction effective mass (which has reached its minimum value, $m_t$, i.e., most of the electrons populate the unprimed valleys) cannot decrease any further. In addition to the reduction of the average conduction effective mass, the greater separation between unprimed and primed subbands also produces a decrease in the intervalley scattering rate between non-equivalent valleys. Both facts, reduction of the conduction effective mass and reduction of the intervalley scattering rate imply an increase in electron mobility.

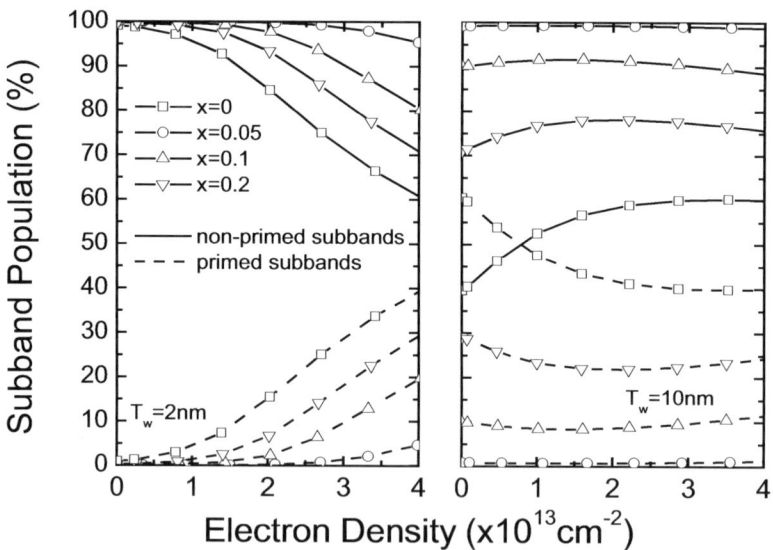

**Fig. 5.30.** Sub-band occupancy (percentage of total inversion density) for the non-primed and primed subbands for different Si strain. Left: $T_w$=2 nm; right: $T_w$=10 nm. Solid lines: non-primed subbands; dashed lines: primed subbands.

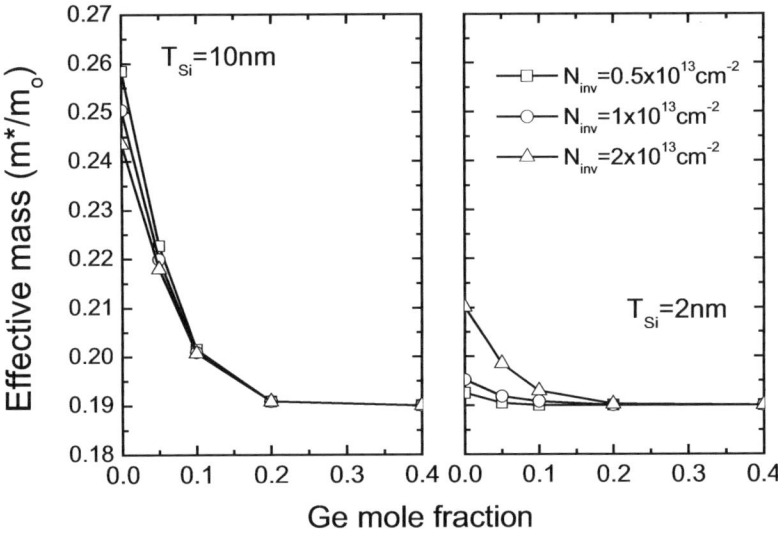

**Fig. 5.31.** Evolution of the conduction effective mass with the strain in a strained DGSOI transistor for two values of the silicon thickness and three values of the electron density.

Figure 5.32 shows the electron mobility in strained DGSOI transistors for different values of silicon thickness and different values of strain.

From observation of Figure 5.32, the following facts can be deduced:

i) For the thicker sample, electron mobility increases as the strain increases, although the increase tends to saturate for the higher amount of strain. The saturation of the average conduction effective mass shown in Figure 5.31 is the factor responsible for this behavior. This effect is more evident in the thinner sample ($T_w$=3nm) where mobility curves for $x$=0.2 and $x$=0.4 coincide.

ii) If we compare the mobility curves for $T_w$=10nm and $T_w$=3nm, we observe that in the unstrained case (dashed line), the mobility in the thinner sample is higher than that in the thicker silicon layer. However, in the strained devices, the trend is the opposite, *i.e.*, the electron mobility is higher in the thicker sample than in the thinner one.

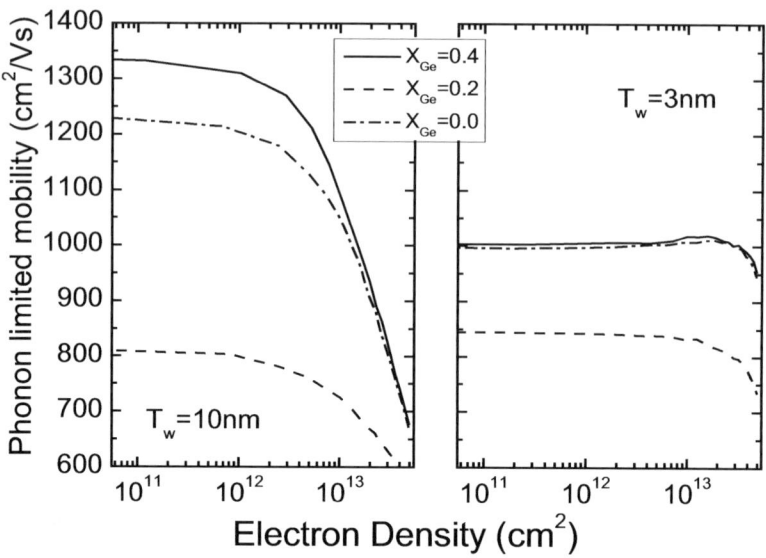

**Fig. 5.32.** Electron mobility in a strained-Si DGSOI MOSFET for two values of the silicon thickness ($T_w$=3nm and $T_w$=10nm) and different strain degrades. Only phonon scattering was considered.

### 5.2.10 Summary

We have studied the behavior of electron mobility in DGSOI transistors and FinFETs, and have compared it to the mobility in single gate SOI and bulk transistors. The role played by volume inversion and the different scattering mechanisms has been analyzed. We can conclude that:

i) Volume inversion plays an important role on the electron mobility. As a consequence, there is a range of silicon thicknesses where electron mobility in ultrathin DGSOI transistors is higher than mobility in conventional bulk MOSFETs.

ii) Surface or interface related scattering mechanism also become of great importance in ultrathin multigate transistors, and as a consequence their effects have to be taken into account.

## 5.3 Silicon multiple-gate nanowires

### 5.3.1 Introduction

The need to scale the active channel region below 30nm requires a silicon body width and thickness of the same dimensions or even lower in order to maintain an acceptable gate electrostatic control of the channel potential. Moreover, it is projected that the size of the transistors will reach 10nm by 2011. [1] For these reduced dimensions, carriers are confined in the directions perpendicular to the transport, such devices being called nanowires.

The potential application of semiconductor nanowire (NW) field-effect transistors (FETs) as potential building blocks for highly downscaled electronic devices with superior performance are attracting considerable attention.[85, 86] However, the term "nanowire" is employed in the bibliography to refer to a different kind of device. With regard to their manufacturing process, it is possible to establish two different approaches, bottom-up and top-down.

The bottom-up approach refers to the methodology that employs chemistry to promote the self-assembly of complex mesoscopic architectures. One of the most important discoveries in recent years has been the growth of single-crystal nanostructured materials at low temperatures using different nanometer-sized metallic nanoparticles (*e.g.* Ni, Au, Fe) as catalysts.[87] A wide variety of semiconductor materials such as Si, Ge, GaAs, GaN and InP [88-95] can be synthesized employing this technique. Different applications, such as laser action, photoluminescence, sensing, pn-junction, and field-effect transistors have already been demonstrated. Currently, the FETs fabricated from Vapor-Liquid-Solid (VLS) grown nanowires may offer better size uniformity than etching for very small diameters (< 5 nm) due to controlled chemical synthesis. However, their performance is not well understood, either experimentally or theoretically.[85]

The top-down group refers to those devices with dimensions in the nanometer range, fabricated using the standard techniques employed for CMOS processing, namely photolithography, thin-film deposition, etching, and metallization to obtain multigate SOI FETs with very small dimensions. The best nanowires demonstrated to date have been fabricated by etching a thin SOI wafer.[96, 97] A study of the electron mobility in this second group of devices (multigate nanowires) will be the main goal of this section.

As has been established in previous chapters, Multiple-Gate SOI MOSFETs are considered an attractive alternative to traditional bulk MOSFETs since they have proved to give better electrostatic control of the channel, allowing a greater reduction of the channel length ($L_g$) while the SCEs are kept under control. Moreover, the use of two, three or even four gates allows a relaxation of the width ($W_{Si}=T_W$) and height ($H_{Si}$) of the silicon fin compared with $L_g$. Under the denomination of multigates FETs, it is possible to find different structures such as FinFETs, Trigates, Gate All-Around (GAA), Pi and Omega-gate MOSFETs[1].

When the dimensions of the semiconductor fin, $W_{Si}$ and $H_{Si}$, see Figure 5.33, reach the nanometer scale, the carriers are confined in two dimensions, in the plane perpendicular to the transport direction. Therefore, the bulk crystal symmetry is not preserved and fundamental magnitudes, such as the density of states, and the band structure, experience important modifications, which will influence the carrier transport properties of these new devices.

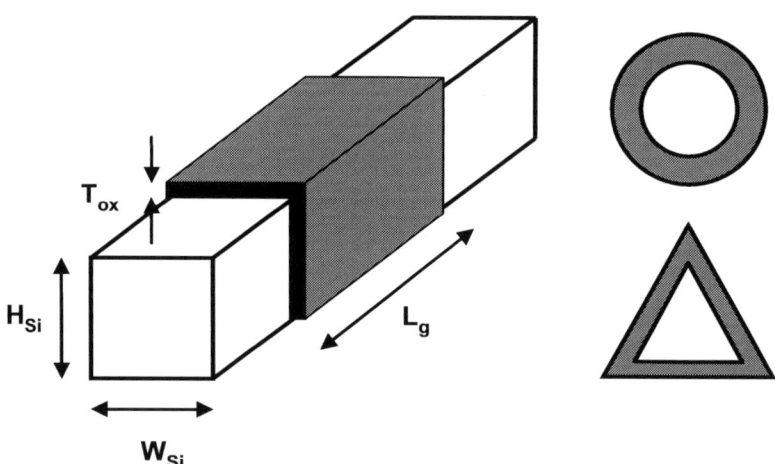

**Fig. 5.33.** Representation of a SiNW where $W_{Si}$ and $H_{Si}$ represent the semiconductor fin width and height respectively, $L_g$ the gate length, and $T_{ox}$ the gate oxide thickness. Different cross sections, such as triangular or circular can be considered in addition to the rectangular one.

---

[1] Detailed information can be found in Chapter 1

It has been demonstrated that band structure effects begin to manifest in silicon nanowires with diameters below 5nm. For higher dimensions, the simple parabolic effective-mass approach with bulk effective-masses is the optimum solution due to its reduced computational cost. Moreover, with very small dimensions (< 5nm) this method can be still used when appropriate tuning parameters are employed.[98]

Therefore, the modified semiconductor band structure should be taken into account when the device dimensions are below the limit of 5nm since other important parameters, such as the threshold voltage and the gate-channel capacitance, suffer considerable modifications.[99]

### 5.3.2 Electrostatic description of Si nanowires

In order to understand in depth the transport properties of these one-dimensional (1D) devices, detailed knowledge of the electron density and the electrostatic potential is necessary[2]. Obviously this requires the solution of the Poisson equation. Two different approximations can be carried out to achieve this goal. On the one hand, whether or not the whole device is considered, the solution of the three-dimensional (3D) Poisson equation for the electrostatic potential has to be carried out.[100-102] On the other hand, if a very long device is considered, this equation can be restricted to a plane perpendicular to the transport direction and the influence of the source and drain contacts neglected. In this case, the 2D Poisson equation must be solved,

$$\nabla(\varepsilon\nabla\phi) = -q\left(p - n + N_D^+ - N_A^-\right) \qquad (5.3)$$

where $\phi$ is the electrostatic potential, $\varepsilon$ is the dielectric constant, $q$ is the electric unit, $n$ and $p$ electron and hole concentrations, $N_D^+$ and $N_A^-$ ionized donor and acceptor concentrations.

The use of the finite element method allows the simulation of different geometries, such as triangular, cylindrical or rectangular cross sections as shown in Figure 5.33.

Due to the reduced dimensions of the devices under study, it is mandatory to include the quantum effects in the simulation, through the self-consistent solution of the Schrödinger equation. The most common

---

[2] More details on the electrostatic of Multiple-Gate FETs can be found in Chapter 4. Here we will focus on specific aspects necessary to study the transport properties of 1D structures.

approximation is the solution of the equation in two dimensions in the channel cross-section. If the whole device is studied, this solution is carried out in an arbitrary number of slices along the device length and then coupled with the corresponding transport equation.[100, 102]

It can be assumed that confinement is produced in the $y$ and $z$ directions and transport in the $x$ direction and, as a first approximation, the effect of source and drain contacts can be neglected. Therefore, the 2D Schrödinger equation can be written as

$$-\frac{\hbar^2}{2}\frac{\partial}{\partial y}\left(\frac{1}{m_y}\frac{\partial \Psi_\upsilon}{\partial y}\right) - \frac{\hbar^2}{2}\frac{\partial}{\partial z}\left(\frac{1}{m_z}\frac{\partial \Psi_\upsilon}{\partial z}\right) + V(y,z)\Psi_\upsilon = E_\upsilon \Psi_\upsilon \quad (5.4)$$

where $m_y$ and $m_z$ are the effective electron masses along the $y$ and $z$ axis respectively, and $\Psi_\upsilon$ the wave function belonging to energy level $E_\upsilon$. In order to self-consistently solve the two equations, different algorithms can be employed. However, the predictor-corrector scheme proposed by Trellakis et al. [103] has been tested by several authors [104-106] on different semiconductor structures with excellent results in every case.

It is also necessary to take into account the six equivalent valleys of the silicon conduction band. Hence, three sets of energy values are calculated according to the different values of the electron effective masses along the confinement directions. Assuming that the transport direction ($x$ axis) is directed along the wire axis, which is aligned with the (100) principal Si lattice direction, the effective masses correspond to those shown in Figure 5.34.

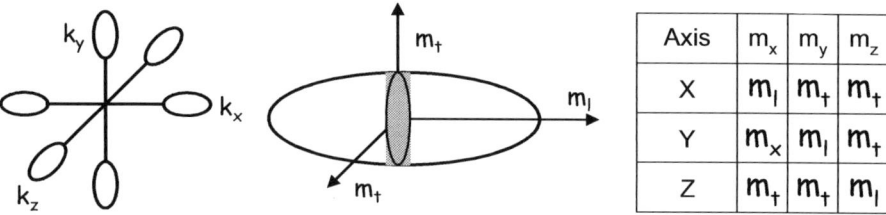

**Fig. 5.34.** Six-fold degenerate conduction band minima in Si (100) with the associated effective masses represented in the table.

Therefore, for silicon-based structures, the quantum electron density is determined by adding the three partial densities corresponding to each direction, where we have different effective masses to consider in the solution of Equation 5.4 and for the calculation of the 2D density of states. All these energies should be gathered and arranged according to their value, starting from the lowest. They are recognized with a subscript υ that refers to the corresponding subband and valley indices. Since we are dealing with a one-dimensional (1D) electron gas, the quantum electron density is obtained from the eigenpairs ($E_υ$, $Ψ_υ$) of Schrödinger's equation as

$$n = \sum_υ N_υ |Ψ_υ|^2 \qquad (5.5)$$

where

$$N_υ = g_V \left(\frac{2m_x k_B T}{\pi^2 \hbar^2}\right)^{\frac{1}{2}} \Im_{-1/2}\left(\frac{E_F - E_υ}{kT}\right) \qquad (5.6)$$

denotes the occupancy of the eigenstate υ, $g_V$ the number of equivalent conduction band valleys ($g_V$ =2), $m_x$ the electron mass along the wire axis, $T$ the temperature, $k_B$ Boltzmann's constant, $E_F$ the Fermi level, $\Im_{-1/2}$ the complete Fermi-Dirac integral of order −1/2 and the remaining symbols have their usual meaning.[107] The number of energy values employed in the simulation should be high enough to capture the occupied levels that contribute significantly to the total electron density. Obviously the higher the lateral size, the higher the number of energy values necessary to satisfy the above condition. However, in such cases, the two-dimensional (2D) confinement disappears and traditional bulk MOSFET behavior emerges.[108]

In order to show the correct operation of the above procedure, it has been applied to a silicon Gate All-Around (GAA) MOSFET, where the semiconductor is completely surrounded by the insulator and the gate contact, as shown in Figure 5.33. In all the calculations, we assumed a square cross-section ($W_{Si} = H_{Si}$), substrate doping of $10^{15} cm^{-3}$, $T_{ox}$ =1nm, and a midgap workfunction metal gate ($\phi_m$ = 4.61eV).

Figures 5.35(a), 5.35(b) and 5.35(c) represent the electron distribution in a silicon GAA with $W_{Si} = H_{Si}$=15nm while Figure 5.35 (d) corresponds to a 4nm lateral size. Figure 5.35(a) was calculated using a classical solution of the structure and an applied gate voltage ($V_G$) equal to 1V. The maximum electron density is located at the Si-SiO$_2$ interface, right in the corners, and its value is clearly overestimated when compared with the corresponding quantum simulation shown in Figure 5.35(b). Figure 5.35(c) was calculated for the same device shown in Figure 5.35(b) but with a gate voltage reduced to 0.25V. It shows how the electrons are spread throughout the whole silicon body with a peak density at the centre of the structure due to the so-called volume inversion effect.[14] Figure 5.35(d) corresponds to a device with reduced silicon fin dimensions ($W_{Si} = H_{Si}$=4nm) and shows that for the same gate voltage as Figure 5.35(b) ($V_G$=1V), the maxima of the electron density is again located at the center of the semiconductor due to the volume inversion effect.

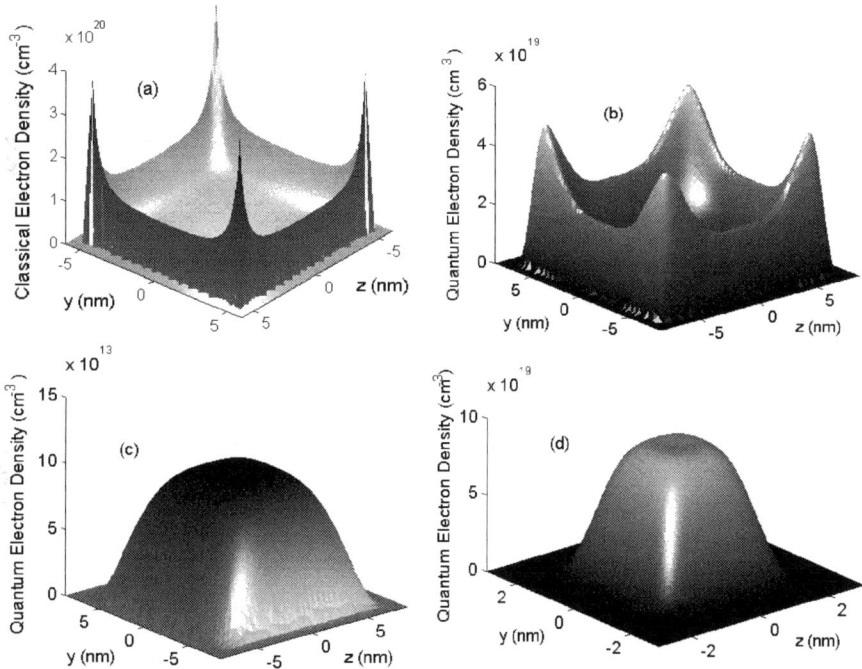

**Fig. 5.35.** Electron density calculated in silicon GAA MOSFETs. In figure (a), the classical electron distribution is shown for an applied gate voltage ($V_G$) of 1V and with $W_{Si}$ =$H_{Si}$ =$15nm$. The peak density is reached at the Si-SiO$_2$ interfaces right in the corners. (b) and (c) show the quantum electron density (QED) when $V_G$=1V and $V_G$ =0.25V respectively and $W_{Si}$ =$H_{Si}$ =15nm. At low $V_G$, the electrons are concentrated in the centre of the silicon fin (volume inversion) and for higher $V_G$, the maximum is shifted towards the interfaces. (d) shows the QED when $V_G$=1V in a device with $W_{Si}$ =$H_{Si}$ =4nm. Again the volume inversion effect is seen, in this case with a higher voltage and reduced lateral size.

### 5.3.3 Electron transport in Si nanowires

Once the electrostatic description of the device has been completed, its electron transport properties can be studied from different approximations such as the Kubo-Greenwood formula [109, 110] modified for 1D transport or the nonequilibrium Green's function (NEGF) formalism.[111, 112]

Several works have employed a full quantum-mechanical treatment to study the transport properties of these devices. Techniques employed for quantum simulation include Nonequilibrium Green's function (NEGF) formalism [111, 112], coupled Schrödinger approaches, the Pauli master equation, simple analytical models and recursive scattering matrices. [101]

However, most have considered the transport as ballistic, neglecting the effect of the scattering processes that occur in these devices. This is because it remains a formidable task to include the microscopic scattering mechanisms in this method.[102] Different approaches to including dissipation in quantum simulations, such as statistical approaches or the addition of an imaginary term to the Hamiltonian can be found in the literature. Here we will briefly mention two of them:

i) The deformation potential theory and the self-consistent Born approximation for obtaining the self-energy functions for the intravalley and intervalley phonon scattering mechanisms [102] have been employed to take into account the effect of the electron-phonon interactions. Interestingly, when the long channel limit is considered, it is possible to obtain an analytical expression for the electron low field mobility that is equal to the Kubo-Greenwood formula modified for a 1D transport.

ii) One of the most popular techniques for including scattering in nanowires is the use of Büttiker probes, which has also been tested in MOSFET simulations. This method has been employed to deal with both elastic (surface roughness scattering and ionized impurity scattering) and inelastic (electron-phonon interaction) scattering mechanisms present in the device. The electron mobility calculated with this method is around 200 $cm^2/Vs$ in the channel.[100]

Furthermore, the use of quantum simulators has demonstrated that the ballistic to diffusive crossover in Si nanowiress occurs at a channel length around 1.5nm, much smaller than previously thought [101], and is far from being a feasible device.

The simulation of long wires (channel length exceeding the carrier mean free path) implies that transport is diffusive, which justifies the use of a semi-classical Monte Carlo (MC) simulation where the quantum effects have been taken into account.[113, 114] This procedure has been quite popular in recent decades and the scattering models have been tested in a large number of semiconductor structures.[42, 115]

However, the Monte Carlo algorithm normally used to calculate the electron transport properties in silicon inversion layers has to be modified to take into account the special characteristics of carrier confinement in two dimensions. Since the 1D density of states presents a large number of peaks [116], the use of a self-scattering mechanism would generate a very large number of events of this kind. To solve this problem, a direct integration method is recommended, following the expression [117]

$$-\ln r = \int_0^t \lambda_v \left[ k_v(t') \right] dt' \tag{5.7}$$

where $r$ is a random number uniformly distributed at the interval [0,1], $t$ is the free flight time, $k_v(t)$ is the momentum in subband $v$ as a function of time, and $\lambda_v(k_v)$ is the scattering rate as a function of the momentum for subband $v$. The total scattering rate has to be calculated beforehand as a function of the different scattering mechanisms considered in the study and tabulated for each subband in uniform energy steps of size $\Delta E$, so the integral can be approximated by a sum

$$-\ln r = \sum_{E=E_i}^{E_f} \lambda_v \left( k_v(E) \right) \Delta t(E) \tag{5.8}$$

with

$$\Delta t(E) = \frac{\hbar}{eF_x} \Delta k_v(E) \text{ and } \Delta k_v(E) = \left| k_v(E \pm \Delta E) - k_v(E) \right| \tag{5.9}$$

It is interesting to remember that in 1D systems, the final scattering state is limited to a backward or forward process from an initial state to a final one, depending on whether or not the scattering event reverses the carrier momentum. Figure 5.36 represents the four fundamental scattering processes in 1D systems when a phonon is emitted ($K_{vv'}^-$) or absorbed ($K_{vv'}^+$). [118]

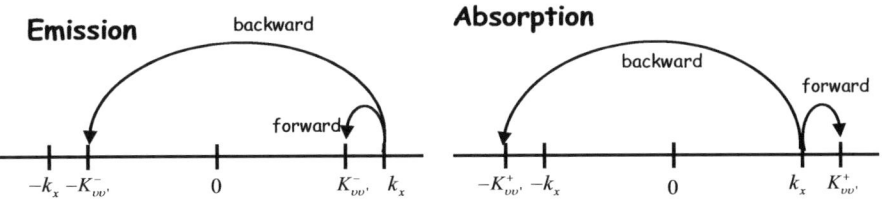

**Fig. 5.36.** Representation of the fundamental scattering processes in 1D systems when a phonon is emitted or absorbed. $K^-_{\upsilon\upsilon'}$ ($K^+_{\upsilon\upsilon'}$) represents the final wave vector after a phonon is emitted (absorbed) and it may or may not reverse the carrier momentum.

Among the different scattering mechanisms which can influence carrier mobility in a SiNW, it has been demonstrated that phonon scattering is the dominant mechanism in mobility degradation in a Si MOSFET under operating conditions at room temperature.[119] Therefore, phonon mobility is dominant in low effective fields and its value is determined by the acoustic and intervalley scattering rates.

The use of elastic approximation in the calculation of electron scattering by acoustic phonons has been shown to be accurate at room temperature.[120] Both intra-subband and inter-subband electron scattering can be considered, using the following expression:

$$\Gamma^{ac}_{\upsilon\upsilon'}(k_x) = \frac{D_{ac}^2 kT}{8\pi\hbar\rho u_l^2} D_{1D}(E) J^{1D}_{\upsilon\upsilon'} \left[ \delta(k_{xf} - k_{\upsilon\upsilon'}) + \delta(k_{xf} + k_{\upsilon\upsilon'}) \right] \quad (5.10)$$

where $\rho$ is the silicon density, $D_{ac}$ is the deformation acoustic potential, $u_l$ the sound velocity in the silicon, and subscripts $\upsilon$ and $\upsilon'$ refer to the initial and final subbands respectively.

The intervalley scattering rate can be expressed as:

$$\Gamma^{int}_{\upsilon\upsilon'}(k_x) = \frac{D_{ij}^2}{16\pi\rho\omega_{ij}} D_{1D}(E) J^{1D}_{\upsilon\upsilon'} \left( N_{ij} + \frac{1}{2} \pm \frac{1}{2} \right) \left[ \delta(k_{xf} - k_{\upsilon\upsilon'}) + \delta(k_{xf} + k_{\upsilon\upsilon'}) \right] \quad (5.11)$$

where $D_{ij}$ is the deformation potential, $\omega_{ij}$ is the intervalley phonon frequency, $N_{ij}$ is the excited phonon number, and subscripts $i$ and $j$ refer to the initial and final valleys respectively.

From (5.10) and (5.11) we can observe that the scattering rates are proportional to the 1D density of states $D_{1D}(E)$ and the overlap integral:

$$J_{vv'}^{1D} = \int_{-\infty}^{\infty}\int_{-\infty}^{\infty}\left|\int \psi_v^*(y,z)e^{\pm i(q_y y + q_z z)}\psi_{v'}(y,z)dydz\right|^2 dq_y dq_z \quad (5.12)$$

In bulk silicon, this integral would reduce to a double delta function. However, in Si nanowires, the lack of translational symmetry in the $y$ and $z$ directions produces a substantially different result, which is responsible for the uncertainty in maintaining the momentum in both directions. The reduction of dimensions ($W_{Si}$, $H_{Si}$) causes a significant increase of this uncertainty, reflected in a rapid growth of the overlap integral.[120, 121] On the other hand, the study of ultrathin-body (UTB) silicon-on-insulator inversion layers has demonstrated the influence of the confinement of acoustic phonons. It has been shown that in order to reproduce experimental mobility curves in UTB SOI inversion layers using a bulk acoustic phonon scattering model, it is necessary to increase the phonon scattering rate.[49-50,53]

Thus, it was concluded that mobility is a strong function of the deformation potential constant ($D_{ac}$). When the typical bulk value of $D_{ac}$ = 9eV is employed in the simulations, the electron mobility reaches very high values and it is necessary to increase $D_{ac}$ to reproduce experimental results in DGSOI transistors.[122] This conclusion was reached employing both a semi-classical and a quantum transport simulator.[102] To be more precise, this value should be fitted with experimental results as a function of the device cross-section. As a reasonable approximation, a value of $D_{ac}$=12.9eV could be used for different dimensions, but this approximation does not accurately treat the effect of confined acoustic phonons, which could have an important influence when very small lateral sizes are analyzed.[123] Other authors have considered similar values such as $D_{ac}$=12.9eV [124] or $D_{ac}$=14.6eV.[24] Therefore, this effect should be taken into account in future works where 2D confinement is evident and phonon quantization more relevant.

The results obtained from MC simulation are shown in Figure 5.37 where the calculated values of phonon-limited mobility are depicted as a function of the gate voltage for three different values of the square cross-section ($W_{Si}$ = $H_{Si}$ = 15nm; 10nm; 5nm). As can be observed, phonon-limited mobility is quite similar for the higher cross sections (15 and 10nm of lateral size). However, if the silicon fin is reduced to 5nm, a significant degradation is found.[114]

Similar results were obtained for cylindrical silicon gated nanowires employing the Kubo-Greenwood formula. In their work, Kotlyar et al. [124] show that phonon-limited mobility is degraded in narrow wires and recovers what is known as MOSFET universal behavior in high surface fields.

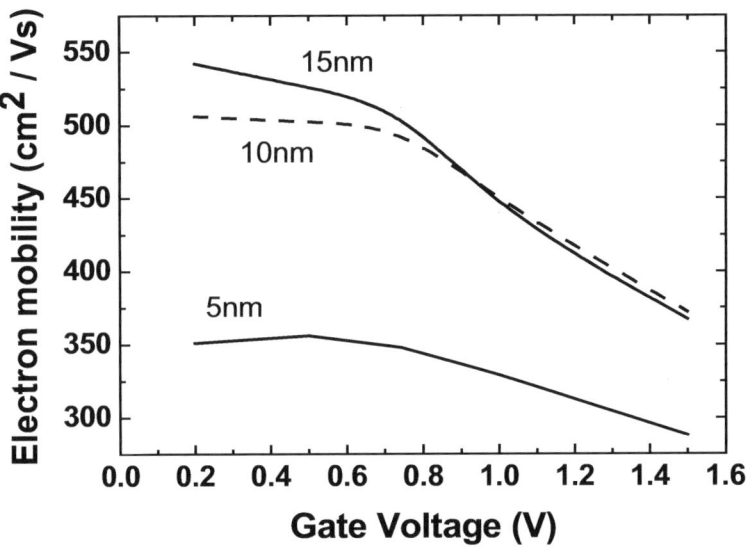

**Fig. 5.37.** Phonon-limited electron mobility in a silicon GAA device as a function of the gate voltage. Three different cross-sections are considered ($W_{Si} = H_{Si}$), 15nm, 10nm, and 5nm.

Analyzing the expressions of the scattering probabilities, Expressions 5.10 and 5.11, it is observed that they are proportional to two factors:

i) The density of states (DOS). The transition from a 3D or 2D DOS to a 1D DOS involves important differences since the two functions show totally different behavior as a function of the electron energy. When the cross-section is big enough, the energy separating the subbands is negligible compared to the thermal energy and the DOS resembles that of a 3D sample. However, for very small cross-sections, the energetic distance between subbands is large and the DOS is that corresponding to a 1D electron gas.[125] The final DOS in 1D conduction channels should be reduced in accordance with the reduced lateral size, and as a consequence, improved carrier mobilities can be expected.

ii) The overlap integral, Expression 5.12, between the initial and final electron wave functions increases when dimensions are reduced.

These two opposite trends will determine the phonon-limited mobility in SiNWs. The overlap integral (5.12) for the ground subband is shown in Figure 5.38 for various fin dimensions and three different values of the inversion charge concentrations (dotted line: $N_{inv}=10^{10}$ cm$^{-2}$; dashed line: $N_{inv}=10^{11}$ cm$^{-2}$; solid line: $N_{inv}=10^{12}$ cm$^{-2}$).

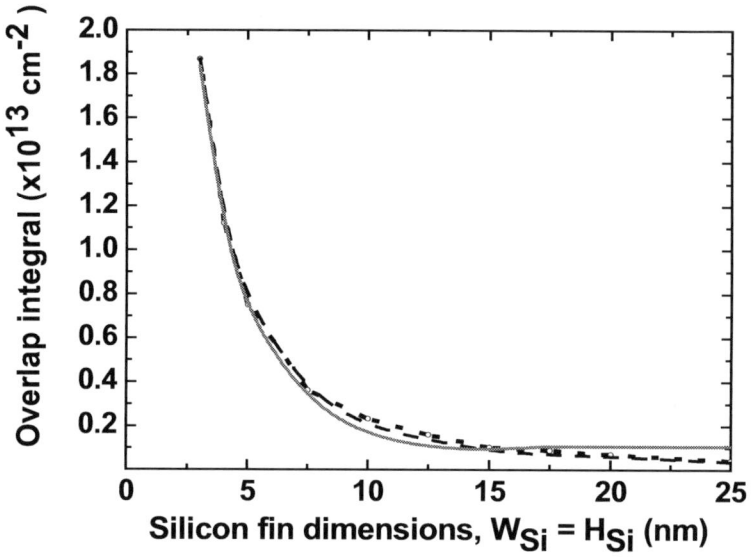

**Fig. 5.38.** Overlap integral for the ground subband as a function of the lateral dimensions of the silicon fin ($W_{Si} = H_{Si}$) for three values of the inversion charge $N_{inv}$ (dotted line: $N_{inv}=10^{10}$ cm$^{-2}$; dashed line: $N_{inv}=10^{11}$ cm$^{-2}$; solid line: $N_{inv}=10^{12}$ cm$^{-2}$).

For the calculation of $N_{inv}$ we first evaluated the total charge per unit length ($N_t$) integrating Expression 5.6 throughout the whole silicon area and then defining $N_{inv} \equiv N_t /[2\cdot(W_{Si}+H_{Si})]$ for the square cross-section or equivalently $N_{inv} \equiv N_t /(2\pi R)$ for the circular one. With this definition, the electron density is normalized for the different perimeters, allowing a fair comparison among all of them.

Two different regions can be distinguished in Figure 5.38. For lateral sizes greater than 10nm, the overlap integral remains almost constant. However, for dimensions below 10nm, a significant increase is seen, due to the geometrical confinement of electrons in a very small region, independently of the applied gate voltage. Phonon-limited electron mobility follows a similar trend, as can be observed in Figure 5.37 and therefore, the electron–phonon wavefunction overlap dominates the density-of-states reduction in the calculation of the phonon scattering rate, causing a considerable reduction of the electron mobility.[114,126]

Similar results were found in cylindrical gated nanowires [124] and have also been previously reported in DGSOI and SGSOI.[21,40,45] However, it is worth highlighting that in these calculations of the form factor in GAA MOSFETs, a range of lateral thickness with a minimum where the phonon scattering rate decreased instead of increasing as happened in DGSOI was not found.[4

### 5.3.4 Surface roughness

An accurate calculation at high effective fields requires the inclusion of surface roughness scattering since its effect is dominant. Intuitively, it could be expected that for very small dimensions, when electrons are completely surrounded by the insulator material, an increasing interaction between carriers and interfaces would produce a mobility degradation. However, various recent works have revealed the opposite behavior.

Ramayya *et al.* [126] have shown that the influence of surface roughness scattering is reduced with decreasing silicon width, due to volume inversion producing an appreciable mobility enhancement at high transverse fields. In their work, the surface roughness scattering (SRS) rate was calculated by assuming exponentially correlated surface roughness [57] and incorporating the wavefunction deformation using Ando's model.[60] As might be expected, the resulting expression is determined mainly by the strength of the field perpendicular to the interface. When the width of the wire is decreased, the electrons move away from the interfaces, causing a reduction in the interface influence. A similar behavior was previously observed in ultrathin double-gate SOI FETs and explained by the onset of volume inversion.[40] Due to the reduced dimensions, carriers are distributed throughout the silicon volume, producing so-called volume inversion. This phenomenon is clearly shown in Figure. 5.35(d) where the electrons are spread throughout the whole semiconductor fin due to the volume inversion effect produced by the geometrical confinement obtained when the dimensions are reduced. The maxima of the carrier density for higher sizes, Figure 5.35(b) and the same applied voltage are shifted next to the interfaces.

In contrast to the previous work, which made use of perturbation theory to compute the SRS rate, Wang et al. [127] employ a full 3D quantum transport simulator of SiNW FETs based on an effective-mass approximation where the microscopic structure of the Si-SiO$_2$ interface roughness is treated directly by using a 3D finite element technique. The main conclusion obtained from this work is that SRS is less important in small-diameter NWs with few modes conducting than in planar MOSFETs with many transverse modes occupied. This result strengthens the theory that SRS is not a limiting factor in electron mobility when device dimensions are reduced.

### 5.3.5 Experimental results and conclusions

Different research groups are currently working on the manufacture and characterization of Si nanowires and a wide variety of experimental results have been presented.

As an example, we could mention the studies published by Lieber's group [128-129], one of the more active in the field, which has shown that the average transconductance and mobility shows substantial advantages for SiNWs obtained from Vapor-Liquid-Solid (VLS) synthesis compared with more conventional multigate FETs. However, a number of issues including device performance, reproducibility, and high quality ohmic contacts must be addressed if such systems are to be implemented in the future.

More recently, electron mobility as high as $\approx 1000 cm^2/Vs$ has been reported for n-channel SiNW FETs made following conventional semiconductor manufacturing techniques, from a p-type SOI wafer with n+ source and drain. Moreover, the authors of this work conclude that as the channel width decreases, the inversion layer mobility of the SiNW increases to approximately twice the mobility of the larger channel-width FETs.[86]

Singh et al. [97] report the fabrication of GAA n– and p– FETs on a SOI wafer with a diameter smaller than 5nm and lengths around 1000nm where the estimated electron and hole mobilities are ~750 and ~325 cm$^2$/Vs at high fields.

These experimental results cover a wide range of mobility values and, to date, a clear explanation is not available for all of them. Various reasons, have been put forward to justify these discrepancies, such as strain in the semiconductor due to the oxidation process or the possible suppression of inter-valley phonon scattering, volume inversion, and reduced surface roughness at high fields.[97]

Three important conclusions can be drawn from the above results:

i) The characteristics of the Si nanowires obtained from chemical synthesis are at the moment similar to those of multigate transistors, indicating no significant improvement in mobility. Moreover, the performance and electrostatics of these new devices (the bottom-up approximation) require further improvement and their scalability still needs to be demonstrated.[130]

ii) Much more theoretical and experimental work is required to further understand the transport properties in multigate NWs where it has been demonstrated that the previous models, which worked properly in inversion layers, fail when applied to these highly confined systems.

iii) The unique properties observed in 1D electron gases can be appreciated only when lateral dimensions are well below 10nm since otherwise their behavior much more closely resembles that found in traditional silicon inversion layers. This conclusion can be relaxed when very low temperatures are applied, as was demonstrated by Colinge *et al.*[131]

## References

1　International Technology Roadmap for Semiconductors. 2006 Edition
2　S.Cristoloveanu and S.S.Li: *Electrical Characterization of Silicon-on-Insulator Materials and Devices* (Kluwer, Boston, 1995)
3　G. C. Celler and S. Cristoloveanu: Frontiers of silicon-on-insulator. J. Appl. Phys. **93**, 4955 (2003).
4　J.P.Colinge: *Silicon-On-Insulator Technology: Materials to VLSI* (Kluwer, Boston, 1991)
5　W. Haensch, E. J. Nowak, R. H. Dennard, P. M. Solomon, A. Bryant, O. H. Dokumaci, A. Kumar, X. Wang, J. B. Johnson, M. V. Fischetti: Silicon CMOS devices beyond scaling. IBM J. Res. & Dev. **50**, 361 (2006)
6　J.A.López-Villanueva, F.Gámiz, J.B.Roldan, Y.Ghailan, J.E.Carceller, and P.Cartujo: Study of the Effects of a Stepped Doping Profile in Short-Channel MOSFET's. IEEE Transactions on Electron Devices **44**, 1425 (1997)

7   S.F. Nelson, K. Ismail, J.O. Chou, and B.S. Meyerson: Room temperature electron mobility in strained Si/SiGe heterostructures. Appl.Phys.Lett., **63**, 367 (1993)
8   Z. Ren, S. Hegde, B. Doris, P. Oldiges, T. Kanarsky, O. Dokumaci, R. Roy, M. Leong, E. C. Jones, and H. P. Wong: An experimental study on transport issues and electrostatics of ultrathin body SOI pMOSFETs, IEEE Electron Device Lett. **23**, 609 (2002).
9   X. Huang, W.-C. Lee, C. Kuo, D. Hisamoto, L. Chang, J. Kedzierski, E. Anderson, H. Takeuchi, Y.-K. Choi, K. Asano, V. Subramanian, T.-J. King, J. Bokor, and C. Hu: Sub 50-nm FinFET: PMOS. *IEDM Tech. Digest*, 67 (1999)
10  D. Hisamoto, T. Kaga, Y. Kawamoto, and E. Takeda: A Fully Depleted Lean-Channel Transistor (DELTA)—A Novel Vertical Ultra Thin SOI MOSFET. *IEDM Tech. Digest* 833 (1989)
11  J. Kedzierski, D. M. Fried, E. J. Nowak, T. Kanarsky, J. H. Rankin, H. Hanafi, W. Natzle, D. Boyd, Y. Zhang, R. A. Roy, J. Newbury, C. Yu, Q. Yang, P. Saunders, C. P. Willets, A. Johnson, S. P. Cole, H. E. Young, N. Carpenter, D. Rakowski, B. A. Rainey, P. E. Cottrell, M. Ieong, and H.-S. P. Wong: High-Performance Symmetric-Gate and CMOS-Compatible Vt Asymmetric-Gate FinFET Devices. *IEDM Tech. Digest*, 437 (2001)
12  T. Park, S. Choi, D. H. Lee, J. R. Yoo, B. C. Lee, J. Y. Kim, C. G. Lee, K. K. Chi, S. H. Hong, S. J. Hynn, Y. G. Shin, J. N. Han, I. S. Park, U. I. Chung, J. T. Moon, E. Yoon, and J. H. Lee: Fabrication of Body-Tied FinFETs (Omega MOSFETs) Using Bulk Si Wafers. *Symp. VLSI Technol.*, 135 (2003).
13  B. Doyle, B. Boyanov, S. Datta, M. Doczy, S. Hareland,B. Jin, J. Kavalieros, T. Linton, R. Rios, and R. Chau,"Tri-Gate Fully-Depleted CMOS Transistors: Fabrication, Design and Layout. *Symp. VLSI Technol.*, 133 (2003)
14  F.Balestra, S.Cristoloveanu, M.Benachir, J.Brini and T.Elewa: " IEEE Electron Device Letters **9**, 410 (1987)
15  D. J. Frank, S. E. Laux, and M. V. Fischetti: Monte Carlo simulation of a 30 nm dual-gate MOSFET: How short can Si go?. *IEDM Tech. Dig.*, 553 (1992)
16  T.Ouisse: Self-consistent quantum-mechanical calculations in ultrathin silicon-on-insulator structures. J.Appl.Phys. **76**, 5989 (1994)
17  B. Majkusiak, T. Janik, and J. Walczak "Semiconductor Thickness Effects in the Double-Gate SOI MOSFET. IEEE Transactions on Electron Devices **45**, 1127 (1998)
18  T. Janik and B. Majkusiak: Analysis of the MOS transistor based on the self-consistent solution to the Schrödinger and Poisson equations and on the local mobility model," IEEE Trans. Electron Devices **45**, 1263 (1998)
19  Y. Taur: An analytical solution to a double-gate MOSFET with undoped body," IEEE Electron Device Lett. **21**, 245 (2000)
20  M.Shoji, Y.Omura and M.Tomizawa: Physical basis and limitation of universal mobility behavior in fully depleted silicon-on-insulator Si inversion layers. J.Appl.Phys. **81**, 786 (1997)

21  M. Shoji and S. Horiguchi: Electronic structures and phonon-limited electron mobility of double-gate silicon-on-insulator Si inversion layers," J. Appl. Phys. **85**, 2722 (1999)
22  D. Esseni, M. Mastrapasqua, C. Fiegna, G.K. Celler, L. Selmi and E. Sangiorgi: An experimental study of low field electron mobility in double-gate, ultra-thin SOI MOSFETs. *IEDM Tech. Digest*, 445 (2001)
23  D. Esseni, A. Abramo L. Selmi and E. Sangiorgi: Study of low field electron transport in ultrathin single and double gate SOI MOSFETs. *IEDM Tech. Digest*, 719 (2002)
24  D. Esseni, M. Mastrapasqua, G.K. Celler, C. Fiegna, L. Selmi, and E. Sangiorgi: An Experimental Study of Mobility Enhacement in Ultrathin SOI transistors operated in Double-Gate Mode. IEEE Transactions on Electron Devices **50**, 802 (2003)
25  D. Esseni, A. Abramo, L. Selmi, and E. Sangiorgi: Physically base Modelling of Low Field Electron Mobility in Ultrathin Single- and Double-Gate SOI n-MOSFETs. IEEE Transactions on Electron Devices **50**, 2445 (2003)
26  G. A. Kathawala, B. Winstead, and U. Ravaioli: Monte Carlo simulations of double-gate MOSFETs," IEEE Trans. Electron Devices **50**, 2467 (2003)
27  Winstead and U. Ravaioli: A quantum correction based on Schrödinger equation applied to Monte Carlo device simulation," IEEE Trans. Electron Devices **50**, 440 (2003)
28  G. A. Kathawala and U. Ravaioli: 3-D Monte Carlo simulations of FinFETs," *IEDM Tech. Dig.*, 683 (2003)
29  J.S. Martin, A. Bournel, and P. Dollfus: On the ballistic transport in nanometer-scaled DG MOSFETs," IEEE Trans. Electron Devices **51**, 1148 (2004)
30  J. Saint-Martin, A. Bournel, F. Monsef, C. Chassat and P. Dollfus: Multi sub-band Monte Carlo simulation of an ultra-thin double gate MOSFET with 2D electron gas. Semicond. Sci. Technol **21**, L29 (2006)
31  V. Sverdlov, E. Ungersboeck, H. Kosina, S. Selberherr: Orientation Dependence of the Low Field Mobility in Double-and Single-gate SOI FETs. *European Solid-State Device Research Conference (ESSDERC)*, 178 (2006)
32  E. Ungersboeck, V. Sverdlov, H. Kosina, S. Selberherr "Electron Inversion Layer Mobility Enhancement by Uniaxial Stress on (001) and (110) Oriented MOSFETs. Proceedings of Simulation of Semiconductor Processes and Devices, *SISPAD-2006*, 43 (2006)
33  L. Ge and J. G. Fossum: Analytical modeling of quantization and volume inversion in thin Si-film DG MOSFETs," IEEE Trans. Electron Devices **49**, 287 (2002)
34  J. G. Fossum, L. Wang, J. Yang, S. Kim, and V. P. Trivedi: Pragmatic design of nanoscale multi-gate CMOS. *IEDM Tech. Dig.*, 613 (2004)
35  J. G. Fossum, L. Ge, and M. Chiang: Speed superiority of scaled double-gate CMOS. IEEE Trans. Electron Devices **49**, 808 (2002)

36 Z. Ren, R. Venugopal, S. Datta, and M. Lundstrom: Examination of design and manufacturing issues in a 10 nm double gate MOSFET using nonequilibrium Green's function simulation. *IEDM Tech. Dig.*, 107 (2001)
37 F. Gámiz, J.B. Roldán, J.A. López-Villanueva, P. Cartujo-Cassinello and J.E. Carceller: Surface roughness at the Si–SiO$_2$ interfaces in fully depleted silicon-on-insulator inversion layers. J.Appl.Phys. **86**, 6854 (1999)
38 F. Gámiz, J.A. López-Villanueva, J. Banqueri, J. Carceller and P. Cartujo: Universality of Electron Mobility Curves in MOSFETs: A Monte Carlo Study. IEEE Trans. Electron Devices **42**, 258 (1995)
39 F. Gámiz, J.A. López-Villanueva, J.A. Jiménez-Tejada, I. Melchor and A. Palma: A Comprehensive Model for Coulomb Scattering in inversion layers. J. Appl. Phys. **75**, 924 (1994)
40 F. Gámiz and M.V. Fischetti: Monte Carlo Simulation of Double Gate Silicon on Insulator Inversion Layers: The Role of Volume Inversion. J. Appl. Phys. **89**, 5487 (2001)
41 M.V. Fischetti, and S.E. Laux: Monte Carlo Study of electron transport in silicon inversion layers. Phys. Rev., **B48**, 2244 (1993)
42 C. Jacoboni and L. Reggiani: The Monte Carlo Method for the solution of charge transport in semiconductors with application to covalent materials. Rev. Mod. Phys. **55**, 645 (1983)
43 P.J. Price: Two-dimensional electron transport in semiconductor layers—Part I: Phonon scattering. Ann. Phys. (N.Y.) **133**, 217 (1981)
44 H. Ezawa: Phonons in a half space. Ann. Phys. (N.Y.) **67**, 438 (1971)
45 F. Gamiz, J.B. Roldán, P. Cartujo-Cassinello, J.E. Carceller, J.A. Lopez-Villanueva, and S. Rodriguez: Electron mobility in extremely thin single-gate silicon-on-insulator inversion layers. J. Appl. Phys. **86**, 6269 (1999)
46 F. Gámiz, J.A. López-Villanueva, J.B. Roldán, J.E. Carceller and P. Cartujo: Monte Carlo Simulation of Electron Transport Properties in Extremely Thin SOI MOSFETs. IEEE Trans. Electr. Dev. **45**, 1122 (1998)
47 F. Gámiz, J.B. Roldán, J.A. López-Villanueva: Phonon-limited electron mobility in ultrathin silicon-on-insulator inversion layers. J.Appl.Phys. **83**, 4802 (1998)
48 C.M. Sotomayor Torres, A. Zwick, F. Poinsotte, J. Groenen, M. Prunilla, J. Ahopelto, A. Mlayah, and V. Paillard: Observations of confined acoustic phonons in silicon membranes. Phys. Stat. Sol. (c) **1**, 2609 (2004)
49 L. Donetti, F. Gámiz, N. Rodriguez, F. Jimenez, and C. Sampedro: Influence of acoustic phonon confinement on electron mobility in ultrathin silicon on insulator layers. Appl. Phys. Lett. **88**, 122108 (2006)
50 L. Donetti, F. Gámiz, J.B. Roldán, and A. Godoy: Acoustic phonon confinement in silicon nanolayers: Effect on electron mobility. J. Appl. Phys. **100**, 013701 (2006)
51 S. Yu, K.W. Kim, M.A. Stroscio, G.J. Iafrate, A. Ballato: Electron acoustic phonon scattering rates in rectangular quantum wires. Phys. Rev, **B50**, 1733 (1994)

52  E. P. Pokatilova, D. L. Nikaa, and A.A. Balandin: Phonon spectrum and group velocities in AlN/GaN/AlN and related heterostructures. Superlattices and Microstructures **33**, 155 (2003)

53  L. Donetti, F.Gamiz, N.Rodriguez, and F.G.Ruiz: Phonon scattering in Si-based nanodevices. Solid-State Electronics **51**, 593 (2007)

54  N. Bannov, V. Aristov, V. Mitin, and M.A. Stroscio: Electron relaxation times due to the deformation potential interaction of electron with confined acoustic phonons in a free standing quantum well. Phys. Rev. **B51**, 9930 (1995)

55  D. Esseni: On the modeling of surface roughness limited mobility in SOI MOSFETs and its correlation to the transistor effective field. IEEE Trans. Electron Devices **51**, 394 (2004)

56  F. Gámiz, J.B. Roldán, P. Cartujo-Cassinello, J.A. López-Villanueva, and P. Cartujo: Role of surface-roughness scattering in double gate silicon-on-insulator inversion layers. J.Appl.Phys **89**, 1764 (2001)

57  S.M. Goodnick, D.K. Ferry, C.W. Wilmsen, Z. Liliental, D. Fathy, O.L. Krivanek: Surface roughness at the Si(100)-SiO2 interface. Phys. Rev. **B32**, 8171 (1985)

58  F. Gámiz, F. Jiménez-Molinos, J.B. Roldan, and P. Cartujo-Cassinello: Coulomb scattering model for ultrathin silicon-on-insulator inversion layers. Appl. Phys. Lett. **80**, 3835 (2002)

59  T.H. Ning and C.T. Sah: Theory of scattering of electrons in a non-degenerate semiconductor surface inversion layer by surface-oxide charges. Phys. Rev. **B6**, 4605 (1972)

60  T. Ando, A.B. Fowler, and F. Stern: Electronic Properties of Two-dimensional systems. Rev. Mod. Phys. **54**, 437 (1982)

61  F. Gámiz: Temperature behaviour of electron mobility in double-gate silicon on insulator transistors. Semicond. Sci. Technol. **19**, 113 (2004)

62  T. Tanaka, K. Suzuki, H. Horie, and T. Sugii: Ultrafast operation of $V_{th}$-adjusted $p^+$-$n^+$ double-gate SOI MOSFET's. IEEE Electron Device Lett. **15**, 386 (1994)

63  K. Suzuki and T. Sugii: Analytical models for n+-p+ double-gate SOI MOSFET's. IEEE Trans. Electron Devices **42**, 1940 (1995)

64  K. Kim and J. G. Fossum: Double-gate CMOS: Symmetrical- versus asymmetrical-gate devices. IEEE Trans. Electron Devices **48**, 294 (2001)

65  J. G. Fossum and Y. Chong, *Proceedings IEEE International SOI Conference*, 107 (1998)

66  Y. Taur: Analytic solutions of charge and capacitance in symmetric and asymmetric double-gate MOSFETs. IEEE Trans. Electron Devices **48**, 2861 (2001)

67  F. Gámiz, J. B. Roldán, A. Godoy, P. Cartujo-Cassinello, and J.E. Carceller: Electron mobility in double gate silicon on insulator transistors: Symmetric-gate versus asymmetric-gate configuration. J. Appl. Phys. **94**, 5732 (2003)

68  L. Chang, M.Ieong, and M.Yang: CMOS Circuit Performance Enhancement by Surface Orientation Optimization. IEEE Trans. Elec. Dev. **51**, 1621 (2004)

69  T. Sato, Y. Takeishi, and H. Hara: Mobility anisotropy of electrons in inversion layers on oxidized silicon surfaces. Phys. Rev. **B4**, 1950 (1971)

70  F. Stern and W.E. Howard: Properties of semiconductor surface inversion layers in the electric quantum limit. Phys.Rev. **163**, 816 (1967)

71  A. Rahman, M.S. Lundstrom, and A.W. Ghosh: Generalized effective-mass approach for n-type metal-oxide-semiconductor field-effect transistors on arbitrarily oriented wafers. J.Appl. Phys. **97**, 053702 (2005)

72  M. Min Yang, E. P. Gusev, Meikei Ieong, O. Gluschenkov, D.C. Boyd, Kevin K. Chan, P.M. Kozlowski, C.P. D'Emic, R.M. Sicina, P.C. Jamison, and A.I. Chou: Performance Dependence of CMOS on Silicon Substrate Orientation for Ultrathin Oxynitride and HfO Gate Dielectrics: IEEE Electron Device Lett., EDL-24, 339 (2003)

73  M.V. Fischetti, D.A. Neumayer, and E.A. Cartier, "Effective electron mobility in Si inversion layers in metal–oxide–semiconductor systems with a high-k insulator: The role of remote phonon scattering. J. Appl. Phys. **90**, 4587 (2001)

74  F. Jiménez-Molinos, F. Gámiz, A. Godoy and J.B. Roldán: Combined influence of Coulomb and remote phonon scattering mechanisms on the electron mobility in SOI-MOSFETs with high-k dielectrics. *Proceedings of Ultimate Integration of Silicon Conference (ULIS)* 2006

75  S.Takagi, J. Koga, and A. Toriumi: Subband structure engineering for performance enhancement of Si MOSFETs. *IEDM Tech. Digest*, 219 (1997)

76  S. Takagi, T. Mizuno, T. Tezuka, N. Sugiyama, T. Numata, K. Usuda, Y. Moriyama, S. Nakaharai, J. Koga, A. Tanabe, N. Hirashita, and T. Maeda: Channel structure design, fabrication and carrier transport properties of strained-Si/SiGe-on-insulator (strained-SOI) MOSFETs. *IEDM Tech. Digest*, 57 (2003)

77  J. Welser, J. L. Hoyt, and J. F. Gibbons: Electron mobility enhancement in strained-Si N-type metal-oxide-semiconductor field-effect transistors. IEEE Electron Dev.Lett., **15**, 100 (1994)

78  M. Rieger and P. Vogl: Electronics band parameters in strained $Si_{1-x}Ge_x$ alloys in $Si_{1-y}Ge_y$ substrates. Phys. Rev. **B 48**, 14276 (1993)

79  J. B. Roldán, F. Gámiz, J. A. Lopez-Villanueva, and J. E. Carceller: A Monte Carlo study on the electron-transport properties of highperformance strained-Si on relaxed $Si_{1-x}Ge_x$ channel MOSFETs. J. Appl. Phys. **80**, 5121 (1996)

80  R. Chau, J. Kavalieros, B. Roberds, A. Murthy, B. Doyle, B. Barlage, M. Dockzy, and R. Arghavani: A 50nm depleted-substrate CMOS transistor. *IEDM Tech. Digest*, 621 (2001)

81  K. Rim, K. Chan, L. Shi, D. Boyd, J. Ott, N. Klymko, F. Cardone, L. Tai, S. Koester, M. Cobb, D. Canaperi, B. To, E. Duch, I. Babich, R. Carruthers, P. Saunders, G. Walker, Y. Zhang, M. Steen, and M. Ieong: Fabrication and mobility characteristics od Ultrathin strained Si directly on insulator (SSDOI) MOSFETs. *IEDM Tech. Dig.*, 47 (2003)

82  F. Gámiz, J. B. Roldán, and A. Godoy: Strained-Si/SiGe-on-insulator inversion layers: The role of strained-Si layer thickness on electron mobility. Appl. Phys. Lett. **80**, 4160 (2002)

83  F. Gámiz, P. Cartujo-Cassinello, J. B. Roldán, and F.Jiménez-Molinos: Electron transport in strained Si inversion layers grown on SiGe-on-insulator substrates. J. Appl. Phys. **92**, 288 (2002)

84  N. Barin, C. Fiegna, and E. Sangiorgi: Analysis of Strained-on-Insulator Double-Gate MOS structures. *IEDM Tech. Dig.*, 169 (2004)

85  H.S.P. Wong: Nanoelectronics: Nanotubes, Nanowires, Molecules, and Novel Concepts. *Proceedings of ESSCIRC*, Grenoble, France (2005)

86  SM Koo, A. Fujiwara, J.P. Han, E.M. Vogel, C.A. Richter, and J.E. Bonevich: High Inversion Current in Silicon Nanowire Field Effect Transistors. Nanoletters **4-11**, 2197 (2004)

87  A. M. Morales, C.M. Lieber: A Laser Ablation Method for the Synthesis of Crystalline Semiconductor Nanowires. Science **279**, 208 (1998)

88  Y. Huang, X. Duan, Y. Cui, L. J. Lauhon, K.H. Kim, C.M. Lieber: Logic gates and computation from assembled nanowire building blocks. Science, **294**, 1313 (2001)

89  Y. Cui, L.J. Lauhon, M.S. Gudiksen, J. Wang, C.M. Lieber: Diameter-controlled synthesis of single-crystal silicon nanowires. Appl. Phys. Lett. **78**, 2214 (2001)

90  D. Wang, Q. Wang, A. Javey, R. Tu, H. Dai, H. Kim, P.C. McIntyre, T. Krishnamohan, K.C. Saraswat: Germanium nanowire field-effect transistors with $SiO_2$ and high-k $HfO_2$ gate dielectrics. Appl. Phys. Lett. **83**, 2432 (2003)

91  X. Duan, Y. Huang, Y. Cui, J. Wang, C.M. Lieber: Indium phosphide nanowires as building blocks for nanoscale electronic and optoelectronic devices. Nature **409**, 66 (2001)

92  X. Duan, J. Wang, C.M. Lieber: Synthesis and optical properties of gallium arsenide nanowires. Appl. Phys. Lett. **76**, 1116 (2000)

93  S. Guha, N.A. Bojarczuk, M.A. L. Johnson, J. F. Schetzina: Selective area metalorganic molecularbeam epitaxy of GaN, and the growth of luminescent microcolumns on $Si/SiO_2$. Appl. Phys. Lett. **75**, 463 (1999)

94  G. Cheng, A. Kolmakov, Y. Zhang, M. Moskovits, R. Munden, M. A. Reed, G. Wang, D. Moses, J.Zhang: Current rectification in a single GaN nanowires with a well-defined p–n junction. Appl. Phys. Lett., 83, 1578 (2003)

95  J.C. Johnson. H.-J. Choi, K.P. Knutsen. R.D. Schaller, P. Yang, R.J. Saykally: Single gallium nitride nanowire lasers. Nature Materials, **1**, 106 (2002)

96  F.L. Yang, H.Y. Chen, F.C. Cheng, C.C. Huang, C.Y. Chang, H.K. Chiu, C.C. Lee, C.C. Chen H.T. Huang, C.J. Chen, H.J. Tao, Y.C. Yeo, M.S. Liang, C. Hu: 5 nm Gate Nanowire FinFET. *Symp. VLSI Technology*, 196 (2004)

97  N. Singh, A. Agarwal, L. K. Bera, T. Y. Liow, R. Yang, S. C. Rustagi, C. H. Tung, R. Kumar, G. Q. Lo, N. Balasubramanian, and D.L. Kwong: High-

Performance Fully Depleted Silicon Nanowire (Diameter ≤ 5 nm) Gate-All-Around CMOS Devices. IEEE Electron Device Letters **27**, 383 (2006)

98   J. Wang, A. Rahman, A. Ghosh, G. Klimeck, and M. Lundstrom: On the Validity of the Parabolic Effective-Mass Approximation for the $I-V$ Calculation of Silicon Nanowire Transistors. IEEE Transactions on Electron Devices **52**, 1589 (2005)

99   G.D. Sanders, C.J. Stanton, Y.C. Chang: Theory of transport in silicon quantum wires. Physical Rev. **B48**, 11067 (1993)

100  J. Wang, E. Polizzi, M. Lundstrom: A three-dimensional quantum simulation of silicon nanowire transistors with the effective-mass approximation.. Journal of Applied Physics **96**, 2192 (2004)

101  M.J. Gilbert, R. Akis, D.K. Ferry: Phonon-assisted ballistic to diffusive crossover in silicon nanowire transistors. Journal of Applied Physics **98**, 094303 (2005)

102  S. Jin, Y.J. Park, and H.S. Min: A three-dimensional simulation of quantum transport in silicon nanowires transistor in the presence of electron-phonon interactions. Journal of Applied Physics **99**, 123719 (2006)

103  A. Trellakis, A.T. Galick, A. Pacelli, U. Ravaioli: Iteration scheme for the solution of the two-dimensional Schrödinger-Poisson equations in quantum structures. J. Appl. Phys. **81**, 7880 (1997)

104  G. Curatola, G. Iannaccone: NANOTCAD2D: Two-dimensional code for the simulation of nanoelectronic devices and structures. Computational Materials Science **28**, 342 (2003)

105  A. Trellakis, U. Ravaioli: Lateral scalability limits of silicon conduction channels. J. Appl. Phys. **86**, 3911 (1999)

106  A. Godoy, A. Ruiz-Gallardo, C. Sampedro, and F. Gámiz: Accurate simulation of Multiple-Gate SOI MOSFETs. *Proceedings of the EUROSOI Conference*, Grenoble (2006)

107  A. Trellakis, U. Ravaioli: Computational issues in the simulation of semiconductor quantum wires. Comput. Methods Appl. Mech. Engin. **18**, 437 (2000)

108  J.P. Colinge, J.C. Alderman, W. Xiong, C.R. Cleavelin: Quantum–Mechanical Effects in Trigate SOI MOSFETs. IEEE Transactions on Electron Devices **53**, 1131 (2006)

109  R. Kubo: Statistical-mechanical theory of irreversible processes. I. General theory and simple applications to magnetic and conduction problems. J. Phys. Soc. Jpn. **12**, 570 (1958)

110  D. A. Greenwood: The Boltzmann equation in the theory of electrical conduction in metals. *Proc. Phys. Soc. London*, **71**, 585 (1958)

111  L.P. Keldysh: Diagram technique for non-equilibrium processes. Sov. Phys. JETP **20**, 1018 (1965)

112  L.P. Kadanoff and G. Baym: *Quantum Statistical Mechanics* (Benjamin, New York, 1962)

113  E.B. Ramayya, D. Vasileska, S.M. Goodnick, I. Knezevic: Electron mobility in silicon nanowires. IEEE Transactions on Nanotechnology **6**, 113 (2007)

114 A. Godoy, F. Ruiz, C. Sampedro, F. Gámiz, U. Ravaioli: Calculation of the phonon-limited mobility in silicon Gate All-Around MOSFETs. Solid State Electronics, *Accepted for publication* (2007)
115 M.V. Fischetti, S.E. Laux: Monte Carlo Simulation of Electron Transport in Technologically Significant Semiconductors of the Diamond and Zinc-Blende Structures. Part II: Submicron MOSFETs. IEEE Trans. Electron Devices **38**, 650 (1991)
116 A. Godoy, Z. Yang, U. Ravaioli, F. Gámiz: Effects of nonparabolic bands in quantum wires. J. Appl. Phys **98**, 013702 (2005)
117 D. Jovanovic, J.P. Leburton: Monte Carlo simulation of quasi-one-dimensional systems. In: Hess K, editor. *Monte Carlo device simulation: Full band and beyond*. Kluwer Academic Publishers. 191 (1991)
118 J.P. Leburton: Optic-phonon-limited transport and anomalous carrier cooling in quantum-wire structures. Physical Rev. **B45**, 11022 (1992)
119 S. Takagi, A. Toriumi, M. Iwase, H. Tango: On the universality of inversion layer mobility in Si MOSFET's: Part I—Effects of substrate impurity concentration. IEEE Trans. Electron Devices **41**, 2357 (1994)
120 R. Mickevicius, V. Mitin: Acoustic-phonon scattering in a rectangular quantum wire. Phys. Rev. **B48**, 17194 (1993)
121 F. Gámiz, J.B. Roldan, J.A. Lopez-Villanueva, P. Cartujo-Cassinello, F. Jimenez-Molinos: Monte Carlo simulation of electron mobility in silicon-on-insulator structures. Solid State Electronics **46**, 1715 (2002)
122 F. Gámiz, P. Cartujo-Cassinello, J.B. Roldán, C. Sampedro, A. Godoy: Influence of confined acoustic phonons on the electron mobility in ultrathin silicon-on-insulator layers. *Proceedings of the Twelfth International Symposium on Silicon-on-Insulator Technology and Devices*. The Electrochem Soc, Inc. Pennington, New Jersey; **2005–03**, 39 (2005)
123 S. Yu, K.W. Kim, M.A. Stroscio, G.J. Iafrate, A. Ballato: Electron-acoustic-phonon scattering rates in rectangular quantum wires. Physical Review **B50**, 1733 (1994)
124 R. Kotlyar, B. Obradovic, P. Matagne, M. Stettler, M.D. Giles: Assessment of room-temperature phonon-limited mobility in gated silicon nanowires.. Journal Applied Physics **84**, 5270 (2004)
125 J.P. Colinge, C.A. Colinge, *Physics of Semiconductor Devices*. Norwell, MA: Kluwer (2002)
126 E. B. Ramayya: Modeling of mobility in a rectangular silicon nanowire transistor. *M.S. Thesis*, Arizona State Univ., Tempe (2006)
127 J Wang, E. Polizzi, A. Ghosh, S. Datta, M. Lundstrom: Theoretical investigation of surface roughness scattering in silicon nanowire transistors. Applied Physics Letters **87**, 043101 (2005)
128 Y. Cui, X. Duan, J. Hu, and C. M. Lieber: Doping and electrical transport in silicon nanowires. J. Phys. Chem. B **104**, 5213 (2000)
129 Y. Cui, Z. Zhong, D. Wang, W. U. Wang, and C. M. Lieber: High performance silicon nanowire field effect transistors. Nano Lett. **3**, 149 (2003)

130 R. Chau, S. Datta, M. Doczy, B. Doyle, B. Jin, J. Kavalieros, A. Majumdar, M. Metz, M. Radosavljevic: Benchmarking Nanotechnology for High-Performance and Low-Power Logic Transistor Applications. IEEE Transactions on Nanotechnology **4**, 153 (2005)

131 J.P. Colinge, A.J. Quinn, L. Floyd, G. Redmond, J.C. Alderman, W. Xiong, C.R. Cleavelin, T. Schulz, K. Schruefer, G. Knoblinger, P. Patruno: Low-Temperature Electron Mobility in Trigate SOI MOSFETs. IEEE Electron Device Letters **27**, 120 (2006)

# 6 Radiation Effects in Advanced Single- and Multi-Gate SOI MOSFETs

Véronique Ferlet-Cavrois, Philippe Paillet and Olivier Faynot

## 6.1 A brief history of radiation effects in SOI

Radiation effects have been studied in microelectronic components for several decades for military and space applications.[1] For these hash environments, robust SOI technologies have been developed with a single-gate partially depleted architecture and body contacts to avoid floating-body effects. The main advantage of SOI for transient and single-event effects comes from the small sensitive volume and the complete dielectric isolation of each transistor. The collected charge under irradiation is limited to the transistor sensitive volume and it is significantly reduced compared to bulk. SOI is naturally hardened to latch-up, charge sharing or funneling effects due the presence of the buried oxide that prevents charge collection from the substrate. On the other hand, total dose remained a concern since the active transistor is surrounded by oxide (gate, lateral, buried oxide) where irradiation generated charge can get trapped and modifies the transistor characteristics. However, hardening techniques were developed for both total dose and transient effects resulting in highly hardened SOI devices.

In the 1990's, a new IC reliability concern emerges in large-scale, non-hardened devices because of the terrestrial radiation environment encountered in avionic and ground-level circuits.[2-6] The two main components of the natural environment threatening the reliability of new generation ICs are alpha particles from the telluric activity and neutrons coming from the interaction of cosmic rays with the earth atmosphere. In both cases, the interaction of particle with semiconductor ICs results in local parasitic charge generation. With technology scaling, the reduction of transistor dimensions and supply voltage, a small collected charge is able to induce an upset. For reference, a 256 Mbyte computer without error-correction software might typically encounter one error every three weeks. [7-8] The errors induced by the natural environment are the major source of errors, by several orders of magnitude, compared to other reliability mechanisms like metal electromigration or gate rupture.[9] Hardening design techniques such as error correction codes and redundancy, have been introduced to mitigate the effects of the natural radiation environment. Special care is also taken to control alpha contaminants in processing and packaging materials.

Few years after, in the early 2000's, SOI reached a level of maturity high enough to allow for the commercial production of complex and high-performance circuits such as microprocessors. The quality of SOI substrates has improved to a level such that SOI circuits now enjoy same fabrication yield figures as high as bulk circuits. The preferred device used in those applications is the single-gate, floating-body partially depleted, thin-film silicon transistor. Body contacts are removed since the purpose of these circuits is high speed and high density. The floating-body effect is even used to accelerate gate switching in digital circuits. Compared to its bulk counterpart, and even without body contacts, SOI has been shown to significantly improve the reliability of large-scale highly integrated circuits in the natural radiation environment. Because of the limited charge collection, SOI appears as one solution to the growing concern of semiconductor manufacturers for the ICs reliability.

In the same time, convinced by the potential of SOI to meet the ITRS requirements for the 50 nm – 20 nm nodes, research laboratories are investigating new SOI-based architectures, fully depleted single- or multi-gate devices. For these new architectures, the issue is to ensure a good reliability of very large-scale circuits in the natural terrestrial environment. These nanoscale devices are also evaluated for hash environments such as space applications (satellites) which need complex on-board electronic treatments. The investigated constraints are then total dose for space applications and single-event effects for both space and terrestrial applications.

## 6.2 Total Ionizing dose effects

As far as radiation effects are concerned, the total dose response of SOI devices is more complex than that of bulk silicon devices. Historically, SOI technologies were first developed for radiation-hardened military and space applications, mostly for their intrinsic immunity to latchup induced by heavy ions crossing the device. Since they are surrounded by oxide layers, individual SOI transistors present a series of parasitic structures which can reveal very troublesome when the device is exposed to ionizing radiation such as x-rays. Some of these parasitic structures are also present in bulk devices (*i.e.* those related to lateral isolation oxide) and some are specific to the SOI technology (*i.e.* the buried oxide).

Radiation-induced trapped charge in the buried oxide can increase the leakage current of partially depleted transistors and decrease the threshold voltage and increase the leakage current of fully depleted transistors. Many efforts have been made to reduce these parasitic structures and very high levels of radiation hardness have been achieved. Process techniques that reduce the net amount of radiation-induced positive charges trapped in the buried oxide and device design techniques that mitigate the effects of trapped charges in the buried oxide have been developed to harden SOI devices to similar total-dose levels as their bulk silicon counterparts.

This section will focus on the effects of total-dose irradiation on advanced fully depleted SOI technologies (with a gate length less than 100nm), both on planar single-gate and on 3D multi-gate devices. The comparison of radiation hardness between more traditional SOI technologies, such as single-gate partially depleted and fully depleted devices, is out of the scope of this chapter, but can be found in [10] and references therein.

### 6.2.1 A brief overview of total Ionizing dose effects

Total dose degradation is a cumulative process caused by ionizing radiation, which creates electron-hole pairs in insulating materials. In the absence of an electric field in the oxide, these radiation-induced pairs will recombine very efficiently, leaving only a limited number of free positive and negative charges in the oxide. Unfortunately, most oxides in modern devices are surrounded by metal and (poly)silicon layers, each biased at a different voltage. Therefore an electric field is almost always present in the oxide, and the generated electron-hole pairs will be separated by this electric field. Electrons are swept very efficiently out of the oxide towards the positive electrode, while holes, which have a much smaller mobility, will have a greater probability of being trapped in the oxide at neutral defect sites. For both SOI and bulk-silicon devices, the same types of oxides are typically used for the gate and field oxides (LOCOS or shallow-trench isolation (STI)) and the radiation-induced build-up of charge in these oxides is similar. The main difference between the total dose response of SOI and bulk-silicon devices is radiation-induced charge build-up in the buried oxide. The net radiation-induced trapped charge is predominantly positive. This positive charge can invert the back-channel interface of n-channel SOI transistors and lead to large increases in leakage current of partially depleted and fully depleted transistors.

Fully depleted transistor circuits are very sensitive to radiation-induced oxide charge build-up in the buried oxide and interface trap build-up at the top buried oxide/silicon interface. The top-gate transistor is electrically coupled to the back-gate transistor and charge trapping in the buried oxide directly affects the top-gate transistor characteristics. For example, positive charge trapping in the buried oxide will decrease the top-gate threshold voltage of an n-channel transistor.[11-13]

In floating-body devices, a sudden increase in the radiation-induced leakage current caused by charge trapping is observed in some cases.[8,14,15] It may be partly due to inversion of the back-channel interface and partly due to a "total-dose latch" effect.[16-18] The total-dose latch effect is caused by charge trapping in the buried oxide modulating the body potential. As the body to source barrier height is lowered, excess electrons can be injected into the body region and be collected at the drain. If the electric field near the drain is high enough to cause impact ionization, this can lead to a current runaway condition causing snapback (in SOI technology, snapback is often called single-transistor latch).

### 6.2.2 Advanced single-gate FDSOI devices

Total Ionizing Dose effects in single-gate fully depleted SOI devices have been extensively studied [5,6,8,10-13,19], mostly on devices with a gate length longer than 0.25 µm. Some total-dose results are also available for gate lengths down to 130 nm, but almost no data is currently available for devices with dimensions below 100 nm. In the following, results from reference [20] are summarized to present the first set of data available on such advanced devices, with gate lengths down to 50 nm.

### 6.2.2.1 Description of advanced FDSOI devices

These advanced devices were fabricated at CEA/LETI. They are fully depleted single-gate SOI NMOS transistors (FDSOI) with a gate length varying from 50 nm to 1 µm. They were processed on a standard P-type substrate, with a resistivity of 10 to 20 Ω.cm. These UNIBOND™ SOI substrates have a buried oxide thickness of 145 nm, and a 11 nm-thick silicon film in the channel region. The transistors are processed with a mid-gap TiN gate, and a low silicon film doping level ($\sim 10^{15}$ cm$^{-3}$). The gate oxide thickness of the transistors is 1.7 nm. After process, the actual gate length is shorter than drawn on the mask. From SEM cross-sections, a reduction of about 20 nm can be observed, meaning that the gate drawn at 50 nm has an effective gate length of about 30 nm. In the following, we will refer to the different gate lengths using their value drawn on the mask, starting at 50 nm.

The process used at CEA LETI involves lateral isolation via oxide trenches, aluminum metal layers and oxide overlayer isolation. The tested devices were either floating-body transistors (gate width of 40 µm), or transistors with an external body contact (gate width of 10 µm). The different drain current characteristics are normalized to account for the difference in gate width.

### 6.2.2.2 Front-gate threshold voltage shift

Charge trapping in the gate oxides of today's MOS devices, whether bulk or SOI, is not a first order concern, since their thickness (< 50 Å) is thin enough to limit the direct influence of their total dose degradation on the electrical characteristics of the device. Thicker oxides such as lateral isolation or buried oxides for SOI devices, on the other hand, are a greater concern, because their related parasitic structures are revealed upon charge trapping, and can dominate the I-V characteristics when total dose accumulates. Therefore, the total dose sensitivity of the device will be increased by charge trapping in these thick oxides. In FD SOI devices, charge trapping in the buried oxide can affect the front transistor through a direct coupling effect between the front and back interfaces.

The front-gate threshold voltage shifts of 80 nm FDSOI devices, resulting from charge trapping in the buried oxide, are presented in Fig. 6.1

as a function of total dose. The $I_D$-$V_G$ characteristics are shifted towards more negative voltages as total dose is accumulated in the gate and buried oxides of the transistor.

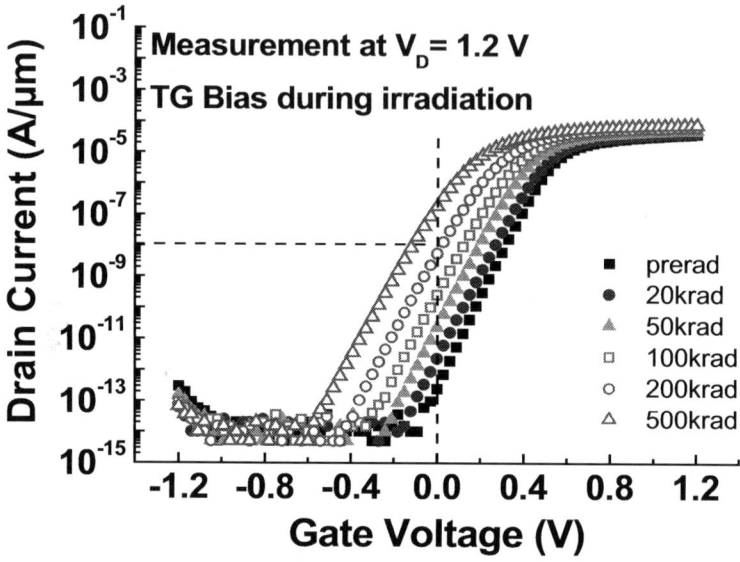

**Fig. 6.1.** Normalized drain current versus front gate voltage, measured at a drain voltage of 1.2V on a 80 nm FDSOI NMOS transistor with external body contact. The device was biased in the OFF-state during irradiation (from [15]).

For a total dose of 500 krad(SiO$_2$) deposited in a 1.8 nm thick oxide, and assuming that 100% of the generated holes are separated and then trapped at the interface, the maximal voltage shift would be of only 60 mV.[15,21] The front gate threshold voltage shifts measured in Fig. 6.2 do obviously not reflect charge trapping in the gate oxide. These shifts are caused by the electrical coupling between the front interface and the charges trapped in the buried oxide.

The bias applied to the device during exposure to radiation is a critical parameter influencing charge trapping. For fully depleted SOI transistors, the worst-case bias is not as well defined as for partially depleted SOI transistors. Similar to the case for partially depleted SOI transistors, for some technologies, the Transmission-Gate (T-G) configuration, (source and drain at $V_{DD}$, other contacts grounded) caused the most radiation-induced charge trapping in the buried oxide.[6] However, for other technologies, the worst-case bias was determined to be the ON bias configuration (gate at $V_{DD}$, other contacts grounded).[9] In these short-gate devices, the T-G bias configuration is considered to be the worst-case bias condition for total-dose effects, as it was the case in devices with larger gate lengths.[10,15] The mechanism causing these differences in bias dependence is still unknown.

The front-gate threshold voltage shifts of TG-biased transistors depend on the gate length of the device. Fig. 6.2 displays the front threshold voltage shift dependence on irradiation dose as a function of transistor gate length. As gate length is decreased, the corresponding threshold voltage shift increases. This effect has been observed already both for PD devices [17] and FD devices.[10] As gate length decreases, source and drain contacts get closer together, and the electric field lines in the BOX, originating from these contacts, converge under the channel region. A more detailed description of charge trapping in FD and PD SOI devices, obtained with numerical simulations using a 3D self-consistent code, can be found in Refs. [15,22]. Differences in trapped charge profiles in the buried oxide are investigated for different bias applied during irradiation.

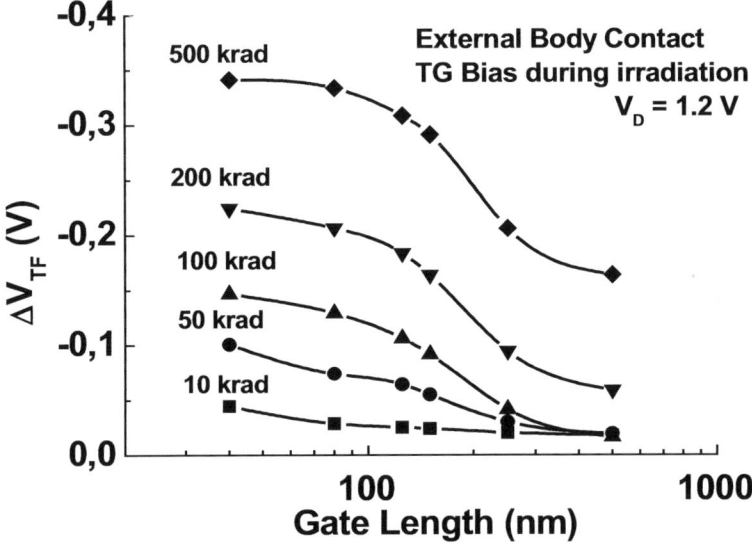

**Fig. 6.2.** Front-gate threshold voltage shifts versus gate length of FDSOI NMOS transistors, biased in the TG-state during irradiation, from Ref. [15].

The important point is that, after a 100 krad($SiO_2$) worst-case irradiation, the characteristics of those devices with external body contact still have a leakage current lower than 0.1 nA/µm. This would be within margins for a low-power technology for most space applications, even though at this point no particular effort was made to optimize the radiation hardness of the technology.

### 6.2.2.3 Single-transistor latch

Charge-trapping effects in advanced fully depleted SOI devices depend greatly on device architecture. The front-gate I-V characteristics are impacted differently for transistors with or without external body contacts. The threshold voltage shifts presented previously were obtained on devices with external body contacts only. Most floating-body devices do exhibit a high-current regime that rapidly dominates their current-voltage characteristics after irradiation. This radiation-induced high-current regime (also called single-transistor latch) has been observed in fully depleted devices with larger gate length.[10-13] Simulations show that the mechanism responsible for the onset of single transistor latch results from the combined effect of small gate length and floating body potential in the intrinsic silicon film.[18,20]

Adding body ties or external body contacts to the device architecture completely suppresses the onset of single-transistor latch. The effects of the radiation-induced charge trapped in the BOX in devices with body contacts are then limited to the expected shift of the threshold voltage, as presented in the previous section. These two different behaviours are illustrated in Fig. 6.3.a, where data plotted with open symbols are obtained on transistors with a grounded external body contact, and data plotted with lines are obtained on floating-body transistors, irradiated and measured with the same bias conditions as their counterparts with external body contacts (open symbols).

In Fig. 6.3a, the first two front-gate *I-V* curves resemble exactly those obtained for devices with an external body contact. As total dose keeps increasing, a high leakage current level is measured in the floating-body device, throughout the negative part of the gate bias sweep. The results obtained for the back gate transistor (Fig. 6.3b) are very similar to the one in Fig. 6.3a, showing the same occurrence of a high-current regime. Thus, the conditions that induce this high current regime on the front-gate characteristics also appear to affect the back-gate conduction, even at high negative back-gate voltages. This observation is specific to these small gate length devices. On 0.25 µm fully depleted devices, this high-current regime or "total-dose latch" was only observed on the front-gate transistor.[10,13]

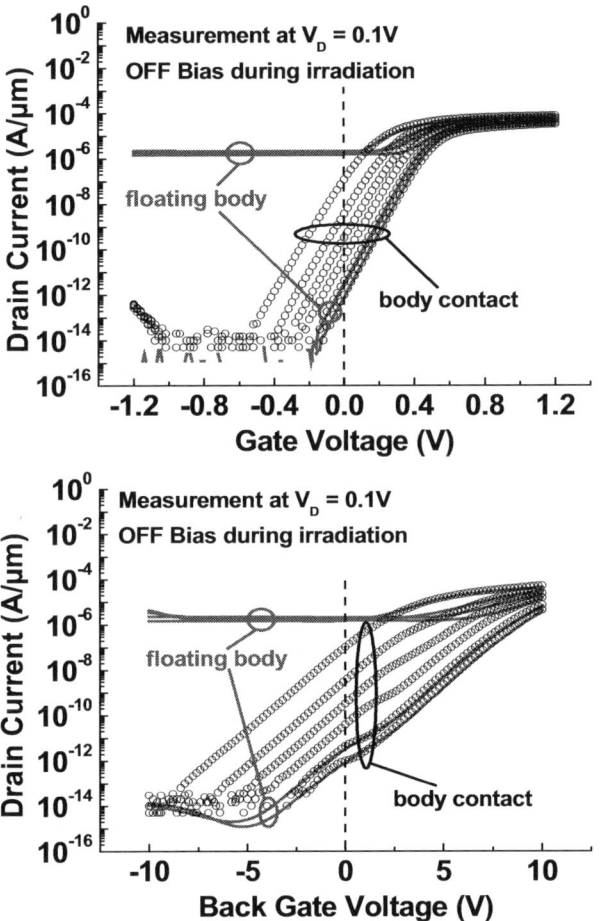

**Fig. 6.3.** Normalized drain current versus front-gate voltage (top) and back-gate voltage (bottom), measured at V = 0.1 V on a 60nm FDSOI NMOS transistor, biased in the OFF-state during irradiation. Data are shown for floating-body devices, at pre-rad, 10, 50, 100, 200, and 500 krad(SiO$_2$), from Ref. [15].

The occurrence of a high-current regime in floating-body devices measured at nominal drain voltage can be expected in short-channel transistors. The data presented in Fig. 6.3 were measured at a much lower drain voltage ($V_D$=0.1V), and the high level of drain current is observed throughout most of the gate voltage sweep. This is the first experimental evidence of single-transistor latch at such a low drain voltage, well below the drain voltage required to initiate impact ionization. before these results were published, the onset of single-transistor latch was usually thought to be maintained, if not induced, by impact ionization in the silicon film. The description of all the precise mechanisms which contribute to its occurrence is still a matter of debate, and is clearly out of the scope of this chapter.

In summary, floating-body fully depleted devices with very short gates and undoped body are sensitive to single-transistor latch, even at low drain voltage, which is new and unexpected compared to devices from older generations. The combined effect of a small gate length with intrinsic floating body results in a strong influence of charge trapping in the BOX on the potential of the body. The increase in floating-body potential, induced by total-dose effects, is responsible for the onset of a high-current regime, without the need for impact ionization. The use of body doping is a way to mitigate this sensitivity to single-transistor latch. Devices with external body contact, on the other hand, are not sensitive to single-transistor latch, even for very small gate lengths. They exhibit an intrinsic resistance to total-dose irradiation, at levels typical of most space applications.

### 6.2.3 Advanced multi-gate devices

Total Ionizing Dose effects in multi-gate, fully depleted SOI devices have not been extensively studied yet, mostly because these devices are still at an early stage of development, thus not readily available for this kind of investigation. The first papers related to total ionizing dose experiments on multi-gate devices were the works of Colinge *et al.* in 1993 [23], and Francis *et al.* in 1994.[24] In reference [23], the authors investigate the total dose sensitivity of Gate-All-Around devices, fabricated using a 3µm technology on SIMOX buried oxide.

The irradiation results of [23] are the first ever reported on multi-gate devices, they are recalled here in Fig. 6.4. Most of the threshold voltage shift induced by ionizing radiation in these GAA devices is shown to saturate rapidly after a total dose of 324 krad(Si), to reach values around 1.5V to 2V. Compared to the threshold voltage shifts obtained in the single-gate FD SOI devices, presented in section 6-3-2-2, this value of 2V can seem to be large. It is not surprising however, since these GAA devices have a 50 nm thick gate oxide, which was thin at the time, but is now much thicker than gate oxides of today's devices. The other total dose experiments [24] performed on GAA-based inverters, clearly evidenced the necessity of a radiation hardened design for SRAM cells to overcome the charge trapping effects in the relatively thick oxides used for this 3-μm generation technology.

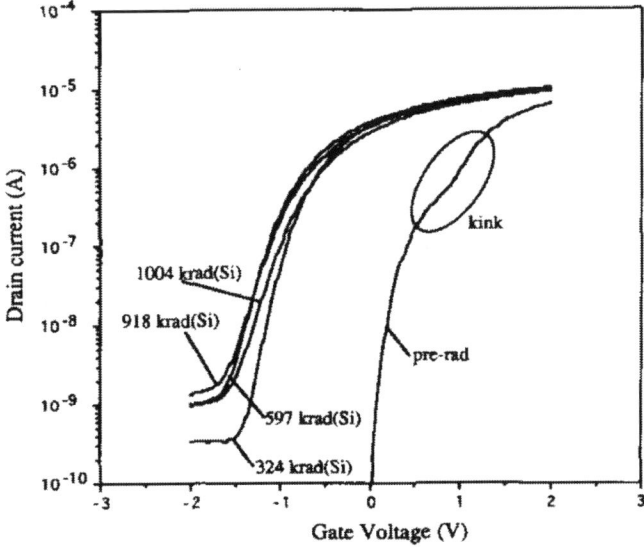

**Fig. 6.4.** Experimental log $I_D(V_G)$ curves of an n-channel GAA device exposed to different irradiation doses. $V_G = 3$ V during $^{60}$Co irradiation, and $V_{DS} = 0.5$ V during parameter extraction. W/L = 3 μm/3 μm, from Ref. [23].

Since the pioneering work of Colinge *et al.*, other experimental data were published recently in 2006, on more advanced devices.[25-27] Another recent work presents only simulations of total-dose effects based on FD SOI devices operating in double-gate mode, with no radiation data on actual devices.[28] In the following, the total-dose results from references [25] and [27] are summarized to present the first set of experimental data available on such advanced non-planar triple-gate transistors.

### 6.2.3.1 Devices and process description

The triple-gate devices used in reference [25] were fabricated at the CEA-LETI on standard P-type Unibond® substrates with a BOX thickness of 100 nm.[29] The silicon body was overetched to form Ω-shaped gates, as schematically described in Fig. 6.5. Successive sacrificial oxidations were performed to suppress the defects on the lateral sides of the silicon finger and to round its corners. The final silicon thickness ($T_{Si}$) is approximately 25 nm.

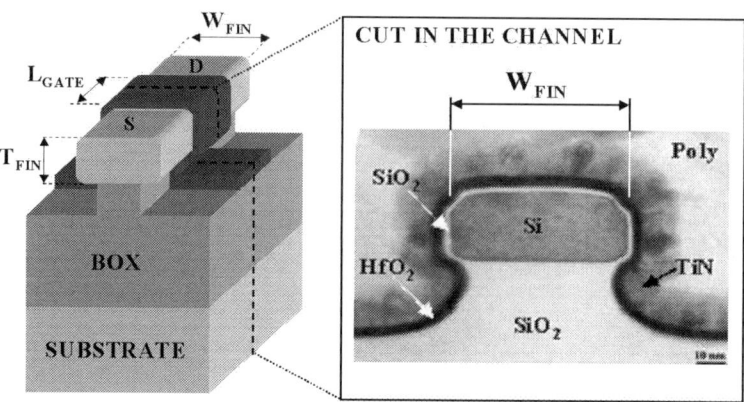

**Fig. 6.5.** Schematic configuration of the Ω-FET. The TEM cross-section of the Ω-FET represents a cut in the channel.[29]

The front-gate oxide was processed in two steps: a thin silicon dioxide layer was grown before the high-k dielectric deposition, with a final equivalent oxide thickness of about 2 nm. The channel is left undoped (~ $10^{15}$ cm$^{-3}$) and a mid-gap TiN metal gate completes the gate stack, resulting in a front-channel threshold voltage $V_{TF}$ of about 0.56 V. The silicon fingers (referred to as "fins") and the gates were patterned using e-beam lithography. The resulting silicon fin widths ($W_{Si}$) vary from 40 nm to 10 μm and the gate lengths from 50 nm to 10 μm. From SEM cross-sections, a reduction of about 20 nm can be observed between the gate drawn on the mask and the final physical gate length. For instance, a 50 nm gate length drawn on the mask corresponds to a physical gate length of about 30 nm.

Devices used in reference [27] were also fabricated on standard Unibond$^{TM}$ substrates. The BOX thickness is 200 nm. The silicon film was doped using boron, resulting in channel doping concentration of 6 x $10^{17}$ cm$^{-3}$. Lithography and RIE are used to etch the silicon film and define 50 nm-wide silicon fins. The devices underwent a hydrogen annealing step to smooth the silicon surfaces, round fin corners and to thin the fins.[30] The radius of curvature at the top and bottom of the fins was measured in TEM cross sections and is 5 nm and 2 nm, respectively. The final fin width and height obtained by this process is 45 nm and 82 nm, respectively. A 2-nm gate oxide was then grown by wet oxidation in an AMAT ISSG reactor. Polysilicon was then deposited and doped N-type by phosphorus ion implantation. Arsenic was implanted to form sources and drains and titanium silicide was formed on sources and drains to reduce parasitic resistance. Classical aluminum/silicon metallization was used to complete the process. The main difference with devices from CEA-LETI is that the minimum printed gate length of these devices is 150 nm and each device consists of 10 fins operating in parallel. A cross section TEM view of a fin is shown in Fig. 6.6.

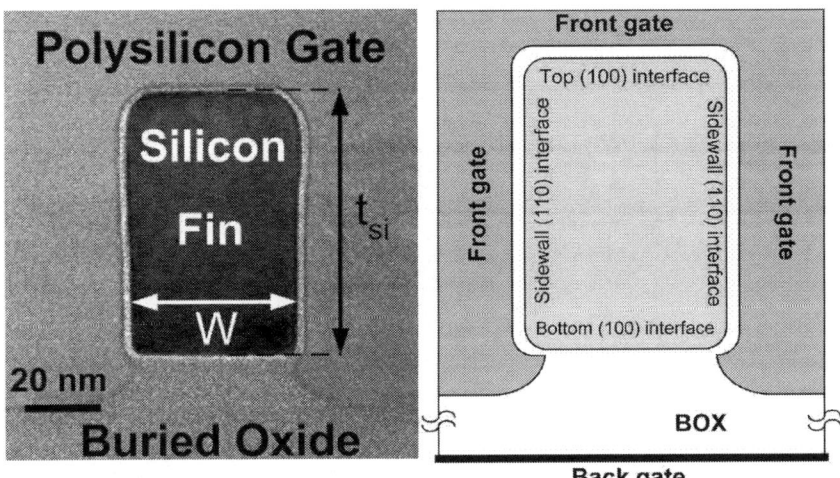

**Fig. 6.6.** (A) TEM cross section of a trigate SOI MOSFET. The fin width is 45 nm and the fin height is 82 nm. Gate oxide thickness is 2 nm. (B). Schematic cross section showing the crystal orientation of the different Si/SiO$_2$ interfaces. Copyright© 2006 IEEE.

### 6.2.3.2 Front-gate threshold voltage shift

Irradiation experiments carried out on omega-gate and triple gate MOSFETs show that the threshold voltage shift induced by the creation of positive charges in the BOX increases with the width of the device fins, as illustrated in Fig. 6.7 from Ref. [25]. As shown in Fig. 6.7, when the fin is very wide, the device operates like a "regular", single-gate, fully depleted SOI device. When the device is very narrow, the presence of the lateral gates in the BOX acts like a virtual back gate. The effect of this virtual gate shields the bottom of the silicon fins from the electric field created by the charges in the BOX. The trapped charge in the BOX then has a very limited impact on the characteristics of the transistor. As a result, narrow devices behave almost like Gate-All-Around MOSFETs in which the charges in the BOX have no influence on the device characteristics. More detailed simulations, using a self-consistent code dedicated to total dose effects in MOS devices, show that the very good total dose tolerance of Ω-FETs is in fact intrinsic, and due to a combined influence of the lateral gates. First, by penetrating into the BOX, they repel the electric field lines originating from source and drain contacts towards the volume of the BOX, instead of towards the channel as in a single-gate device. Therefore, charge trapping is reduced under the channel region, since radiation-generated holes are drifting to and trapped in a region where they can less impact the device electrical response. This modification of the trapped charge profile combines to the previously mentioned "virtual gate" effect, which reduces the influence of the trapped charge even further.

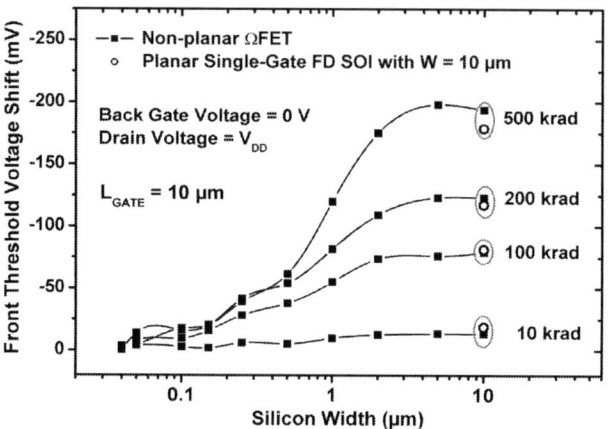

**Fig. 6.7.** Front-gate threshold voltage shift $\Delta V_{T,F}$ versus silicon fin width for non-planar $\Omega$-FETs (black squares), and planar single-gate FD SOI transistors (open circles), from Ref. [25].

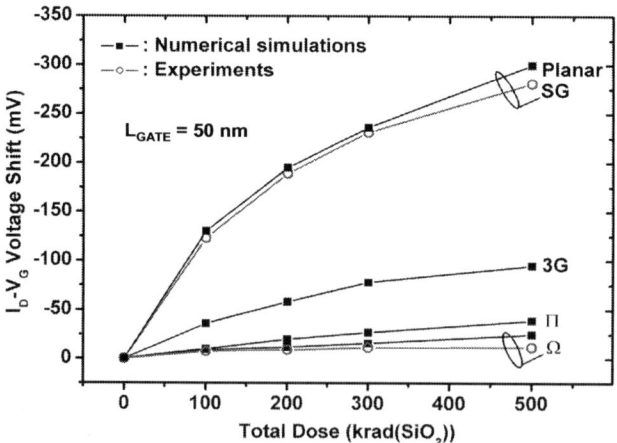

**Fig. 6.8.** Simulated voltage shift (black squares) on different MOSFET architectures: planar SG, Triple-Gate, Π-Gate and Ω-gate FETs. Experimental data obtained under similar irradiation condition are added as open circles. Data from Ref. [25].

When comparing planar and non-planar devices, the Ω-FET architecture is the most efficient in improving the radiation tolerance of advanced 3D devices. As shown in Fig. 6.8 from Ref. [25], all the simulated multi-gate devices exhibit either a very limited or no threshold voltage shift after exposure to ionizing radiation. These simulations are in perfect agreement with the experimental data currently available.

In conclusion, the total-dose effects investigated in scaled non-planar multi-gate devices clearly evidence the high ionizing dose tolerance of narrow scaled Ω-FETs. This tolerance is intrinsic to the geometry of the device, and benefits from the efficient control over the potential in the fin from the lateral gates. On the one hand it decreases the influence of the radiation-induced charge trapped in the buried oxide on the electric properties of the device (so called "virtual gate" effect). On the other hand, it also modifies the profile of charge trapping in the BOX. Holes are trapped near the lateral gates in the bulk of the BOX instead of under the channel region, less impacting the device electrical response. From existing data and simulations, the Ω-FET architecture appears to be the one with the highest intrinsic radiation tolerance of all advanced 3D devices investigated so far.

## 6.3 Single-event effects

The sensitive volume for charge collection in SOI devices is much smaller than for bulk silicon devices, making SOI devices potentially much harder to single-event upset (SEU). However, bipolar amplification caused by floating-body effects can significantly reduce the SEU hardness of SOI devices. Body ties are used to reduce floating-body effects and improve SEU hardness. SOI ICs are completely immune to classic four-layer p–n–p–n single-event latchup; however, floating-body effects make SOI ICs susceptible to single-event snapback (single-transistor latch). The sensitive volume for dose-rate effects is typically two orders of magnitude lower in SOI devices than in bulk-silicon devices. Using body ties to reduce bipolar amplification, much higher dose-rate upset levels can be achieved for SOI devices than for bulk-silicon devices.

### 6.3.1 Background

The first paper showing single-event effects in multi-gate devices was published in 1995.[31] The irradiated devices were 1k SRAMs based on 3-μm gate length Gate-All-Around (GAA) transistors irradiated by heavy ions. A very high threshold LET of about 100 MeV cm$^2$/mg was measured when the SOI/GAA SRAMs were operated at low supply voltage (1.9 V), which is the worst case for single-event effects. The saturated cross section was found to be of $3 \times 10^{-8}$ cm$^2$/bit, corresponding to the sensitive surface of the body region (enclosed in the GAA structure) of the OFF-state transistors.

This paper [31] was also the first to suggest an analytic expression for the minimum charge necessary to induce an upset (*i.e.* the critical charge). The upset mechanism was explained by taking into account the temporal response of the SRAM cell (see Fig. 6.9a). In an SRAM cell, when an OFF-state transistor is struck by an ionizing particle, a transient current is collected at the drain. The cell upset then results from the competition between the feedback effect on the opposite inverter causing the cell upset, and the recovery transient current through the ON-state load transistor mitigating the upset mechanism. In the considered SOI/GAA SRAMs, the critical charge is high (in the pC range) consistent with the 3-µm generation. Because of the small sensitive volume (delimited by the GAA structure), it resulted in a very high threshold LET (100 MeV cm$^2$/mg).

For present-day multi-gate devices the orders of magnitude of radiation effects have to be re-considered. Fig. 6.9b shows the simulated upset of an SRAM cell for a 65 nm partially depleted SOI technology. In this case, the critical charge, obtained from the transient current at the threshold LET 1.5 MeVcm$^2$/mg, is exactly 1 fC. The critical charge for a 50-nm generation device is then typically lower than 1 fC. Embedded or stand-alone SRAMs are one of the most sensitive parts in a digital system. Fig. 6.10a shows that the SEU rate for 1 Mbit SRAMs is about constant over generations. However, system complexity has increased considerably with technology scaling, and computers and complex system (like computer clusters) have now huge SRAM capability.

Another reliability issue has appeared recently, called the "single-event transient" (SET).[32] Irradiation induces parasitic transients that propagate in digital logic and can get latched in a memory element and then generate an erroneous information. Fig. 6.10b shows an example of SET propagation and latch in a 0.18 µm bulk technology. In Fig. 6.10a, the effect of SET ("non-SRAM SER") is shown to increase with generation scaling. It seems the SET sensitivity might reach the SEU sensitivity for future generations. In this case, mitigation techniques will have to be adopted for both radiation effects in memory and logic treatment parts.

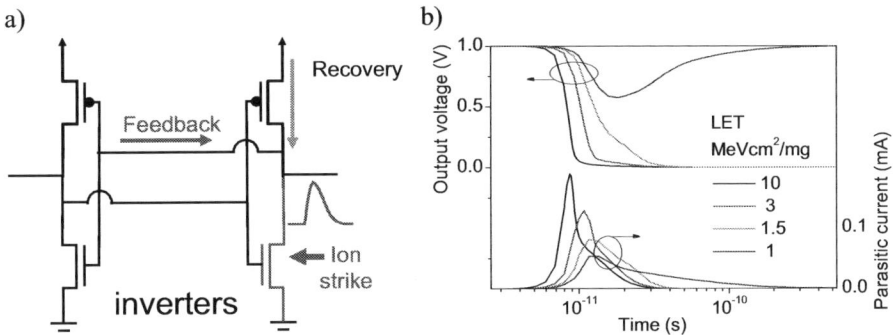

**Fig. 6.9.** (a) Schematic of the upset mechanism in an SRAM cell.[8,33] (b) Simulation of the SRAM cell output voltage and the parasitic current crossing the hit OFF-state NMOS transistor at different LET.

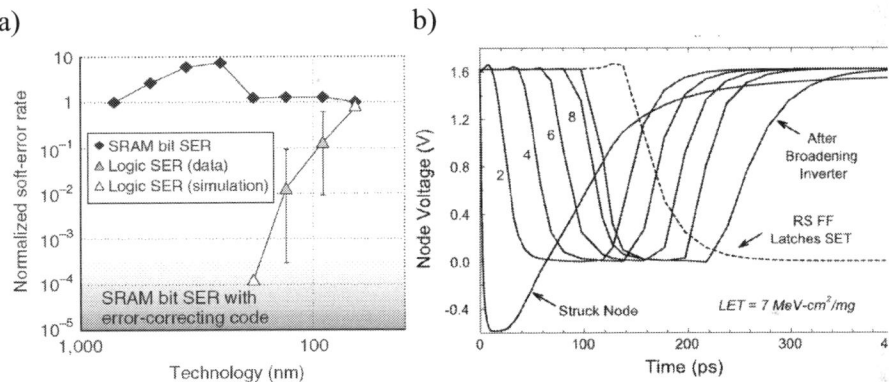

**Fig. 6.10.** (a) Simulated logic SER (SET effects) as a function of scaling (black diamonds) and measured data for the 0.13 μm node, and measured SRAM SEU rate (white diamonds).[6,34] The SET and SEU rates are expressed in FIT/Mbit (1 error for $10^9$ hours in a 1Mbit SRAM). (b) Simulation of single-event transient (SET) propagation in a 10-inverter delay chain in 0.18 μm bulk technology.[35]

The investigation of single-event effects in new devices is still under progress. First, nanoscale devices are developed in research laboratories mostly at the device level. Irradiation can thus only be performed on devices and then extrapolated by simulation and analysis to the circuit level. Transient effects in single transistors are difficult to measure since the transistor response is very fast (in the 10-50ps range) and the collected charge is small. A specific high-bandwidth experimental set-up must be developed, and the measured signal is often at the limit of the system capability. The fast transistor response is due to its small dimensions and its reduced sensitive silicon volume. The measured quantities are then orders of magnitude faster and smaller than for micrometers-devices.

The simulation of single nanoscale transistors also faces difficulties. New physical effects, (quantization of carrier energy, tunneling current and ballistic operation) must be taken into account for a proper simulation and modeling of radiation effects in these new devices. The involved physical phenomena still need to be validated in the specific unusual range of device operation of under irradiation.

Since complex circuits (SRAMs, DSP) are not available in nano-scale technologies yet, their response under irradiation is extrapolated from single devices using mixed-mode simulations. This can be achieved with TCAD tools for total dose (X-rays or $^{60}$Co) or heavy ions that induce direct ionization. However, some particles, such as high-energy protons or neutrons encountered in space or terrestrial environments, deposit charge by indirect ionization through nuclear interactions that generate a wide population of ionizing recoils. The analysis of future nano-scale circuits in such environments then require a multi-scale simulation with first the calculation of the deposited charge, then the device response, and eventually the circuit response.

In the following, we will present the state-of-the-art of single-event effects in advanced single and multi-gate transistors. We will first show measured and simulated responses of single and multi-gate transistors of the 50-nm generation and then investigate the effect of technology scaling on device sensitivity

### 6.3.2 Effect of ion track diameter in nanoscale devices

The ion deposited charge $Q_{dep}$ in SOI planar devices is calculated from the ion LET (Linear Energy Transfer) and the silicon film thickness $t_{Si}$ (Fig. 6.11). The ion generates charges along its track in the sensitive transistor volume, and the track diameter is assumed to be small compared to the transistor dimensions. The angle of incidence $\theta$ of the striking ion can be used during tests to increase the deposited charge in long gate length transistors. The ion LET is usually expressed in MeVcm$^2$/mg in hardness assurance test methods, and the conversion factor in silicon devices is that an ion of 100 MeV cm$^2$/mg generates 1 fC/nm along its track length. Also detailed in Fig. 6.11 is the principle of the parasitic bipolar amplification. The floating body reacts as the bipolar base, and it results in the amplification of the deposited charge. The bipolar gain $\beta$ is calculated as the ratio of the charge collected at the drain electrode to the charge generated in the transistor body.[36]

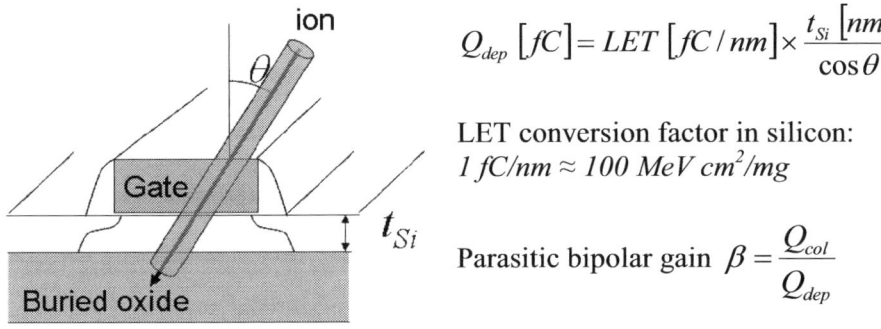

$$Q_{dep}\,[fC] = LET\,[fC/nm] \times \frac{t_{Si}\,[nm]}{\cos\theta}$$

LET conversion factor in silicon:
$1\,fC/nm \approx 100\,MeV\,cm^2/mg$

Parasitic bipolar gain $\beta = \dfrac{Q_{col}}{Q_{dep}}$

**Fig. 6.11.** Radial charge density of ion tracks in silicon for low- and high-energy ions with same LET, compared to new generation nanoscale devices.[37]

As far as nanoscale devices are concerned, Fig. 6.12 clearly shows that the notion of ion LET is no longer appropriate to describe charge generation.

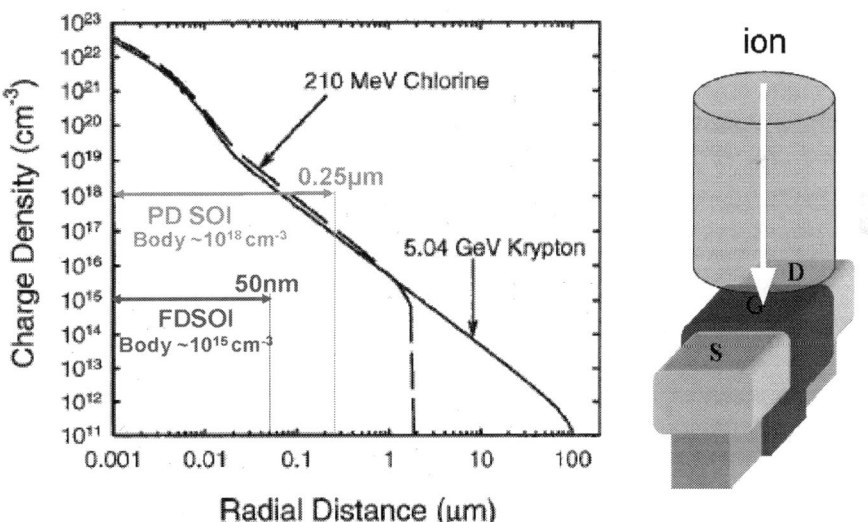

**Fig. 6.12.** Left: Radial charge density of ion tracks in silicon for low- and high-energy ions with same LET (about 10 MeVcm²/mg), compared to new generation nanoscale devices. [37]; Right: Schematics of the track of an ion striking a MuGFET.

The radial profile of the ion track, which varies with both the ion type and its energy, has to be taken into account in a 3D simulation. Furthermore, contrary to old generation partially depleted devices, the body of nanoscale transistors is lightly doped (near intrinsic) with a low recombination rate. The main consequence is that even the low charge density at the edges of the ion track must be taken into account to calculate the generated charge. The advantage of the ion track being larger than the transistor dimension is that part of the ion-induced charge is not collected by nanoscale transistors. The drawback, however, is that in high-density integrated circuits transistors are very close together, and a single ion track can generate charges in two neighbouring transistors, inducing multiple-bit upset (MBU). This ion track effect was already observed in partially depleted devices. This is shown in Fig. 6.13 which shows the gain measured in partially depleted NMOS transistors processed with three different SOI technologies (0.25μm, 130nm and 70nm). For each technology, the gain is calculated from the ratio of the charge collected in a short gate length transistor compared to that collected in a long gate length transistor.

An example of gain extraction is given in Fig. 6.13a for the 70-nm technology. The right axis displays the statistics of measured transients versus the collected charge for two SOI NMOS transistors of the 70 nm PD SOI technology, with a gate length of 70 nm and 10 μm respectively. The longer-gate length transistor (10 μm) does not exhibit bipolar amplification, while the nominal gate-length transistor (70 μm) is assumed to have the maximum gain that can be observed with the technology. The bipolar gain is then calculated as the ratio of the respective collected charges. The bipolar gain is close to 1 for low values of collected charge, *i.e.* no bipolar amplification is observed when the ion hits in less sensitive area such as the vicinity of the source.[38] The bipolar gain reaches its maximum value (of ~ 2) when the ion hits the device at its most sensitive area, the body region close to the drain junction.

Fig. 6.13b compares the bipolar amplification measured in three SOI PD devices for a medium energy ion, ($^{48}$Ca, 6.2 MeV/a) in full symbols. As expected from the statistic nature of the ion strike position, a distribution of bipolar gain is also observed for all three SOI devices. Indeed, depending on the ion strike location, bipolar amplification and charge collection efficiencies vary greatly. The highest value of the bipolar gain, representative of ion strikes near the drain junction of the device, is shown to significantly decrease with device scaling. This effect of the track radius was predicted [39,40], and recent numerical simulations showed the same trend in SOI technology scaling.[35] However, these results are the first experimental evidence of the effect of technology scaling for a given ion track.

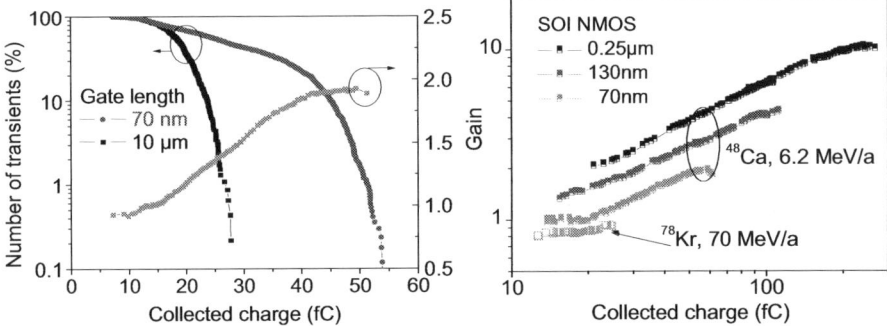

**Fig. 6.13.** (a) Statistics of the measured drain collected charge for 70 nm or 10 μm gate length SOI NMOS transistors, processed with a partially depleted 70 nm technology. The bipolar amplification (gain, right axis) of the 70 nm gate length device is plotted versus its collected charge. (b) Bipolar amplification measured in 0.25-μm, 130-nm and 70-nm gate length partially depleted SOI transistors. The floating-body transistors were irradiated with either medium energy ions (6.25-MeV/a $^{48}$Ca, LET ~ 15 MeV-cm$^2$/mg) in full symbols and high energy ions (70 MeV/a $^{78}$Kr ion LET ~ 10 MeV/(mg.cm$^2$)) in open symbols.[41]

The same experiment was also conducted with a higher energy ion ($^{78}$Kr, 70 MeV/a, instead of $^{48}$Ca, 6.2 MeV/a) with a close value of LET (10 instead of 15 MeV cm$^2$/mg). Data for 70 nm devices are plotted with open squares in Fig. 6.13b. Clearly, no bipolar gain is now observed (maximum gain ~ 1) when this device is struck by an ion of higher energy, and the gain distribution even decreasess below unity. This result can be explained by the fact that the high-energy ion track radius is probably larger than that of an ion with same LET, but much lower energy. For a very short gate length device (such as this 70 nm PDSOI), the channel becomes then actually narrower than the ion track itself. Part of the ion-generated charge is therefore deposited outside of the sensitive area, in highly doped or insulating regions where it cannot be collected. For devices with longer gate length (130 nm, 0.25 μm), the effect is not as clearly visible, which means that the energy of ions should be taken into account only for technology generations below 100-nm.

Beyond geometry considerations resulting from the comparison of the ion track and the transistor dimensions, the electrical response of nanoscale transistors is also modified because of the shape of the ion track. In Ref. [42], device simulations were conducted in 45-nm Double-Gate transistors irradiated with a gaussian-shape ion track in normal incidence striking in the middle of the gate. Different LETs were simulated with the same ion track characteristic radius (Gaussian sigma 14 nm). Fig. 6.14 shows the charge density in the ion track, from the middle of the gate to the drain region. It also shows the electric field with a peak at the drain junction. The electric field peak magnitude is high for low ion LETs (Fig. 6.14a), but collapses completely for high LETs (Fig. 6.14b) when the ion generated charge density at the drain junction reaches the drain doping concentration (approximately $10^{20}$ cm$^{-3}$). Because of the electric field collapse, the impact ionization at the body-drain junction also considerably decreases and does no longer contribute to the parasitic bipolar amplification.

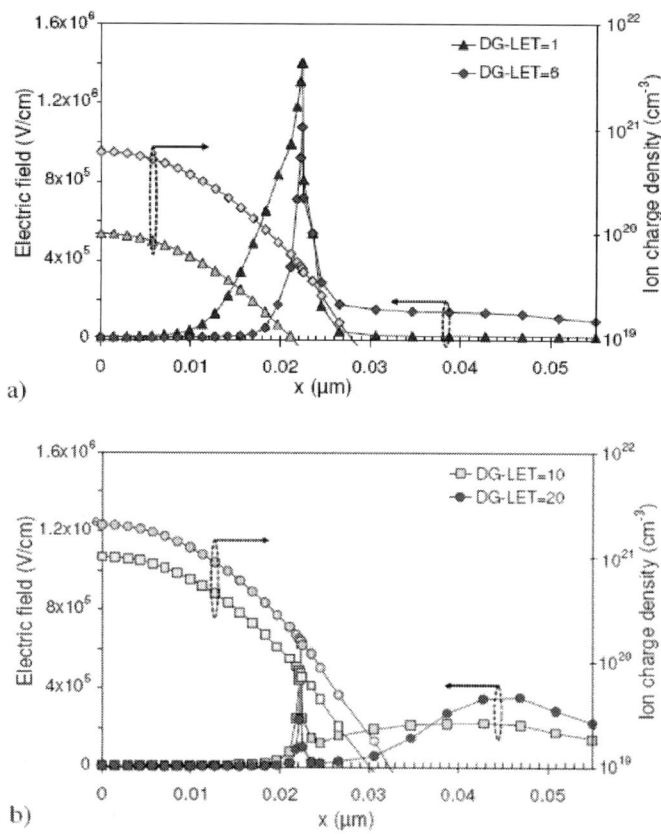

**Fig. 6.14.** Device simulation a Double-Gate 45-nm gate length NMOS transistor irradiated in normal incidence by an ion striking in the middle of the gate (at $x = 0$ μm) at low (a) and high LET (b). The electric field (left axis) and the ion track profile (right axis) are plotted for a cut in the silicon body in the gate to drain direction. The drain junction is located at $x$=22.5 nm.[42]

Two features are required to observe this particular behavior: the track diameter needs to be large enough compared to the transistor gate length in order to create a conduction path between the source and the drain, and the charge generated needs to be large enough to collapse the electric field at the body-to-drain junction. The result of the electric field collapse is that the bipolar gain decreases from 2 at 1 MeV cm²/mg to 1 at 10 MeV cm²/mg. At higher LET, the deposited charge is collected without amplification.

Another effect that must be taken into account in very thin film transistors is fluctuations in the energy deposition.[39, 40] Two ions with the same energy traveling the same path length in a thin silicon film will not create exactly the same number of electron-hole pairs. This is primarily due to collision statistics (*i.e.* energy-loss straggling) of the incident ion with the surrounding silicon. In very thin silicon layers (10 nm and below), there are only few interactions creating energetic electrons (called δ-rays) which generate electron-hole pairs. For a $^{78}$Kr ion at 10MeV/amu (LET 30 MeV cm²/mg) commonly used for device testing, the straggling effect in 10nm thin silicon layers is close to 15%. This uncertainty in the deposited charge increases for lighter ions and thinner silicon films. The main effect for nanoscale devices will be some fluctuations in their single-event response, both during testing and their lifetime in radiation environment.

### 6.3.3 Transient measurements on single-gate and FinFET SOI transistors

The first transient signals measured on 50-nm fully depleted SOI transistors were obtained using a pulsed laser and a high frequency (50-GHz) measurement set-up.[43] The laser setup delivers optical pulses centered at 590-nm (2.1 eV) with a pulse duration of 1-ps and a 12 kHz repetition rate.[44-46] The laser beam is focused with a ×100 microscope objective resulting in a 1.1-μm Gaussian spot on the irradiated device. The transistor is irradiated in the OFF-state at a constant voltage of 1 V through a bias-T, and the transient signal is recorded on a 20-GHz sampling scope. The principle behind sampling scope operation is the following: recorded transients are formed by averaging multiple (typically 4000) single-strike transients measured at the laser repetition rate. Such measurement method is possible using a laser setup which benefits from very stable characteristics (hit location, repetition rate and laser pulse energy). Large-bandwidth measurements can then be achieved, which helps detecting small signals with a high sensitivity.

Fig. 6.15a shows the transient signals obtained on 50-nm single-gate fully depleted transistors, with or without body contacts, at a pulse energy of 80 pJ. The main feature of these signals is that they are particularly short, about 35 ps at half maximum width and less than 70 ps at the signal base (10% of the peak). The 10-90% rise time is 18 ps, which indicates that the signal measurement is limited by the oscilloscope 20-GHz bandwidth. It is thus most likely that the actual transients are shorter than the signals measured on the scope. The total transient current duration of these 50nm fully depleted SOI devices (<70 ps) is much shorter than that of partially depleted 50-nm and 0.25 μm SOI transistors which display a significant transient tail (floating-body effect) resulting in a long signal base duration (280 ps and 170 ps, respectively).[43, 47] Further transient measurements were achieved two years later [48] with an improved system bandwidth, and signal FWHM of 25 ps (instead of 35 ps) were measured on the same single-gate fully depleted technology.

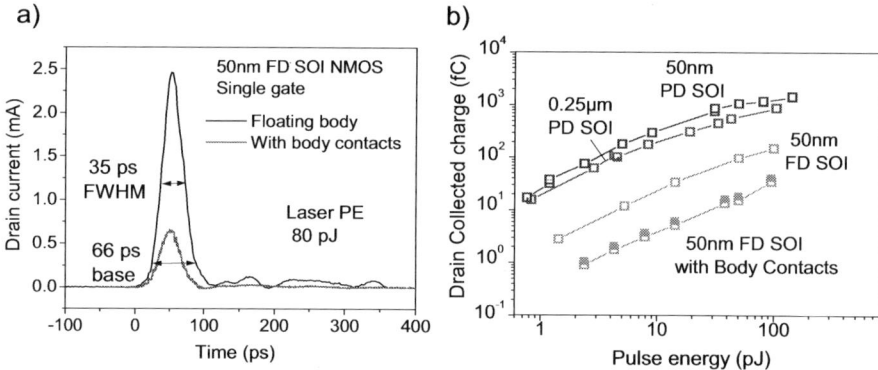

**Fig. 6.15.** (a) Transient drain current obtained on 50-nm fully depleted SOI transistors irradiated by a 1ps-pulse focused laser beam with a laser pulse energy (laser PE) of 80pJ. The tested transistors have either a floating body or body contacts. Transistors are irradiated in the OFF-state, with the drain biased at the nominal supply voltage, 1 V; all other electrodes are grounded. (b) Collected charge measured on three types of SOI NMOS transistors processed using different technologies as a function of laser pulse energy. The transistors have a floating body (open symbols), except one of them labeled "50 nm FD SOI with body contacts" (full symbols).[43]

This short transient duration in single-gate FD SOI devices is particularly interesting for SET hardening. Hardening methods are often based on delay elements with a typical delay time of about 1 ns in bulk devices.[49] With the 50-nm SOI technology, a delay time of 70 ps would be enough to ensure the hardening of the designs.

The second important point is that the floating body transistor delivers a larger signal than the transistor with external body contacts. Contrarily to the signal width and amplitude, the collected charge does not depend on the system bandwidth. It can be measured by integrating the transient current, and it represents an intrinsic characteristic of the transistor response. A parasitic bipolar gain of 6 is calculated from the charge ratio between the two transistors. The presence of body contacts on these fully depleted single-gate transistors is very efficient to reduce the signal amplitude. Even if the silicon film is thin (11 nm), the body doping is very low (nearly intrinsic) and the generated charges easily diffuse over several micrometers, from the laser spot to the body contacts.

Fig. 6.15b shows the collected charge at a function of the laser pulse energy for the 50-nm single-gate fully depleted transistors (with and without body contacts), and compares it to the charge collected in 0.25 µm and 50 nm partially depleted SOI transistors. The charge collected in the fully depleted transistor is much lower than that of partially depleted devices, mainly because of the thinner silicon film and then the smaller body volume. The thin and small body volume in fully depleted devices is the key parameter to fast response and reduced collected charge.

The transient response of fully depleted single-gate transistors was also measured under heavy-ion irradiation. Fig. 6.16a shows three examples of transient currents chosen among the worst-case strike positions. Contrarily to laser-induced transients measured with a sampling scope, the heavy ion measurements were achieved with a high-performance single-shot oscilloscope (12-GHz bandwidth). The reason is that, depending on the ion strike location, the transient amplitude and shape can be very different. Furthermore, a high number of ion strikes would be necessary to get an averaged transient with a sampling scope, which is difficult to get in a reasonable irradiation time for such small dimension transistors. The measurement bandwidth is then lower than for laser measurements, but ion irradiation is necessary to quantify the transistor response in a space environment.

The measured transistors have a 100-nm gate length, and are processed with the same 50-nm single-gate fully depleted technology as those used for laser experiments. They were irradiated with 10MeV/amu $^{78}$Kr beam (LET 30 MeV cm$^2$/mg). The transistors had to be tilted at 60° to increase the deposited charge (by a factor of 2) and get a measurable signal. At 0°, no signal could be measured. Transistors with body contacts (and no bipolar amplification) could not be measured either, even at 60°, because the signal amplitude was too small to be detected by the single-shot scope. The transients in Fig. 6.16a have a very short duration (55-ps FWHM, 100-ps base width) and do not exhibit any tail, meaning that no floating body effect is measurable at this time scale, which is consistent with laser irradiation. A negative rebound is even observed following the positive current transient because of the limited bandwidth and a possible impedance mismatch in the transistor packaging.

The charge generated in the silicon film is calculated from the heavy ion LET (30 MeVcm$^2$/mg) and the silicon film thickness ($t_{Si}$ = 11 nm). At an incidence of $\theta = 60°$, the ion path length in the silicon film is $cos\theta \times t_{si}$ = 22 nm and the generated charge is 6.6 fC. The ion straggling effect (deposited charge uncertainty) is approximately 10-15% [8, 39], which is still acceptable in our experimental conditions. In the same way, the ion track density profile induces some variations of the deposited energy which are estimated at 25% of the *average* deposited charge. This will have important incidence in the radiation response of complex circuits. However, for the *worst-case* response of our irradiated transistor, the ion hits well in the middle of the transistor and the deposited charge is close to its maximum value (6.6-fC) as calculated before from the ion LET.

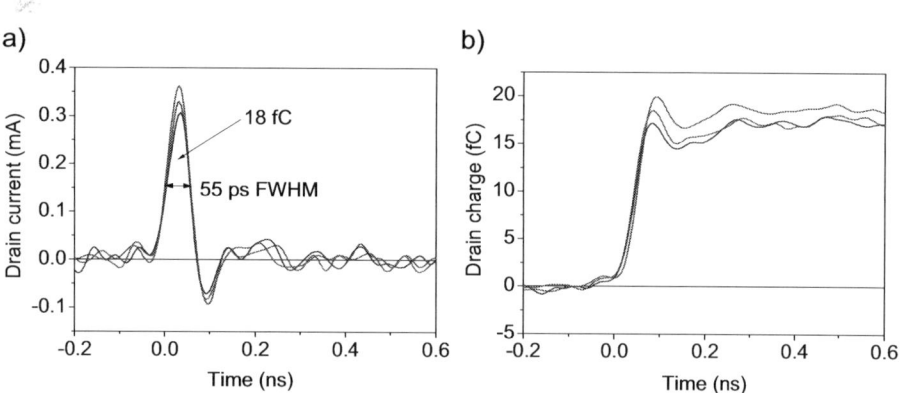

**Fig. 6.16.** Transient current (a) and charge (b) measured for the worst-case strikes in the 10 nm gate length floating-body fully depleted SOI transistor. The transistor was irradiated with a 10MeV/amu $^{78}$Kr beam (LET 30 MeV cm$^2$/mg) with an angle of incidence of 60°.[50]

As for laser transients, the collected charge is an intrinsic data of the transistor response. Fig. 6.16b shows the charge versus time obtained by integration of the transient current. The parasitic bipolar gain can be calculated by the ratio of the collected charge (18 fC) on the maximum deposited charge (6.6 fC). Once again, we consider only the *worst-case* deposited and collected charges. The gain is found to be 2.7 for these 100-nm gate length floating-body FD transistors. This gain value is consistent with the gain of 6.6 found with pulsed laser irradiation performed on the shorter gate length (50-nm) transistors fabricated in the same technology.[43] The bipolar gain has an inverse gate length (*1/L*) behavior [36,51] which brings about a reduction of the bipolar gain as gate length decreases.

FinFET transistors irradiated with the same heavy ion beam ($^{78}$Kr, 10MeV/amu, LET = 30 MeV cm$^2$/mg), delivers a larger transient signal than single-gate transistors (Fig. 6.17a). FinFETs were irradiated in normal incidence, and the signal could be measured without tilting the device as was necessary for single-gate transistors. The signal amplitude and the transient width are larger (FWHM 70ps) than that for single-gate transistors. A tail can be detected on the drain transient, especially for the shorter-gate length transistor (60 nm). This current tail has an obvious effect on the collected charge. Contrary to the single-gate transistor which shows charge saturation at 0.2 ns, the charge continues to increase in the FinFETs up to 1 ns for the shortest-gate device. The effect is less obvious, but still present, in the longest gate transistor (100 nm).

The transient tail is due to a more pronounced floating-body effect in the tested FinFET than in the single-gate fully depleted transistors. The collected charge at 0.2 ns due to the prompt component is higher (about twice if we compare the 100 nm gate length transistors), and the charge still increases up to 0.6-1ns. The silicon volume in the tested FinFET is thicker and larger (silicon film thickness 25 nm, fin width 50 nm) than for the single-gate transistor (silicon film thickness 11 nm). A higher amount of charge is then deposited in the FinFET body, and this deposited charge is slower to evacuate before the $\Omega$-shape gate recovers the potential control in the silicon body. Moreover, for the tested FinFETs, the active transistor (gate length, silicon film) has been scaled to the dimensions of a 50 nm generation technology, except for the low-doped source and drain regions (150 nm) because of layout and process constraints at this research stage. The large resistive access region to the active transistor induces a "de-biasing" effect under irradiation when a large current flows between the source and drain contacts. On the other hand, the single-gate transistors benefit from a fully-scaled, high-performance technology with narrow low-doped source and drain regions of only 30 nm.

The observed differences on these two technologies show that the key parameter for a hardened architecture is first to reduce the silicon volume. Despite the single-gate architecture, which does not ensure the same body control as for a multi-gate technology, the smaller body volume induces a reduced transient response, *i.e.* lower collected charge and very short transient duration. A second point that has to be mentioned is that during its evolution from research to production, technology architectures are optimized and their response under irradiation can change significantly.

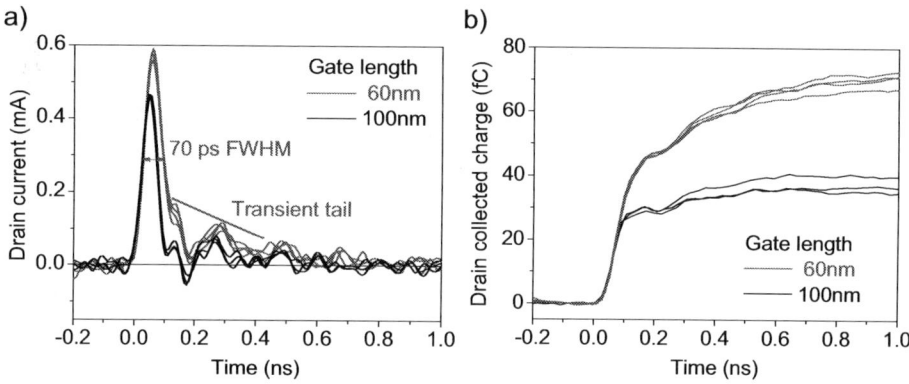

**Fig. 6.17.** Transient current (a) and charge (b) measured for the worst-case strikes in the 60 nm and 100 nm gate length FinFET transistors. The transistors were irradiated with a 10MeV/amu $^{78}$Kr beam (LET 30 MeV cm$^2$/mg) at normal incidence (0°).

In order to study the influence of the gate architecture on the transient response, it is important to compare devices with the same body dimensions. This was done by simulation in Fig. 6.18 for 32 nm gate length NMOS transistors with double-date, triple-gate, Ω-Gate, and surrounding-gate architectures.[52] The transistors have the same rectangular intrinsic body region (32 nm×10 nm×10 nm) and abrupt source and drain junctions. Fig. 6.18 shows that, when the equivalent number of gates increases (from double-gate to surrounding-gate), the amplitude of the transient response is reduced. This behavior is consistent with a better body control for a higher number of gates as already observed with static sub-threshold and OFF-state current. These simulations also confirm that the higher collected charge measured on irradiated FinFET transistors is only due to the larger silicon volume, and not to the gate architecture.

The same simulations were performed by taking into account the quantum confinement of carriers using the Density-Gradient model.[53] The quantum effect in very thin body volume also contributes to decrease the transient current, again consistently with static characteristics. In the performed simulations, the transient response is very short, on the order of 10 ps, with a limited transient tail, which suggests that the body potential is efficiently controlled in these very thin SOI devices.

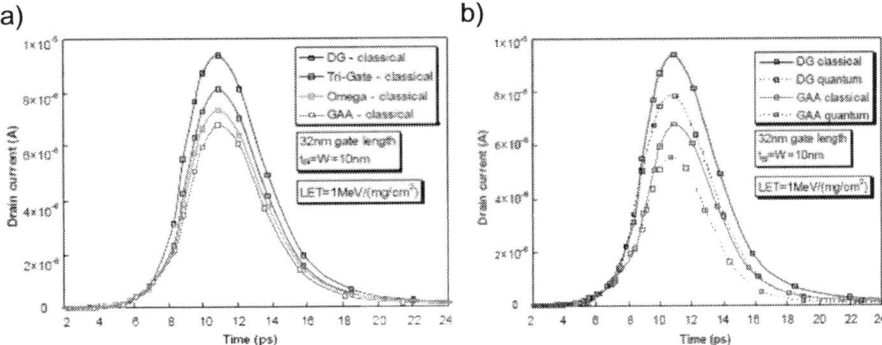

**Fig. 6.18.** Drain current transients induced by an ion strike at normal incidence in the middle of the silicon film. The ion track generation has a time gaussian shape centered at 10ps (sigma 2ps) and a spatial track radius of 14 nm. The current is calculated in both classical and quantum cases. a) Classical simulations; b) classical and quantum simulations.[52]

### 6.3.4 Scaling effects

In order to evaluate the sensitivity of future 50nm SOI devices, mixed-mode simulations were performed for both an SRAM cell and on inverter chains.[54] The simulated devices are the 50 nm single-gate fully depleted MOSFETs. These devices are representative of the technologies that will be in production within a short future, and device simulations could be validated by transient measurements on single transistors. The transistors are of the floating-body type (without body contacts). The struck transistor is described in 3D and the others are simulated in 2D (with a width scaling factor) to get an accurate response of the device under irradiation.

Fig. 6.19a shows the threshold LET for SEU and SET effects in the 50 nm SOI technology. Our results are compared to previous simulations showing the impact of generation scaling on bulk and partially depleted SOI with body contacts.[35] The first interesting point is that the threshold LET for SEU and SET have similar values for the 50 nm generation, meaning that logic inverter chains are becoming potentially as sensitive to single-event effects as SRAM cells. Secondly, the very thin-film, 50 nm technology compares favourably to partially depleted SOI. The 50 nm threshold LET for both SEU and SET is equal to 2 MeVcm$^2$/mg and does not experience a drastic decrease compared to the 0.1 μm threshold LET. Even though our 50 nm transistors have a floating body (with parasitic bipolar amplification), the very thin film is about ten times thinner than for partially depleted, which is an obvious advantage for radiation hardness.[55] The 50 nm critical charge is low, but the generated charge in the thin sensitive volume is also very small.

In Fig. 6.19b, the minimum width of voltage transients propagating in inverter chains is plotted as a function of technology generation. This critical transient width is an important parameter for SET hardening techniques based on temporal filtering. Shorter transients can be more easily filtered than large transients. The minimum transient width has been found at 15 ps by mixed-mode simulation in the 50 nm SOI technology. The simulations confirm the measurements on single 50-nm generation transistors, which produce the shortest transients ever observed. The very thin-film SOI architecture is then an interesting architecture for single-event effects reduction.

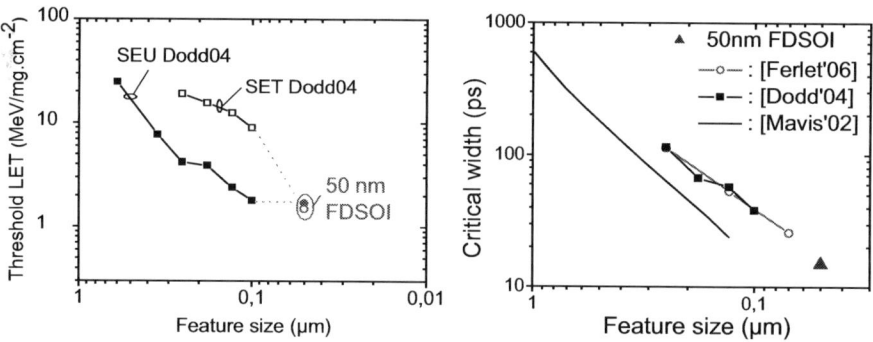

**Fig. 6.19.** Left: Simulated SET threshold LET for unattenuated transient propagation and SEU threshold LET as a function of scaling for SOI CMOS technologies.[35] Right: Critical transient width for unattenuated propagation as a function of technology scaling. Our results (red points) are superimposed on circuit simulations from [56] and mixed-mode simulations from [57].

## References

1. J.L. Leray, E. Dupont-Nivet, J.F. Péré, Y.M. Coïc, M. Raffaelli: CMOS/SOI Hardening at 100 Mrad($SiO_2$). IEEE Transactions on Nuclear Science **37-6**, 2013 (1990)
2. T.C. May and M.H. Woods: Alpha-particle-induced soft errors in dynamic memories. IEEE Transactions on Electron Devices **26**, 2 (1979)
3. A. Taber, E. Normand: Single Event Upset in Avionics. IEEE Transactions on Nuclear Science **40-2**, 120 (1993)
4. E. Normand: Single-Event Effects in Avionics. IEEE Transactions on Nuclear Science **43-2**, 461 (1996)
5. J.F. Ziegler, H.W. Curtis, H.P. Muhlfeld, C. J. Montrose, B. Chin, M. Nicewicz, C.A. Russell, W.Y. Yang, L.B. Freeman, P. Hosier, L.E. LaFave, J.L. Walsh, J.M. Orro, G.J. Unger, J.M. Ross, T.J. O'Gorman, B. Messina, T.D. Sullivan, A.J. Sykes, H. Yourke, T.A. Enger, V. Tolat, T.S. Scott, A. H. Taber, R.J. Sussman, W.A. Klein, and C.W. Wahaus: IBM experiments in soft fails in computer electronics (1978–1994). IBM J. Res. Develop. **40-1**, 3 (1996)
6. R. Baumann: The impact of technology scaling on soft error rate performance and limits to the efficacy of error correction. T*echnical Digest of IEDM*, 329 (2002)
7. P.E. Dodd, M.R. Shaneyfelt, J.R. Schwank, G. L. Hash: Neutron-induced soft errors, latchup, and comparison of SER test methods for SRAM technologies. *Technical Digest of IEDM*, 333 (2002)
8. P.E. Dodd, L.W. Massengill: Basic Mechanisms and Modeling of Single-Event Upset in Digital Microelectronics. IEEE Transactions on Nuclear Science **50-3**, 583 (2003)
9. P. Roche, and G. Gasiot: Impacts of Front-End and Middle-End Process Modifications on Terrestrial Soft Error Rate. IEEE Transactions on Device and Materials Reliability **5-3**, 382 (2005)
10. J.R. Schwank, V. Ferlet-Cavrois, M.R. Shaneyfelt, P. Paillet, and P.E. Dodd: Radiation Effects in SOI Technologies. IEEE Transactions on Nuclear Science **50-3**, 522 (2003)
11. W.C. Jenkins, S.T. Liu: Radiation Response of Fully-depleted MOS Transistors Fabricated in SIMOX. IEEE Transactions on Nuclear Science **41-6**, 2317 (1994)
12. V. Ferlet-Cavrois, O. Musseau, J. L. Leray, J. L. Pelloie, C. Raynaud: Total Dose Effects on a Fully-Depleted SOI NMOSFET and Its Lateral Parasitic Transistor. IEEE Transactions on Electron Devices **44-6**, 965 (1997)
13. V. Ferlet-Cavrois, G. Gasiot, C. Marcandella, C. D'hose, O. Flament, O. Faynot, J. du Port de Pontcharra, and C. Raynaud: Insights on the transient response of full and partially depleted SOI technologies under heavy ion and dose rate irradiations. IEEE Transactions on Nuclear Science **49-6**, 2948 (2002)

14  J.R. Schwank, M.R. Shaneyfelt, P.E. Dodd, J.A. Burns, C.L. Keast, and P.W. Wyatt: New insights into fully-depleted SOI transistor response after total-dose irradiation. IEEE Transactions on Nuclear Science **47-6**, 604 (2000)
15  O. Flament, A. Torres, V. Ferlet-Cavrois: Bias dependence of FD transistor response to total dose irradiation. IEEE Transactions on Nuclear Science **50-6**, 2316 (2003)
16  F.T. Brady, H.L. Hughes, P.J. McMarr, and B. Mrstik: Total-dose hardening of SIMOX buried oxides for fully-depleted devices in rad tolerant applications. IEEE Transactions on Nuclear Science **43-6**, 2646 (1996)
17  F.T. Brady, R. Brown, L. Rockett, and J. Vasquez: Development of a radiation tolerant 1MSRAM on fully-depleted SOI. IEEE Transactions on Nuclear Science **45-6**, 2436 (1998)
18  V. Ferlet-Cavrois, S. Quoizola, O. Musseau, O. Flament, and J.L. Leray: Total dose latch in short channel NMOS/SOI transistors. IEEE Transactions on Nuclear Science **45-6**, 2458 (1998)
19  V. Ferlet-Cavrois, G. Gasiot, C. Marcandella, C. D'hose, O. Flament, O. Faynot, J. d. P. d. Pontcharra, and C. Raynaud: Insights on the transient response of full and partially depleted SOI technologies under heavy ion and dose rate irradiations. IEEE Transactions on Nuclear Science **49-6**, 2948 (2002)
20  P. Paillet, M. Gaillardin, V. Ferlet-Cavrois, O. Faynot, C. Jahan, L. Tosti and S. Cristoloveanu: Total Ionizing Dose Effects on Deca-Nanometer Fully Depleted SOI Devices. IEEE Transactions on Nuclear Science **52-6**, 2345 (2005)
21  P. Paillet, J.R. Schwank, M.R. Shaneyfelt, V. Ferlet-Cavrois, R.A. Loemker, O. Flament and E.W. Blackmore: Comparison of charge yield in MOS devices for different radiation sources. IEEE Transactions on Nuclear Science **49-6**, 2656 (2002)
22  V. Ferlet-Cavrois, T. Colladant, P. Paillet, J.L. Leray, O. Musseau, J. R. Schwank, M. R. Shaneyfelt, J. L. Pelloie, and J. du Port de Poncharra: Worst-case bias during total dose irradiation of SOI transistors. IEEE Transactions on Nuclear Science **47-6**, 2183 (2000)
23  J.P. Colinge and A. Terao: Effects of Total-Dose Irradiation on Gate-All-Around (GAA) Devices. IEEE Transactions on Nuclear Science **40-2**, 78 (1993)
24  P. Francis, C. Michel, D. Flandre, J.P. Colinge: Radiation-Hard Design for SOI MOS Inverters. IEEE Transactions on Nuclear Science **41-2**, 402 (1994)
25  M. Gaillardin, P. Paillet, V. Ferlet-Cavrois, O. Faynot, C. Jahan, and S. Cristoloveanu: Total Ionizing Dose Effects on Triple-Gate FETs. IEEE Transactions on Nuclear Science **53-6**, 3158 (2006)
26  X. Wu, P.-C.H. Chan, A. Orozco, A. Vazquez, A. Chaudhry, J. P. Colinge: Dose radiation effects in FinFETs. Solid State Electronics. **50**, 287 (2006)
27  J.P. Colinge, A. Orozco, J. Rudee, W. Xiong, C. Rinn Cleavelin, T. Schulz, K. Schrüfer, G. Knoblinger, P. Patruno: Radiation dose effects in trigate SOI MOS transistors. IEEE Transactions on Nuclear Science **53-6**, 3237 (2006)

28  B. Jun, H.D. Xiong, A. L. Sternberg, C.R. Cirba, D.C. Chen, R.D. Schrimpf, D.M. Fleetwood, J.R. Schwank and S. Cristoloveanu: Total-Dose Effects on Double Gate Fully Depleted SOI MOSFETs. IEEE Transactions on Nuclear Science **51-6**, 3767 (2004)

29  C. Jahan, O. Faynot, M. Casse, R. Ritzenthaler, L. Brevard, L. Tosti, X. Garros, C. Vizioz, F. Allain, A.-M. Papon, H. Dansas, F. Martin, M. Vinet, B. Guillaumot, A. Toffoli, B. Giffard, and S. Deleonibus: ΩFETs transistors with TiN metal gate and $HfO_2$ down to 10 nm. *Proceedings of IEEE Symposium on VLSI Technology*, 112 (2005)

30  W. Xiong, G. Gebara, J. Zaman, M. Gostkowski, B. Nguyen, G. Smith, D. Lewis, C.R. Cleavelin, R. Wise, S. Yu, M. Pas, T.J. King, and J.P. Colinge: Improvement of FinFET electrical characteristics by hydrogen annealing. IEEE Transactions on Electron Devices **25-8**, 541 (2004)

31  P. Francis, J.P. Colinge, G. Berger: Temporal analysis of SEU in SOI/GAA SRAMs. IEEE Transactions on Nuclear Science **42-6**, 2127 (1995)

32  S. Buchner, M. Baze, D. Brown, D. McMorrow, and J. Melinger: Comparison of error rates in combinational and sequential logic. IEEE Transactions on Nuclear Science **44-6**, 2209 (1997)

33  J.R. Schwank, P.E. Dodd: Radiation Effects in SOI Microelectronics. *IEEE SOI Conference Short Course* (2001)

34  R. Baumann: Soft errors in advanced computer systems. IEEE Design & Test of Computers **22-3**, 258 (2005)

35  P.E. Dodd, M.R. Shaneyfelt, J.A. Felix, and J.R. Schwank: Production and Propagation of Single-Event Transients in High-Speed Digital Logic ICs. IEEE Transactions on Nuclear Science **51-6**, 3278 (2004)

36  V. Ferlet-Cavrois, G. Gasiot, C. Marcandella, C. D'Hose, O. Flament, O. Faynot, J. Du Port de Pontcharra and C. Raynaud: Insights on the transient response of fully and partially depleted SOI technologies under heavy-ion and dose-rate irradiations. IEEE Transactions on Nuclear Science **49-6**, 2948 (2002)

37  O. Musseau, V. Ferlet-Cavrois, A.B. Campbell, A.R. Knudson, S. Buchner, B. Fischer and M. Schloegl: Technique to measure an ion track profile. IEEE Transactions on Nuclear Science **45-6**, 2563 (1998)

38  D.E. Fulkerson, E.E. Vogt: Prediction of SOI single-event effects using a simple physics-based SPICE model. IEEE Transactions on Nuclear Science **52-6**, 2168 (2005)

39  M.A. Xapsos: Applicability of LET to single events in microelectronic structures. IEEE Transactions on Nuclear Science **39-6**, 1613 (1992)

40  A. Akkerman and J. Barak: Ion-track structure and its effects in small size volumes of silicon. IEEE Transactions on Nuclear Science **49-6**, 3022 (2002)

41  V. Ferlet-Cavrois, P. Paillet, M. Gaillardin, D. Lambert, J. Baggio, J.R. Schwank, G. Vizkelethy, M.R. Shaneyfelt, K. Hirose, E.W. Blackmore, O. Faynot, C. Jahan and L. Tosti: Statistical Analysis of the Charge Collected in SOI and Bulk Devices Under Heavy Ion and Proton Irradiation: Implications for Digital SETs. IEEE Transactions on Nuclear Science **53-6**, 3242 (2006)

42  K. Castellani-Coulie, D. Munteanu, J. L. Autran, V. Ferlet-Cavrois, P. Paillet and J. Baggio: Analysis of 45-nm Multi-Gate Transistors Behavior under Heavy Ion Irradiation by 3-D Device Simulation. IEEE Transactions on Nuclear Science **53-6**, 3265 (2006)

43  V. Ferlet-Cavrois, P. Paillet, D. McMorrow, A. Torres, M. Gaillardin, J. S. Melinger, A. R. Knudson, A. B. Campbell, J. R. Schwank, G. Vizkelethy, M. R. Shaneyfelt, K. Hirose, O. Faynot, C. Jahan and L. Tosti: Direct measurement of transient pulses induced by laser and heavy ion irradiation in deca-nanometer devices. IEEE Transactions on Nuclear Science **52-6**, 2104 (2005)

44  J. S. Melinger, D. McMorrow, A. B. Campbell, S. Buchner, L.H. Tran, A.R. Knudson, W.R. Curtice: Pulsed laser induced single event upset and charge collection measurements as a function of optical penetration depth. J. Appl. Phys. **84-2**, 690 (1998)

45  J.S. Melinger, S. Buchner, D. McMorrow, W.J. Stapor, T.R. Weatherford, A.B. Campbell, H. Eisen: Critical evaluation of pulsed laser method for single-event effects testing and fundamental studies. IEEE Transactions on Nuclear Science **41-6**, 2574 (1994)

46  D. McMorrow, J. S. Melinger, S. Buchner, T. Scott, R.D. Brown, N.F. Haddad: Application of a pulsed laser for evaluation and optimization of SEU-Hard designs. IEEE Transactions on Nuclear Science **47-3**, 559 (2000)

47  O. Musseau, V. Ferlet-Cavrois, J. L. Pelloie, S. Buchner, D. McMorrow, A. B. Campbell: Laser probing of bipolar amplification in 0.25-µm MOS/SOI transistors. IEEE Transactions on Nuclear Science **47-6**, 2196 (2000)

48  M. Gaillardin, P. Paillet, V. Ferlet-Cavrois, *et al.*: Transient Radiation Response of Single- and Multiple-Gate FD SOI Transistors. *Nuclear and Space Radiation Effects Conference* (2007)

49  M. Baze, in *Proc. IEEE NSREC short course*, Monterey CA (2003)

50  D. Munteanu, V. Ferlet-Cavrois, J.L. Autran, P. Paillet, J. Baggio, O. Faynot, C. Jahan and L. Tosti: Investigation of Quantum Effects in Ultra-Thin Body Single- and Double-Gate Devices Submitted to Heavy Ion Irradiation. IEEE Transactions on Nuclear Science **53-6**, 3363 (2006)

51  O. Musseau, J.L. Leray, V. Ferlet-Cavrois, Y.M. Coic, B. Giffard: SEU in SOI SRAMs – A Static Model. IEEE Transactions on Nuclear Science **41-3**, 607 (1994)

52  D. Munteanu, J.L. Autran, K. Castellani-Coulié, V. Ferlet-Cavrois, P. Paillet, J. Baggio: 3D Quantum Numerical Simulation of Single-Event Transients in Multiple-Gate Nanowire MOSFETs. IEEE Trans. Nucl. Sci. (2007)

53  A. Wettstein, A. Schenk, W. Fichtner: Quantum device-simulation with Density-Gradient model. IEEE IEEE Transactions on Electron Devices **48**, 279 (2001)

54  V. Ferlet-Cavrois, P. Paillet, M. Gaillardin *et al.*: New Measurement Techniques of Voltage Transients in Inverters and Chains of Inverters under Pulsed Laser Irradiation. *Nuclear and Space Radiation Effects Conference* (2007)

55  P. Oldiges, K. Bernstein, D. Heidel, B. Klaasen, E. Cannon, R. Dennard, H. Tang, M. Ieong, and H-S.P. Wong: Soft-Error Rate Scaling for Emerging SO1 Technology Options. *Proceedings of IEEE Symposium on VLSI Technology*, 46 (2002)

56  D. G. Mavis, P. H. Eaton: Soft error rate mitigation techniques for modern microcircuits. *Proc. Int. Reliability Physics Symp.*, 216 (2002)

57  P. E. Dodd, M. R. Shaneyfelt, J. A. Felix, J. R. Schwank: Production and propagation of single-event transients in high-speed digital logic ICs. IEEE Transactions on Nuclear Science **51-6**, 3278 (2004)

# 7 Multi-Gate MOSFET Circuit Design

Gerhard Knoblinger, Michael Fulde and Christian Pacha

## 7.1 Introduction

For a circuit designer the Multi-Gate MOSFET (MuGFET) is a disruptive device architecture because the third dimension is explicitly exploited to reduce short-channel effects and to limit the increase of leakage currents in CMOS technologies beyond the 45nm node. Moreover, new gate-stack materials are introduced to fabricate enhancement-type nFET and pFET devices, which is a prerequisite for digital and analog CMOS circuit design. In this chapter we will therefore analyze the current achievements in MuGFET device technology in regard to various, circuit-relevant figures of merit, specific layout and design aspects, as well as the requirements for a future CMOS technology platform to provide System-on-a-Chip (SoC) integration capability.

The basic idea behind this chapter is to describe the interrelationship between the MuGFET device properties and elementary digital and analog circuits, such as CMOS logic gates, SRAM cells, reference circuits, operational amplifiers, and mixed-signal building blocks. This approach is motivated by the observation that a cost-efficient, heterogeneous SoC integration is a key factor in modern IC design. Typical examples are GSM/EDGE baseband processors for cellular phones [1], low-power multimedia processors [2] and ultra-low-power IC's for wireless sensor networks and ambient intelligent applications.[3] From a technical point of view, common feature of these SoC applications is that they are all operated in an active and leakage power-limited environment. The prospect that MuGFET devices offer reduced leakage currents and improved low-voltage performance compared to planar bulk devices on the one hand and the challenges caused when leaving the evolutionary scaling path of planar CMOS on the other hand motivates an early circuit investigation in close cooperation with technology development.

## 7.2 Digital Circuit Design

Along with the introduction of sub-100nm technology nodes a significant change of digital circuit design paradigms can be observed. The rapidly increasing subthreshold and gate leakage currents result in nearly constant threshold voltages on the order of $V_{TSAT}$=300-400mV and a slow-down of gate oxide thickness reduction for logic core devices in low-power technologies. Performance gains of 30%-50% due to pure technology shrinking as achieved in technology nodes above the 130nm node are no longer feasible. This has initiated the recent introduction of various strain techniques for mobility increase into planar CMOS technologies.[4-5] For the same reason, high-k/metal gate stacks are investigated.[6] However, the reduction of supply voltages, when introducing a new low power CMOS technology generation beyond the 90nm generation, seems to be limited to about 100mV compared to the previous node to facilitate at least a moderate increase of dynamic performance of 10-15%.

Dynamic voltage scaling, a well-established circuit technique to adapt active power and performance to temporal varying system performance requirements, becomes less efficient in terms of energy reduction since the operating windows of supply voltages is reduced to $V_{DD}$=1.0 and 1.4V.[1-2] Dimension scaling of about ×0.7 per minimum feature size and increase of transistor density close to ×2 per generation is, therefore, the only remaining parameter that continues the ideal scaling trend of previous years. In regard to MuGFET technology, this puts an intense pressure on lithography and etching to print circuit-relevant multi-fin device configurations with an area efficiency comparable or even better than planar bulk CMOS. In addition to the three-dimensional device architecture a further challenge results from novel gate stack materials. In this section we will, therefore, discuss the basic device figures of merit and design issues for digital circuits and SRAMs. Experimental results of prototype circuits fabricated with relaxed 130nm and 65nm dimensions are presented and analyzed in regard to large scaling integrated circuit performance, low-power design and SoC compatibility.

### 7.2.1 Impact of device performance on digital circuit design

For analog and digital circuit design the fabrication of nFET and pFET enhancement type multi-fin MuGFET devices (Fig. 7.1) is a key requirement to be provided by the technology. Enhancement-type devices, *i.e.* nFET transistors with positive threshold voltages and pFET transistors with negative threshold voltages, are required to build a CMOS logic family with input and output voltage compatibility. Thus, the output signals of a logic gate, usually $V_{DD}$ and $V_{SS}$, can be directly used to switch on and off the transistors of the subsequent circuit stage without introducing a level shifter.

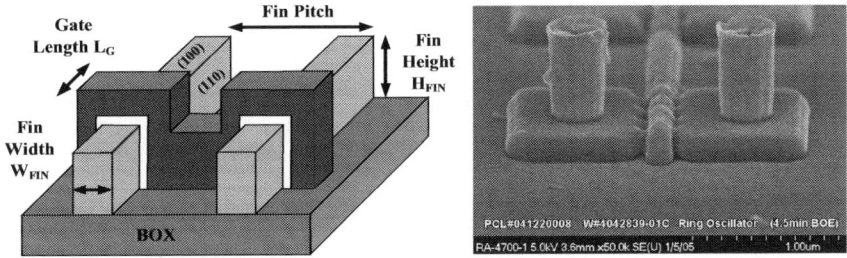

**Fig. 7.1.** Cross section of a multi-fin MuGFET and SEM picture of a 4-fin nFET. Copyright© 2006 IEEE.

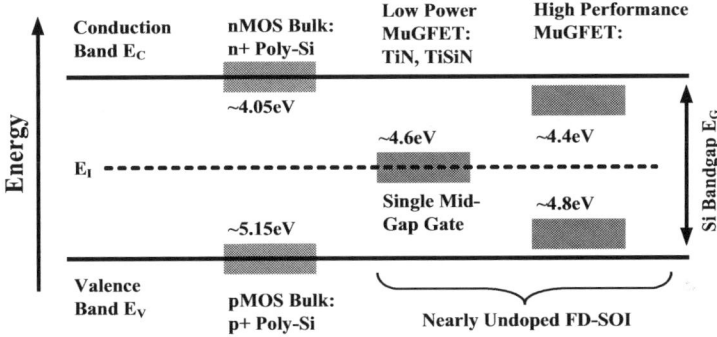

**Fig. 7.2.** Overview of gate workfunction requirements for planar bulk CMOS, low power MuGFET, and high performance MuGFET technologies. Copyright© 2006 IEEE.

In planar CMOS technology, nFET and pFET devices are implemented using n-type and p-type doped poly-Si. Here, the dual $n^+/p^+$ gate work functions align with the conduction and valence band edges so that threshold voltages of planar nFET and pFET devices are defined by channel and halo doping. However, for MuGFETs the situation is different, since channel doping for $V_T$ adjustment is not a viable solution for very thin fins, and therefore, metal gates with appropriate work functions are the first choice. Considering typical industrial requirements to provide low power CMOS compatible devices with threshold voltages around 300-400mV, gate electrode materials with midgap metal gate workfunction are required. Typical candidates are TiN or TiSiN (Fig. 7.2). In contrast to this, high performance CMOS technologies typically contain

CMOS devices with lower threshold voltages of 200mV, which can be in principle provided by appropriate gate workfunction tuning.

**Table 7.1.** Projection of MuGFET device dimensions for Low-Standby-Power technologies.

|  | 65nm | 45nm | 32nm | 22nm | 16nm |
|---|---|---|---|---|---|
| $L_G$ [nm] | 55-50 | 45-38 | 35-30 | 25-22 | 18-16 |
| $W_{FIN}=L_G/1.5$ [nm] | 36-33 | 30-25 | 23-20 | 16-14 | 12-10 |
| $H_{FIN}=2\ W_{FIN}$ [nm] | 72-66 | 60-50 | 46-40 | 32-28 | 24-20 |
| $W_{EFF}$ [nm] TriGate | 180-165 | 150-125 | 115-100 | 80-70 | 60-50 |
| $W_{EFF}$ [nm] FinFET | 144-132 | 120-100 | 92-80 | 64-56 | 48-40 |

From the viewpoint of practical circuit design, the opportunity to adapt the driving strength of a transistor in a special circuit topology is of key relevance. While this is simply done by varying the transistor width $W$ in planar bulk or SOI technology, MuGFET technology only allows a modification of multi-fin device structures in discrete steps by adding or removing one or several fins. Fig. 7.1 shows such a multi-fin device structure with the fin pitch $P_{FIN}$ as essential figure of merit for the integration density. In regard to electrostatic integrity of the device thin fins following the rule of thumb of $L_G(min) \geq 1.5\ W_{FIN}$ are mandatory. Considering that fin aspect ratios of $H_{FIN}/W_{FIN}=2$ are manufacturable, this results in small fin pitch requirements of about $P_{FIN}=100nm<W_{EFF}$ for a 32nm Low Standby Power CMOS technology in order to achieve equal or better transistor density than planar CMOS technology (Table 7.1). Here, $W_{EFF}$ is the effective transistor width of a TriGate or MuGFET device, respectively, while the fin pitch $P_{FIN}$ is equivalent to the silicon ground area consumption. At the device level, a 120nm fin pitch has been demonstrated.[14]

Considerations on the fin sidewall orientations as well as the impact of the gate oxide thickness are described in more detail in Section 1.5. The device dimensions in Table 7.1 also show that the minimum effective transistor width to gate length ratio is on the order of $W_{EFF}/L_G=3.2$ for TriGate devices and $W_{EFF}/L_G=2.7$ for MuGFETs. While this lower limit of transistor width discretization is no major issue for analog and digital design, SRAM design needs a certain attention to achieve sufficient read and write stability.

**Table 7.2.** Basic MuGFET device parameters. Copyright© 2007 IEEE.

| Gate material/dielectric | 5nm TiN midgap metal on HfSiON |
|---|---|
| Gate lengths | $L_G$ = 55nm and 75 nm (complex circuits) |
| Fin widths, height | $W_{FIN}$ = 30 nm, $H_{FIN}$ = 60 nm |
| Transistor width per fin | $W_{EFF} = 2 \cdot H_{FIN} + W_{FIN}$ = 150 nm |
| Threshold voltages | $V_{TSATN}=V_{TSATP}$=330mV |
| On currents (n/pFET) | $I_{ON}$=404/317 µA/µm at $V_{DD}$=1V |
| Off currents (n/pFET) | $I_{OFF}$=25/36 pA/µm at $V_{DD}$=1V |
| Drain-induced Barrier lowering | DIBL=15mV/V |
| Subthreshold slope | S=63mV/decade |

Most of the circuits in this chapter are fabricated in a low-power compatible MuGFET technology with focus on small leakage currents. Main device parameters and the corresponding transistor $I_D$-$V_{GS}$ input characteristics are shown in Tables 7.2 and Fig. 7.3.[7] Both, nFET and pFET devices have symmetric threshold voltages, ideal subthreshold slope of $S$=63mV/dec and an excellent low drain-induced barrier lowering (DIBL) of 15mV/V. Due to the (110) sidewall orientation, high hole mobilities are achieved resulting in large pFET on currents of about 78% of nFET currents. This device property is particularly beneficial since CMOS logic becomes more symmetric in terms of nFET/pFET drive currents.

In regard to circuit performance, it is interesting to analyze the device switching behavior. This is illustrated for the relevant operating regions of digital circuits in the $I_D$-$V_{DS}$ characteristics of an n-type MuGFET and a planar nFET in Fig. 7.4 and in Table 7.3. For same off-currents and gate length, the better short-channel effect control, *i.e.* a smaller DIBL and steeper subthreshold slope, of MuGFETs causes larger switching currents in the four operating regions and hence a better circuit speed. While DIBL and subthreshold slope are usually assessed in regard to leakage current, these results show their relevance for performance, especially in transistor stack configurations (regions III and IV). Note that these intrinsic advantages of MuGFETs become not visible if a simple $CV/I_{ON}$ delay metrics is used for device performance evaluation.

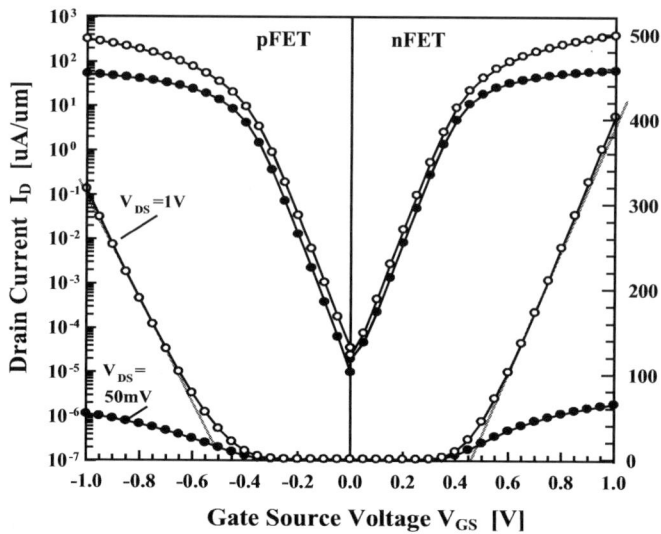

**Fig. 7.3.** $I_D$-$V_{GS}$ characteristics of MuGFETs with symmetrical nFET/pFET $V_T$'s. Copyright© 2007 IEEE.

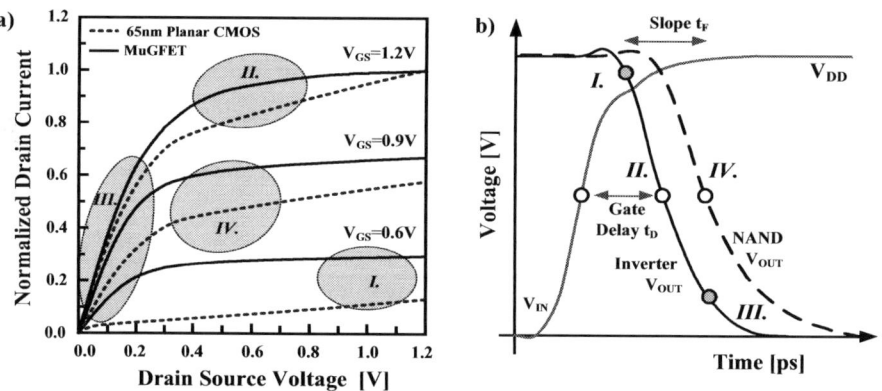

**Fig. 7.4.** Normalized drain currents of 65nm planar bulk and MuGFET nFETs with comparable channel length and off-currents (a) and illustration of relevant regions for inverter and NAND circuit operation (b) Copyright© 2007 IEEE.

Dynamic circuit performance is explored using simple CMOS inverter based ring oscillators. Fig. 7.5 shows the evolution of the trade-off between performance and leakage currents for various MuGFET CMOS technologies.[8] The starting point is CMOS MuGFET circuits fabricated using a conventional n/p-type Poly-Si Gate process with 70nm gate length and 50nm fin width. Obviously, the circuits are too slow and even leakage current exceeds typical the Low-Standby-Power Technology requirements of 0.1-1nA/stage.

**Table 7.3.** Relevant operating regions of digital circuits and impact of MuGFET device properties on performance.

| Region | Circuit Topology | MuGFET Feature | Performance Impact |
|---|---|---|---|
| I. | Inverter, $V_{IN} < V_{DD}/2$ | Steeper Subthreshold slope | Larger initial gate overdrive |
| II. | Inverter switching point: $V_{IN} = V_{OUT} = V_{DD}/2$ | Lower DIBL, larger effective drive current | Faster gate delay |
| III. | Inverter: $V_{out} < 0.1\ V_{DD}$ Stacked bottom devices | Larger linear current due to lower $V_{TLIN}$ | Faster output slope Lower source resistance |
| IV. | Stacked top device: $V_{OUT} \approx V_{DD}/2, V_S > 0V$ | Larger drive current | Faster stack delays |

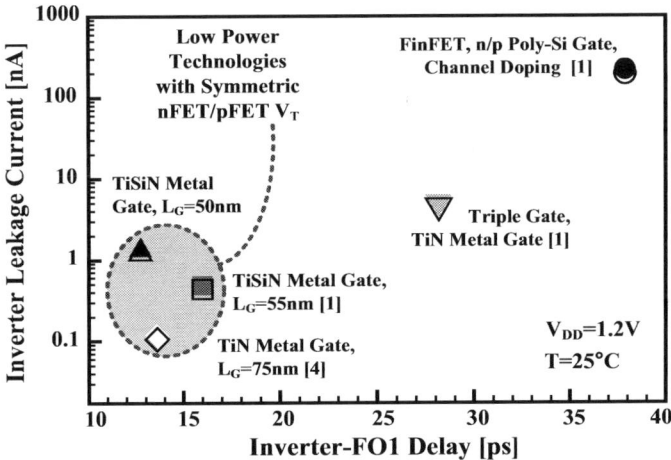

**Fig. 7.5.** Evolution of INV-FO1 inverter delay and leakage current for different MuGFET technology options starting from doped FinFET devices with n+/p+ Poly-Si gates to Triple Gate FETs with various single midgap metal gate options. Copyright© 2007 IEEE.

The introduction of a metal gate process together with the Triple-Gate device topology improves the nFET on-current by 50%. This reduces the leakage current by ×80 and the inverter delay to 28ps. However, the non-symmetrical work function of the TiN metal gate in these first working metal-gate MuGFET CMOS circuits results in a large threshold voltage of -600mV and limits pFET-dependent 0-1 logic transitions. In subsequent optimization steps, competitive digital performance of 13-15ps inverter delays has been achieved by providing symmetric nFET and pFET threshold voltages of $V_T$=300-350mV.[7,9] Here, gate lengths are comparable to 90nm and 65nm low-power CMOS technologies. Circuit performance of better than 10ps per stage is achievable for shorter gate length (40nm to 45nm) at the price of a dramatic increase in leakage current, since the fin widths are not been scaled down to the 25nm-20nm regime. Therefore, short-channel effect control is degraded.

The scaling of MuGFET circuit performance can be predicted using the semi-empirical Alpha-power MOS model.[10] Fig. 7.6 shows that MuGFETs can meet 32nm technology node requirements by scaling the gate length and fin width, adding conventional gate nitridation and by applying strain techniques.[11-12] Extrapolating existing ring oscillator delay of 15ps at $V_{DD}$=1V, this study indicates the feasibility of sub-10ps INV-FO1 delays. When scaling down MuGFET device dimensions it is assumed that electrostatic device integrity is not degraded, *i.e.* $L_G/W_{FIN}$ =2.

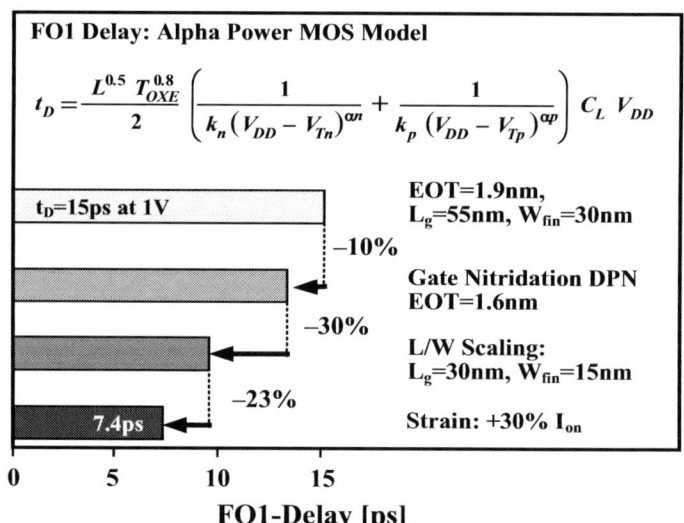

Fig. 7.6. Gate-delay scaling study for 32nm Low Standby Power Technology based on MuGFET performance of 15ps at 1V and Alpha-Power MOS model. Copyright© 2007 IEEE.

Besides pure digital inverter performance, an essential step for MuGFET circuit design is the demonstration of full CMOS logic functionality. Therefore, Fig. 7.7 shows the gate delays of various ring oscillators composed of NAND and NOR gates with different fan-ins.[13] All ring oscillator stages have a fan-out of 2. This measurement is a key step towards large-scale digital circuits. Typical static CMOS logic behavior is observed since the gate delay increases for larger fan-in, *i.e.* the inverter is faster than the NAND2 and NAND3 gates with stacked transistor configurations.

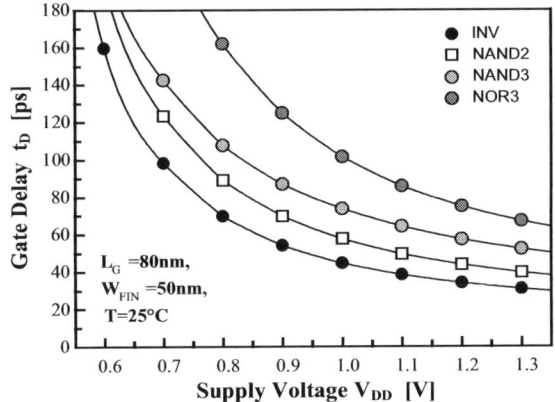

**Fig. 7.7.** Demonstration of full CMOS logic functionality for doped MuGFET devices with $n^+/p^+$ Poly-Si gates, ring oscillator stages with a fan-out FO=2. Copyright© 2006 IEEE.

The additional dynamic performance benefit of MuGFETs results from excellent stacked device performance and is already illustrated for the $I_D$-$V_{DS}$ output characteristics in Fig. 7.3. A detailed comparison of NANDs and NORs delays with different stack heights (fan-in) of for metal gate based MuGFETs is shown in Fig. 7.8a. The delay penalty of complex CMOS gates is significantly lower than for typical 130 and 65nm CMOS technologies (Fig. 7.8b). As a result, the delay penalty of a MuGFET NAND5 is lower than for a NAND3 in 65nm bulk CMOS.

Three effects explain this major advantage of MuGFETs with good SCE's (Fig. 7.8c): Firstly, near ideal subthreshold slope allows for lower $V_T$'s at constant $I_{OFF}$. The gate overdrive $V_{DD}$-$V_T$ is increased and the drain current $I_D$ at $V_{GS}$<$V_{DD}$ rises disproportional to $I_{ON}$ at $V_{GS}$=$V_{DD}$. Secondly, a low DIBL effect is beneficial for $I_D$ at reduced drain-source voltage $V_{DS}$. Since the effective $V_{GS}$ and $V_{DSs}$ are significantly lower in NAND/NOR gates than in inverters a device with lower DIBL and high linear current $I_{DLIN}$ results in better stack performance. Thirdly, the removal of the auto-reverse biasing with SOI technologies yields in higher stack performance. However, due to small body effects in sub-100nm technologies latter effect has minor impact. Due to the performance loss caused by high fan-ins, logic gates with more than 4 nFETs or 3 pFETs are typically not used in sub-100nm bulk technologies. Since higher fan-ins can be used with

MuGFETs, the same functionality can be implemented with less logic gates, resulting in an additional area reduction in more complex circuits.

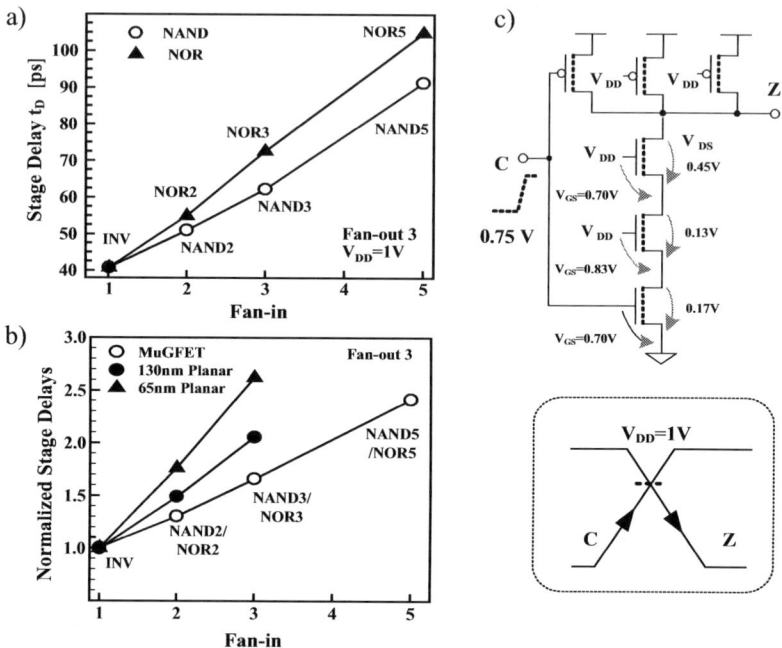

**Fig. 7.8.** Stage delays of complex MuGFET CMOS gates (a) and comparison of average stage delays to planar CMOS gates (b) for increasing fan-in. MuGFETs are based on single-midgap metal gate with symmetric threshold voltages. A snapshot of a NAND3 operation during the rising input transition is shown in (c). Copyright© 2007 IEEE.

### 7.2.2 Large-scale digital circuits

After exploring simple ring oscillator circuits with a complexity of about 300-500 devices, the next step is the demonstration of large-scale integration capability using more product-typical circuits. This also offers the opportunity to assess low voltage behavior and low power design techniques, which play a significant role in today's circuit design. The assumption is that, despite of the excellent short channel behavior of MuGFETs, today's power reduction techniques, such as clock gating, mixed-$V_T$ design, and combined voltage and frequency scaling will still be required in future MuGFET CMOS circuits to meet low power product requirements. Therefore, there is need to investigate the efficiency of these low power techniques for MuGFET circuits.

To close the gap between simple ring oscillators and complex digital CMOS circuits such as microprocessor or DSP cores, a digital test macro composed of product-representative 4-stage pipeline structures and 10k devices was fabricated.[8] Fig. 7.9 shows the schematic of an inline measurable 4-stage pipeline in a loop structure, which is composed of 20 stages of CMOS logic between set/reset capable edge-triggered master-slave flip-flops (FF). Logic depth, functionality, and gate loads are chosen to replicate the critical paths of low power microprocessor cores in a generic way. The external clock signal is buffered and distributed by a small clock tree within the circuit. Basic idea behind this test structure is to detect the maximum clock frequency by generating a setup time violation at one of the FFs if the clock period is smaller than

$$T_{min} = t_{CLK-Q} + t_{SU} + \sum_{i=1}^{20} t_i \qquad (7.1)$$

With $t_{CLK-Q}$, $t_{SU}$, and $t_i$ being FF clock-Q delay, setup time, and logic gate delay, this is achieved by resetting FF2-FF4 to logic-0 and inserting a logic-1 into FF1 via the start signal. On the rising clock edge, the inserted logic-1 starts to circulate within the loop and indicates correct pipeline functionality. Therefore, the whole pipeline is operated in a ring-oscillator mode at $f_{CLK}/4$. This is observed by monitoring the outputs QO and CLKO on an oscilloscope. A setup time violation prevents the sampling of the circulating logic-1 at the receiving FF. This immediately interrupts the oscillation so that the failure point can be detected by a constant QO signal.

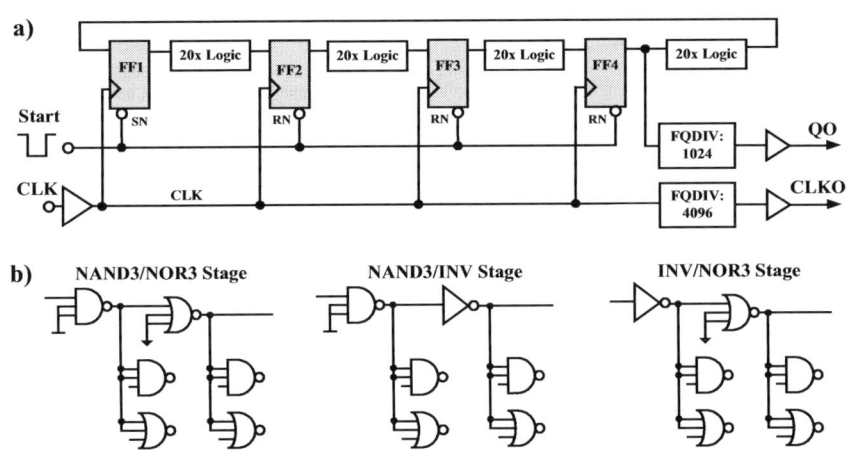

Fig. 7.9. Large-scale MuGFET circuit demonstrator (a) composed of a four stage pipeline configuration. Three different paths are implemented with loaded NAND3/NOR3, NAND3/IN, INV/NOR3 (b) functionalities. Copyright© 2007 IEEE.

The complete test macro comprises three types of pipeline structures with 10 cascaded NAND3/NOR3, NAND3/INV, and INV/NOR3 logic stages. Fig. 7.10 shows SEM pictures of the NAND3/NOR3 stage. Together with clock buffers, frequency dividers, and I/O-drivers a path structure contains about 3500 devices and results in an overall complexity of about 10k devices for the whole macro. For early circuit exploration in new CMOS technologies, the proposed test macro is particularly useful due to its simplicity and capability to characterize performance and power dissipation. The test macro is fabricated using the low power MuGFET process described in Table 7.2.

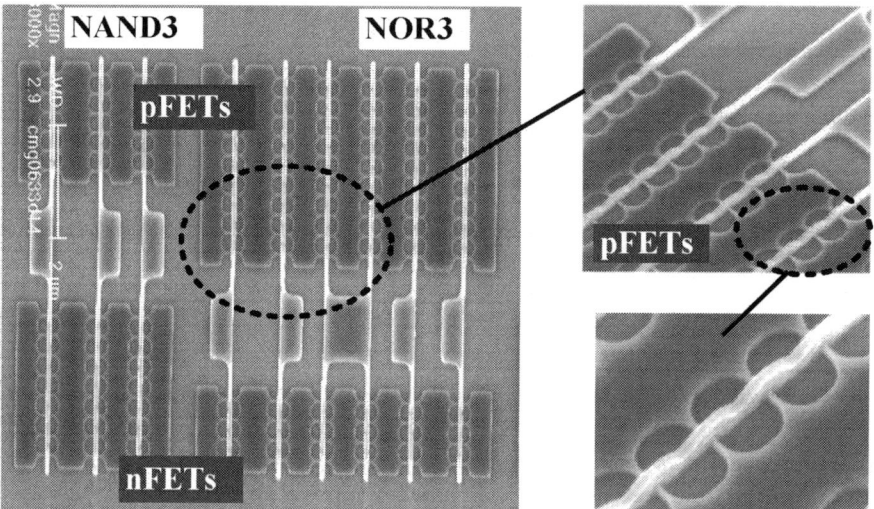

**Fig. 7.10.** SEM photograph of the NAND3/NOR3 combination. Copyright© 2007 IEEE.

Fig. 7.11 shows the measured clock frequencies of the three paths for supply voltages down to the subthreshold region at $V_{DD}$=0.3V. Clock frequencies of $f_{CLK}$=370MHz for the NAND3/NOR3 path, followed by $f_{CLK}$=430MHz for the INV/NOR3 path, and $f_{CLK}$=505MHz for the NAND3/INV path are observed at $V_{DD}$=1.2V. This demonstrates that MuGFET technology complies with low-power product speed requirements of 300MHz to 500MHz in typical critical paths structures. The speed difference between the p-type dominated INV/NOR3 and n-type dominated INV/NAND3 paths of 15% is significantly smaller than in planar CMOS circuits with up to 60% due to the higher hole mobility in p-type MuGFETs with (110) sidewall orientations. This device property is particularly beneficial since CMOS logic becomes more symmetric in terms of nFET/pFET drive currents.

To get detailed insight into the relevance of these measurements, voltage scaling of MuGFET circuits is compared to 90nm, 65nm, and 45nm planar CMOS circuits with similar topologies (Fig. 7.12). The measured performance is normalized, *i.e.* 100% performance corresponds to nominal $V_{DD}$ for each technology. MuGFET circuits show an excellent low voltage performance and outperform planar CMOS in this region. The reason for this superior voltage scalability is again the small DIBL and good short-channel effect control.

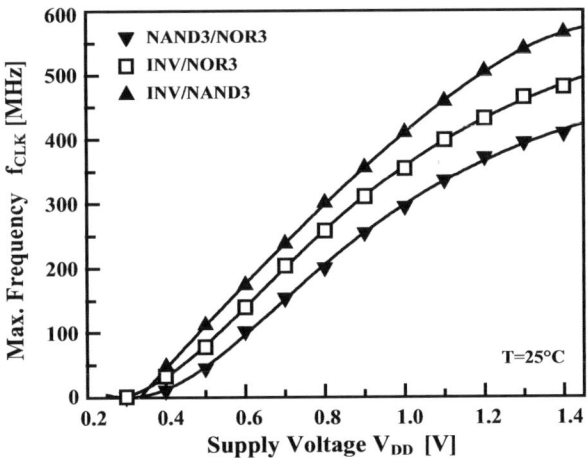

**Fig. 7.11.** Measured MuGFET large-scale circuit performance of the three different critical paths shown in Fig. 7.4. Copyright© 2007 IEEE.

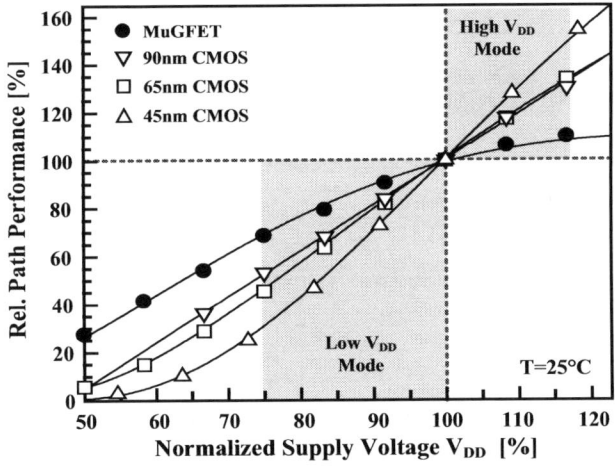

**Fig. 7.12.** Voltage scaling efficiency of MuGFET, 45nm and 65nm planar bulk CMOS circuits. Copyright© 2007 IEEE.

When increasing $V_{DD}$ to 110% of the nominal value, as usually done in mobile products to facilitate temporarily high performance modes, MuGFET clock frequency shows a much stronger saturating behavior compared to planar CMOS technologies. The reason is that drain currents of MuGFETs increase only linear with the gate overdrive $V_{GS}$-$V_T$ as shown in Fig. 7.3. This is in contrast to the nonlinear increase of drain current in planar CMOS devices following the α-power MOS model $I_D \propto (V_{GS}-V_T)^\alpha$ with α≈1.3. The linear current dependence of MuGFETs is attributed to the large parasitic source/drain resistance of 500Ω/μm. The resulting intrinsic voltage drop in the source reduces the effective gate overdrive. The parasitic resistance will be reduced if 30nm to 50nm selective epitaxial silicon growth is applied in the source/drain regions.[14]

### 7.2.3 Leakage-performance trade off and energy dissipation

Motivated by demand for low active and leakage power dissipation, the NAND3/NOR3 path is further investigated under three operating modes at full data activity of 25%, with 0% data activity where FFs and clock distribution are switching, and in clock gating mode with disabled clock $f_{CLK}$=0MHz. For all three operating modes the current is measured and plotted against the maximum clock frequency to analyze the trade-offs between performance and power dissipation (Fig. 7.13). Since FFs belong to the most complex cell types within a digital standard cell library and hence are much larger in terms of transistor numbers compared to combinatorial logic gates, the measured ×7.4 difference between 0% and 25% data activity is reasonable.

However, the difference between the leakage current and the current for 0% data activity when only clocks and flip-flops are dissipating active power is 70x only. Considering that excellent short-channel behavior and low sub-threshold related source currents down to 30pA/μm are obtained, an explanation for this considerable high current during standby mode is the on-state gate leakage current due to the newly introduced HfSiON-based gate stack on MuGFETs which still has improvement potential in terms of thickness homogeneity and interface states in circuits with large integration density.

Due to the relevance of voltage scaling for leakage reduction, we now compare the leakage-performance trade offs of MuGFET and planar CMOS circuits for $V_{DD}$=0.4V to 1.4V (Fig 7.14). Performance and leakage currents at nominal supply voltages are marked by the dashed lines. Fig. 7.14 shows that MuGFET circuits exhibit a steeper slope and a strong leakage increase above $V_{DD}$=1V than the planar CMOS circuits. Again, this is caused by the on-state gate leakage of the TiN/HfSiON gate stack. In contrast to this, the leakage current of the planar CMOS circuits is dominated by subthreshold current with a typically weaker $V_{DD}$ dependence.

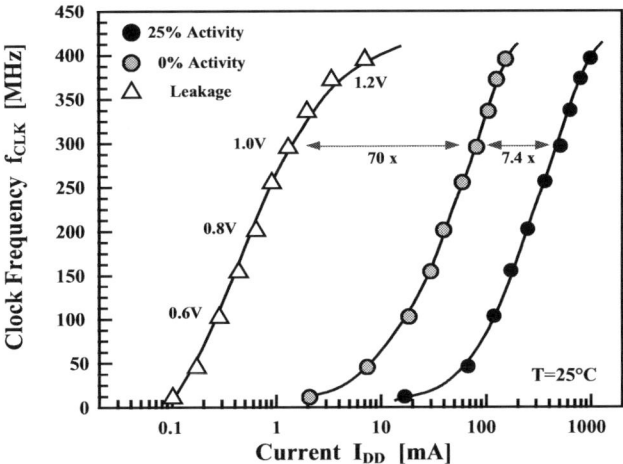

**Fig. 7.13.** Trade-offs between circuit performance and current dissipation of the NAND3/NOR3 path for three operating modes: active with 25% data activity in the logic, 0% data activity, and leakage state with disabled clock.

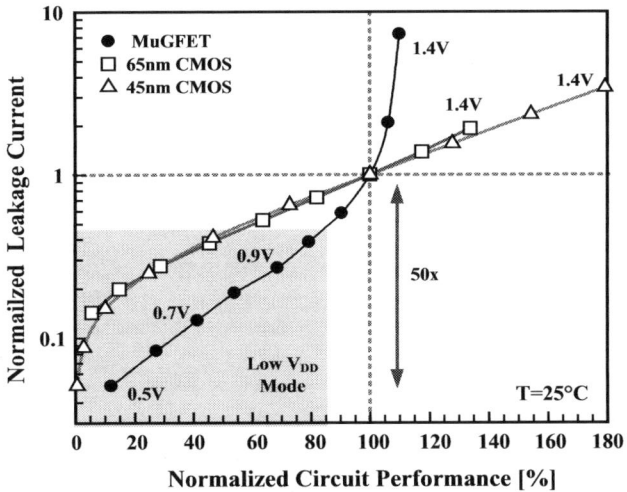

**Fig. 7.14.** Comparison of leakage-performance trade offs of MuGFET and planar CMOS circuits for a wide range of supply voltages. Copyright© 2007 IEEE.

In addition to leakage-performance trade-off, the efficiency of dynamic voltage scaling is of key importance for low power in MuGFET circuit designs. In modern product designs this technique is frequently applied to temporarily reduce capacitive energy consumption for all applications where maximum circuit performance is not required. If a new CMOS technology or process option is introduced for low voltage circuit design two main questions arise:

- Where is the most energy-efficient operating point of product-typical circuits?
- How does the gain in energy at low supply voltages affect circuit performance compared to nominal operating point?

Therefore, the energy dissipation per operation of the fabricated circuits is derived from the current measurements shown in Fig. 7.13 for full circuit activity of 25%, disabling of the logic-1 propagation within the pipeline to emulate 0% data activity, and $f_{CLK}=0$ to emulate clock gating. Fig. 7.15 shows the results for the NAND3/NOR3 pipeline operated at maximum possible frequency for each voltage level according to Fig. 7.11. Energy dissipation for a lower core activity of 5% is calculated from the measurements and according to:

$$E_{TOT} = \eta V_{DD}^2 (\gamma C_{LOG} + C_{CLK}) + I_{LEAK} V_{DD} T_{CLK} \qquad (7.2)$$

where $\eta$ is the clock gating factor ($0 \leq \eta \leq 1$) and factor $\gamma$ is the logic activity of the pipeline structure ($0 \leq \gamma \leq 0.5$). $C_{LOG}$ and $C_{CLK}$ are the total capacitances of logic gates and the clocked devices in FFs and clock tree. For room temperature, the ratio of total energy to leakage energy is ×100. Minimum leakage energy occurs at $V_{DD}=0.6V$. For $V_{DD}<0.6V$ and especially in the subthreshold region, leakage contribution increases due to the disproportional rise of the clock period $T_{CLK}$ in Eqn. (7.2). For 5% activity and maximum speed, Fig. 7.15 indicates a ×5 total energy reduction when scaling $V_{DD}$ from 1.2V to 0.6V.

So far energy dissipation has been discussed for room temperature only. However, mobile SoC products are operated at about 85°C junction temperature, where MuGFET off-current increases by about ×7. In this case, the relative contribution of the leakage energy increases. To estimate this effect, total energy dissipation at 85°C is calculated for a low data activity of 5% (Fig. 7.16).

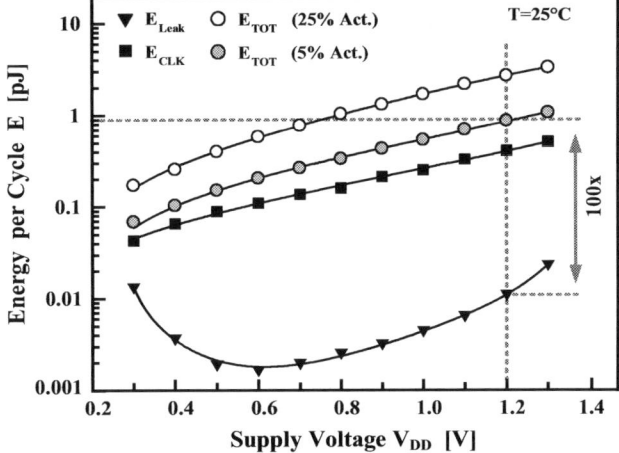

**Fig. 7.15.** Energy dissipation of the NAND3/NOR3 path at maximum speed for two activities without clock gating. Copyright© 2007 IEEE.

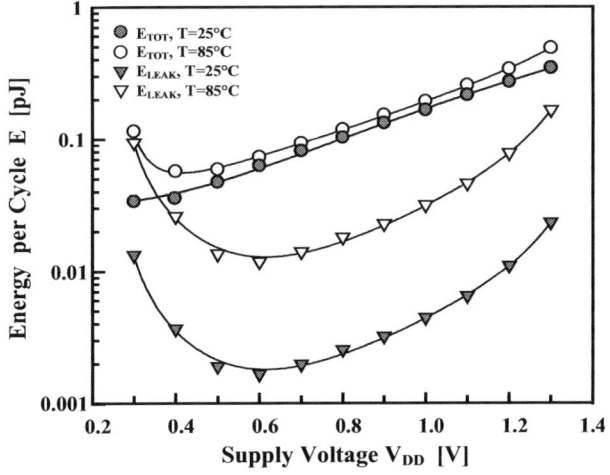

**Fig. 7.16.** Energy dissipation of the NAND3/NOR3 path at maximum speed at 25°C and 85°C for 5% data activity and 30% clock gating.

In addition a clock gating efficiency of 30% is assumed, which reduces clock buffer and flip-flop dynamic energy consumption. Due to the increased leakage contribution, it becomes obvious, that for this low activity scenario the minimum energy per clock cycle is in the supply voltage range between 0.4V to 0.5V. The general effect is well known in circuit design and is the reason why optimum supply voltages of SRAM circuits with even lower activity are at about 1V.[15] Mixed-$V_T$ integration is a favored low power circuit technique to reduce leakage energy during active mode. The feasibility of this circuit technique will be discussed in Section 7.2.4.

Besides energy per operation, the energy-delay product $E \times t_D$ of a logic stage is a useful figure of merit to consider the penalty of strong speed reduction at low $V_{DD}$. Fig. 7.17a shows the average switching energies, gate delays and the $E \times t_D$ product obtained from the MuGFET NAND3/NOR3 pipeline. According to this metric, the most efficient voltage range in terms of energy and performance is between 0.7-0.9V. Below 0.6V the delay is highly sensitive to external and internal supply voltage variations. Setting a low voltage limit of 0.6V leaves a sufficient margin of 100mV to avoid delay increase larger than ×2 and thus ensuring robust circuit operation. Fig. 7.17b compares the normalized $E \times t_D$ products of MuGFET and planar CMOS circuits. The minimum $E \times t_D$ product for MuGFET circuits is 300mV and 400mV lower compared to 65nm and 45nm CMOS technology, respectively. Again, the advantage of MuGFETs is due to the superior voltage scalability originating from the small DIBL and subthreshold slope.

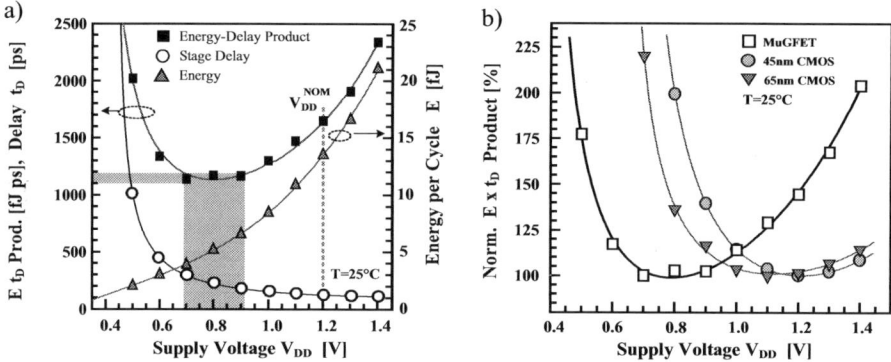

**Fig. 7.17.** Energy-delay product, average energy per clock cycle, and average stage delay of the NAND3/NOR3 path (a) and comparison of energy-delay products of MuGFET, 45nm and 65nm planar bulk CMOS circuits. Copyright© 2007 IEEE.

A conclusion of the investigations of these large-scale circuits is that MuGFETs offer an opportunity to reach the 1V supply voltage barrier for 300MHz critical path performance. For applications with about 100MHz performance 0.6V seems to be a reasonable lower limit. Even though energy efficiency for extremely low voltage operation below $0.6V \approx 2 \times V_T$, degrades, this supply voltage region is attractive for applications with very low performance requirements of 10kHz to 1MHz, *e.g.* data processing in wireless sensor nodes and mobile phone standby modes. Assuming that a gate length to fin width ratio of $L_G/W_{FIN}=1.5$ is achieved for 32nm and

22nm feature sizes, a significant performance benefit of MuGFET-based CMOS circuits compared to planar CMOS circuits of about ×2 is feasible in these operating region.

### 7.2.4 Multi-$V_T$ devices and mixed-$V_T$ circuits

Multi-$V_T$ device integration is a key requirement to implement mixed-$V_T$ circuits and power gating for leakage reduction (Fig. 7.18a). Adjusting channel doping is not a viable solution for thin fin widths to generate different $V_T$'s. Moreover, multiple metal gate workfunction integration is expensive in terms of process complexity. An alternative solution is the intentional design of slightly different fin widths in combination with gate length tuning to provide low-$V_T$ and high-$V_T$ devices (Fig. 7.18b). The device physical background of this approach is the MuGFET gate length to fin width ratio $L_G/W_{FIN}$. Since this ratio quantifies the capability to suppress the penetration of the electrical drain field into the fin region, it can be used to adjust the trade-off between source leakage current and drive current.

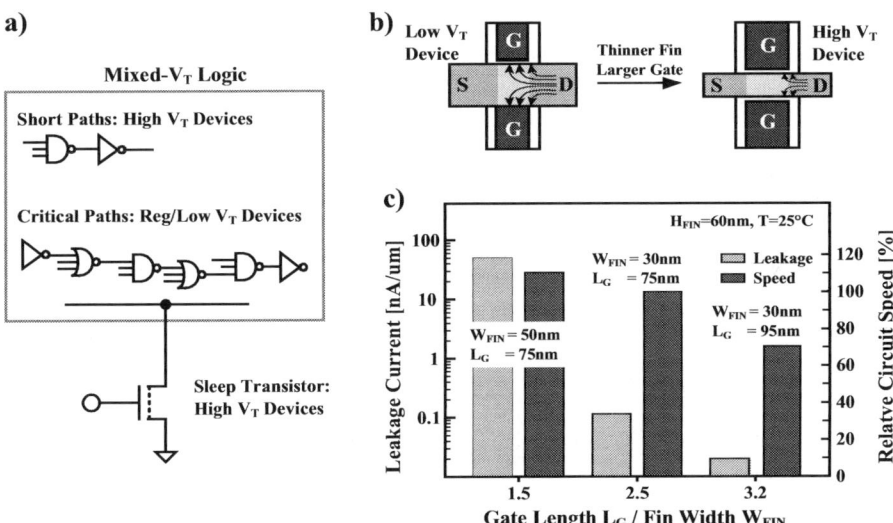

**Fig. 7.18.** Mixed-$V_T$ design (a), implementation of different device types using MuGFETs with different fin widths and gate lengths in a single midgap metal gate technology (b), and impact on circuit performance and leakage to provide mixed-$V_T$ capability (c). Copyright© 2007 IEEE.

In Fig. 7.18c the reg-$V_T$ devices are tall triple gate FETs with $L_G$=95nm and $L_G/W_{FIN}$=2.5, which is sufficient to eliminate short-channel effects. Increasing the initial $L_G/W_{FIN}$ ratio of the reg-$V_T$ device by 28% to 3.2 decreases the off current by ×5 at the penalty of 25% speed loss. Using these n-type MuGFETs as sleep transistors to separate the circuit temporarily from supply lines, and assuming 5% of the total transistor width for the sleep transistors, shows that ×100 leakage reduction during sleep modes is achievable. This value is comparable to planar CMOS designs based on reg-$V_T$/high-$V_T$ combination. For a typical mixed-$V_T$ design with 20% reg-$V_T$ circuits in critical paths and 80% high-$V_T$ circuits for the remaining logic, larger $L_G$ reduces active-mode leakage by ×3.

For wide fins with nearly rectangular cross section and $L_G/W_{FIN}$=1.5 leakage increases by ×600 to 55nA/μm. Considering the minor speed improvement of 15% over the reg-$V_T$ core device with tall fins, fin width increase is less efficient to provide low-$V_T$ devices in a cost-sensitive low power MuGFET technology. However, the leakage current value of 55nA/μm still fulfills microprocessor requirements. Maintaining the $L_G/W_{FIN}$=1.5 scaling ratio and then reducing gate length result in a significant performance increase as shown on device level for high performance CMOS devices.[16]

### 7.2.5 High-temperature circuit operation

Many SoC applications, especially for automotive and industrial markets, require circuit operation at elevated temperatures. Fully depleted SOI circuits have been demonstrated at 275-300°C, while partially depleted SOI circuits are limited to 225°C. The operation of bulk silicon circuits is limited to a temperature of approximately 200°C. Fig. 7.19 shows that metal-gate based MuGFET circuits are operating even at temperatures of 300°C.[17]

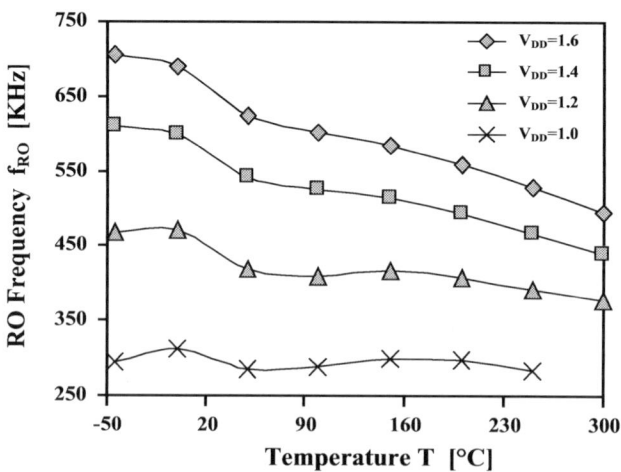

**Fig. 7.19.** MuGFET ring oscillator speed *vs.* temperature. Copyright© 2006 IEEE.

At higher supply voltages, performance reduces with temperature due to carrier mobility reduction, but at about $V_{DD}$=1V, the performance reduction is flat since the threshold voltage $V_T$ is also decreasing. These initial results indicate that MuGFET CMOS technology is also an interesting opportunity for future high temperature applications and not limited to low power products.

### 7.2.6 SRAM design

SRAM cells (Fig. 7.20a) are traditionally one of the most critical circuit components in a SoC technology platform and considerable effort is spent to reduce cell sizes, achieve robust read and write operation as well as voltage scalability. In sub-100nm CMOS, the degradation of read and write stability at low operating voltages typically defines the minimum supply voltage of the complete IC, if a cost-efficient single supply voltage architecture is chosen. In planar CMOS SRAM design, device dimensions W/$L_G$ of the pull-down nFET (PD), access nFET (AC) and pull-up pFET (PU) device can be adjusted continuously to optimize cell stability.

Usually, PD devices have the largest W/$L_G$ ratio and are chosen to be 1.3-2× stronger in terms of drive currents as the AC devices. The current ratio $I_{PD}/I_{AC}$, reflecting the geometric device dimension, is known as beta ratio to describe the static read stability of the cell. A sufficient beta ratio assures that the stored 0 is maintained on the internal storage nodes SN or /SN if the word lines WL are open and not overwritten by the pre-charged bit lines BL during the voltage divider configuration. This is achieved by using minimum transistor width and increase gate length for the AC devices. While high-density cells are designed with beta ratios of less than 1.5, dedicated low voltage cells are typically larger and have beta ratios of 1.5-2. In MuGFET SRAM design, the opportunity to vary the transistor width continuously does obviously not exist in minimum sized cells, where each device consists of a single-fin transistor (Fig. 7.20b). Therefore, the remaining degree of freedom to weaken the access devices is to use a slightly larger gate length. This approach is explored in an SRAM design shown in Fig. 7.21.[7] The measured static noise margin of 185mV for $V_{DD}$=1V is acceptable.

A further issue is the write stability. In section 7.2.1 we have seen that the current drive of pFET devices can reach up to 80% that of an nFET. While this renders CMOS logic more symmetrical, there can be an issue when writing a logic-0 into the SRAM cell, if the pFET pull up path is too strong. Again, slightly increased gate length is helpful to weaken the pFET access devices. MuGFET device tuning by locally adjusting the fin widths is a further technique but requires very precise control of the fin dimensions.

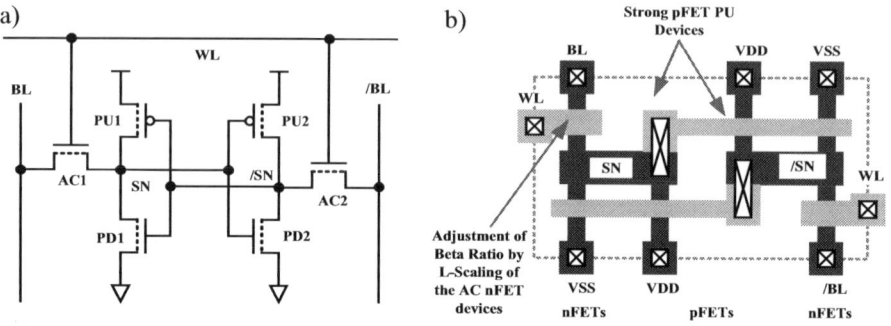

**Fig. 7.20.** SRAM schematic (a) and MuGFET SRAM layout (b).

Theoretical circuit studies show that static noise margin can be also improved by using 45°C rotated fins within the cell. Depending on the orientation of the whole memory array, either drive currents of PD devices are increased or the AC and PU devices are weakened due to different mobilities along the fin sidewalls.[18] If a cell area increase is acceptable, an additional solution is to use two-fin pull-down devices. The smallest MuGFET SRAM cell reported so far has a cell size of 0.274 µm$^2$.[19] However, at the time this book is being written, dynamic SRAM operation is still to be demonstrated and requires higher yield and co-integration of CMOS logic for peripheral circuits together with SRAM core cells.

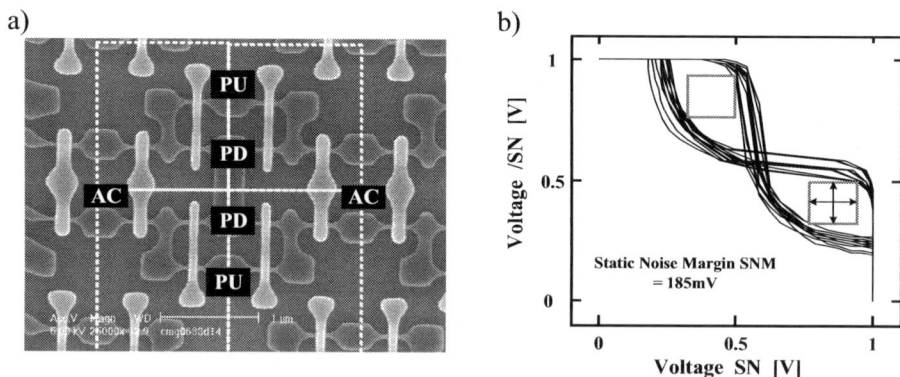

**Fig. 7.21.** MuGFET SRAM SEM microphotography (a) and SRAM butterfly curves with word line at 1V (b). [7] Copyright© 2007 IEEE.

## 7.3 Analog Circuit Design

The ongoing technology scaling causes some severe challenges for analog and RF circuit design using standard planar digital CMOS technologies for example in system on a chip (SoC) solutions. While transistors are getting faster, their analog properties degrade from node to node. One of the most critical issues is the degrading intrinsic gain (defined as $g_m/g_{ds}$) of the devices due to strong pocket implants and growing impact of short channel effects. For technology nodes below 65nm single digit values are expected for $g_m/g_{ds}$. But also increasing flicker noise, variability and mismatch represent serious issues and demand smart solutions from analog and RF designers to maintain or enhance circuit performance. The introduction of multi-gate device architectures (MuGFET) and novel materials (high-k, metal gate) will significantly change analog and RF device properties and hence will also impact circuit design.

In this chapter first the most important analog and RF device figures of merit for single- and multi-gate transistors are compared. In addition new device and material specific parasitic effects are shown. Based on device properties some multi-gate circuit design challenges and prospects are explained, beginning with basic analog building blocks up to mixed-signal and RF circuits. If not explicitly stated, all multi-gate device and circuit measurements are carried out on a MuGFET technology as described in Ref. [7].

### 7.3.1 Device figures of merit and technology related design issues

#### 7.3.1.1 Transconductance

The transconductance $g_m$ describes how efficient a small voltage signal at the transistor gate is converted into a drain current signal. The available transconductance limits the bandwidth of operational amplifiers as well as the maximum frequency of oscillators. Figure 7.22 shows the gate length dependence of $g_m$ for n-type MuGFET and bulk MOSFET in comparable technologies (in terms of minimum feature size). The $g_m$ of the MuGFET is slightly lower than that of the bulk FET mainly due to the high parasitic source/drain resistances. The situation gets worse for shorter channel lengths and increasing overdrive voltage $V_{GS}$-$V_T$.

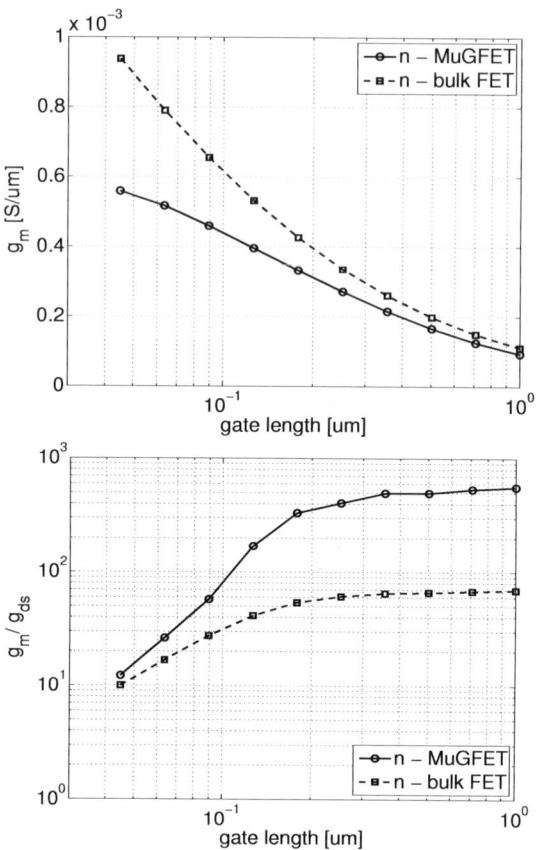

**Fig. 7.22.** $g_m$ (top) and $g_m/g_{ds}$ (bottom) of n-type MuGFET and bulk MOSFET at $V_{GS}=V_T+200$mV and $V_{DS}=1$V.

#### 7.3.1.2 Intrinsic transistor gain

The intrinsic transistor gain is defined as $g_m/g_{ds}$ and relates the effectiveness of the transistor as controlled current source in regard to the output resistance. The $g_m/g_{ds}$ ratio limits the open-loop gain of operational amplifiers for a specified bandwidth as well as the resolution of analog-to-digital converters. Figure 7.22 compares $g_m/g_{ds}$ of bulk and MuGFET devices. Even at short gate lengths the MuGFET features an improved gain, although its $g_m$ is slightly lower. The main reason for the high gain of the MuGFET is the very low output conductance due to the undoped fins without any pocket implants.[20] Additionally, the excellent short channel behavior helps to reduce the output conductance. From analog perspective, the high intrinsic gain is a strong argument to use multi-gate devices in future technology nodes, as it overcomes one of the most critical scaling issues in planar bulk CMOS. A quantitative estimation of the resulting benefits for analog circuits can be seen later in this chapter.

### 7.3.1.3 Matching behavior

The amount of transistor parameter mismatch of nominal identical devices defines the resolution-speed-power trade-off in analog and mixed-signal circuits to a great extent.[21] In deep sub-micron CMOS technologies there are many sources of transistor parameter mismatch such as $V_T$ or current factor $\mu C_{ox}$. Fluctuations of channel and gate doping concentrations, oxide charges and surface roughness contribute to local variations of device parameters. The dominant effect in today's CMOS technologies is the $V_T$ mismatch due to the statistical variation of channel dopants. It can be expressed as

$$\sigma_{V_T} = \frac{A_{VT}}{\sqrt{W \cdot L}} \tag{7.3}$$

with a matching constant $A_{VT}$ proportional to the oxide thickness and the doping level [22]:

$$A_{VT} \propto t_{ox} \sqrt[4]{N_D} \tag{7.4}$$

Undoped, metal-gate based MuGFETs therefore outperform doped bulk devices in terms of matching as illustrated in Fig. 7.23.

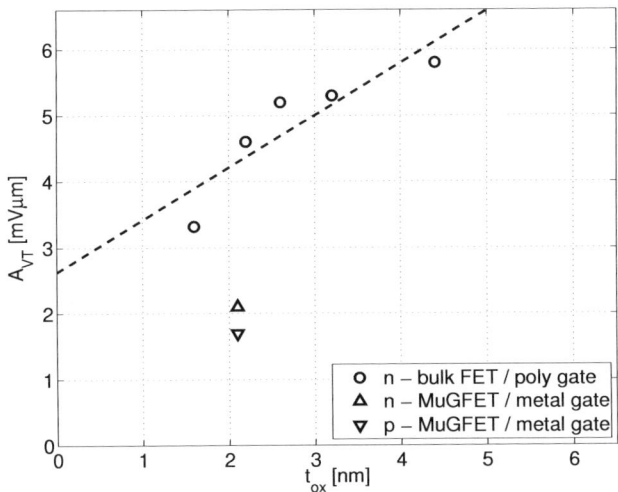

**Fig. 7.23.** Matching performance for bulk and MuGFET devices.[23]

On the other hand, new mechanisms come into play for MuGFETs with scaled dimensions. For very narrow fins, variations of source/drain resistance, gate misalignment, line edge and fin sidewall roughness degrade the matching performance again.[18-19] Although not related to the multi-gate device architecture, also the introduction of high-k dielectrics may worsen the matching behavior.[20] However the impact of these effects can be reduced by technology optimization whereas the limiting $V_T$ mismatch in case of planar MOSFETs is just a statistical effect.

### 7.3.1.4 Flicker noise

Similar to matching, noise is a limiting factor for the area-power trade-off in analog circuit design. The noise sources in single- and multi-gate devices do not differ in general terms. However there are new or other effects that contribute to the individual noise mechanisms, especially in case of flicker noise. Following similar effects as described above, also flicker noise is increased for narrow fins due to the high density of states at the non-perfect surface (110 crystal orientation and roughness) of the fin sidewalls. The introduction of high-k dielectrics will further degrade noise performance (Fig. 7.24).[25]

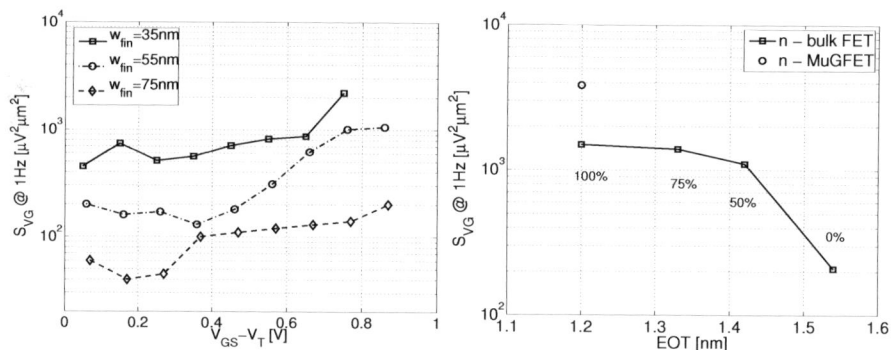

**Fig. 7.24.** Flicker noise *vs.* overdrive voltage (top) and Hf content of the dielectric (bottom). Copyright© 2007 IEEE.

Flicker noise is measured for planar nFETs with varying Hf content in the dielectric, going from SiON to pure $HfO_2$. Replacing SiON with HfSiON (50% Hf) increases the flicker noise by more than a factor of 5. For pure $HfO_2$ flicker noise is increased by another 35%. MuGFETs (fin width = 20nm, TiN gate, undoped channel) show the same trend (Fig. 7.24 shows only $HfO_2$), but with an overall increased noise level. This is attributed to additional noise sources at the source/drain resistances and the low interface quality of the rough fin sidewalls and dielectric. Especially for narrow fins the flicker noise is higher compared to planar, in this case about 2.5 times.

### 7.3.1.5 Transit and maximum oscillation frequency

The transit frequency $f_t$ as well as the maximum oscillation frequency $f_{max}$ are key figures of merit for RF circuit design. $f_t$ and $f_{max}$ define the unity gain frequencies for current and power respectively. Both are limited by the relation of achievable transconductance versus parasitics as the gate-drain and gate-source capacitance $C_{gd/gs}$ or in case of $f_{max}$ the gate resistance as shown in equation 7.5:

$$f_T = \frac{g_m}{2\pi(C_{gd}+C_{gs})} \quad ; \quad f_{max} = \sqrt{\frac{f_t}{8\pi R_{gate} C_{gd}}} \qquad (7.5)$$

Compared to planar devices $f_t$ and $f_{max}$ are significantly decreased due to the low $g_m$, the high source/drain resistances and the high $C_{gd/gs}$. The high $C_{gd/gs}$ per unit width can be explained by the additional fringing capacitance of the source/drain contact landing pads to the gate region in between the fins that does not contribute to the channel width. Figure 7.25 compares $f_t$ and $f_{max}$ of bulk and MuGFET devices.

Selective epitaxial growth (SEG) is a possible solution to reduce source/drain parasitics and will be a key achievement to enhance the RF performance of MuGFETs.[14] For applications below 10GHz, nevertheless, the actual achievable $f_t$ and $f_{max}$ of MuGFETs are sufficiently high.[20]

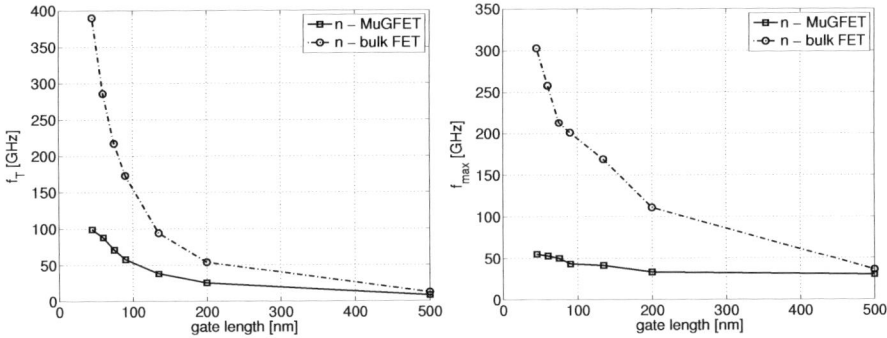

**Fig. 7.25.** $f_t$ (top) and $f_{max}$ (bottom) for n-type bulk and MuGFET devices.

### 7.3.1.6 Self-heating

The dissipated power of active devices leads to local temperature increases that influence important transistor parameters as $V_T$ or mobility. Self-heating is not a multi-gate specific effect but occurs pronounced in SOI technologies such as MuGFET due to the low thermal conductivity of the surrounding BOX and the small silicon volume. Scaling down the volume of the silicon fin due to the scaling of the gate length and the fin width will even more increase the effect.[26] Figure 7.26 shows pulsed output characteristics at different temperatures compared to DC measurement at room temperature. For low current and voltage values the DC curve fits the room temperature pulsed measurement, whereas with increasing power dissipation self-heating shifts the curve to the 75°C pulsed curve. Measurements and simulations show that self-heating can degrade the $I_{ON}$ of MuGFETs by 10% with a time constant in the range of ten to hundred nanoseconds.[27] Self-heating can be included in circuit simulation with a simple equivalent circuit as shown in Fig. 7.26. The power dissipation of the device is measured and applied to a thermal RC network creating the time-dependent local temperature that is fed back to the actual device. Transient drain current variations may have serious impact on analog circuits as shown later in this chapter.[26]

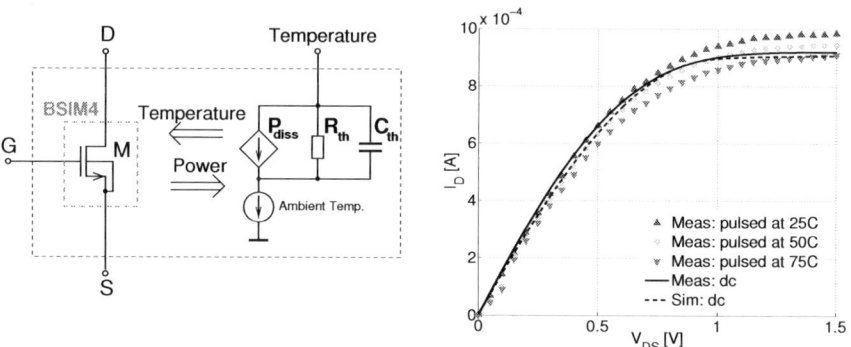

**Fig. 7.26.** Equivalent circuit for self-heating and comparison of (pulsed) measurements and simulation. Copyright© 2007 IEEE.

### 7.3.1.7 Charge trapping in high-k dielectrics

Another effect that is non-specific to multi-gate devices is charge trapping in high-k dielectrics. High-k is seen as important part of sub 45nm single- or multi-gate CMOS technologies, therefore the related issues for the design of analog and mixed signal circuits are discussed here.

Depending on many technology parameters such as the material of the dielectric or gate electrode, the interface quality, bias and temperature, charges can be trapped in preexisting states or defects in the high-k material. This effect leads to a dynamic shift of $V_T$.[28] Typical $V_T$ shifts are in the range of some mV up to hundreds of mV, while time constants in the range from μs to ms are reported.[28] Under appropriate bias conditions detrapping of the charges is possible, revealing hysteresis effects in the IV characteristics (Fig. 7.27). Similar to self-heating, this effect can be modeled in first order by means of an equivalent circuit as shown in Fig. 7.27.[26]

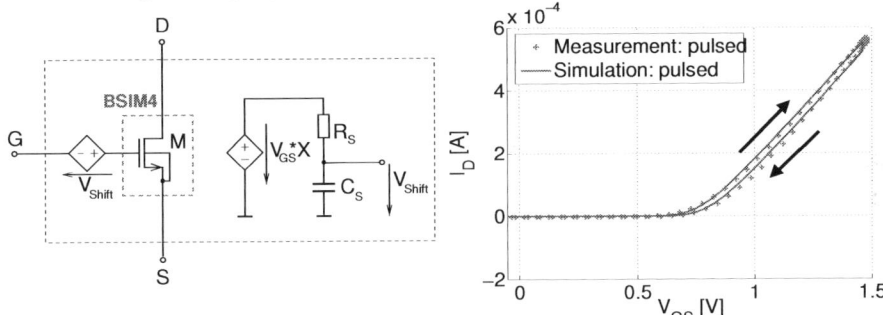

**Fig. 7.27.** Equivalent circuit for charge trapping and comparison of (pulsed) measurements and simulation. Copyright© 2007 IEEE.

### 7.3.2 Design of analog building blocks

This section illustrates some MuGFET specific features in the design of basic analog building blocks. Important differences compared to standard planar designs are discussed.

#### 7.3.2.1 $V_T$-based current reference circuit

The aim of a current reference is to provide a constant current that is insensitive against variations of the supply voltage $V_{DD}$. A common solution is shown in Fig. 7.28.[29] The p-type current mirror M3 – M4 forces the currents to be equal, while the absolute value of the current is determined by $R_S$, $V_{T1}$ and the current factor $K'_1$ (ignoring λ):

$$I_{OUT} = \frac{V_{T1}}{R_S} + \frac{1}{K'_1 R_S^2} + \frac{1}{R_S}\sqrt{\frac{2V_{T1}}{K'_1 R_S} + \frac{1}{K'^2_1 R_S^2}} \tag{7.6}$$

To ensure a proper operating point a start-up circuit is necessary (not shown in Fig. 7.28).

The design of this circuit with multi-gate devices does not imply big surprises coming form planar bulk technologies. Due to the low $g_{ds}$ and the good matching behavior, the requirements on channel length and device area are relaxed compared to planar. Fig. 7.28 shows the measured performance of a MuGFET implementation of the reference. The output current at a fixed output voltage is measured for varying values of $V_{DD}$. Quantitatively the $V_{DD}$ dependence can be expressed using a resistor $R_{DD}$ which is defined as:

$$R_{DD} = \frac{\partial V_{DD}}{\partial I_{OUT}}\bigg|_{V_{OUT}=const.} \quad (7.7)$$

The example in Fig. 7.27 yields a $R_{DD}$ of about 5MΩ.

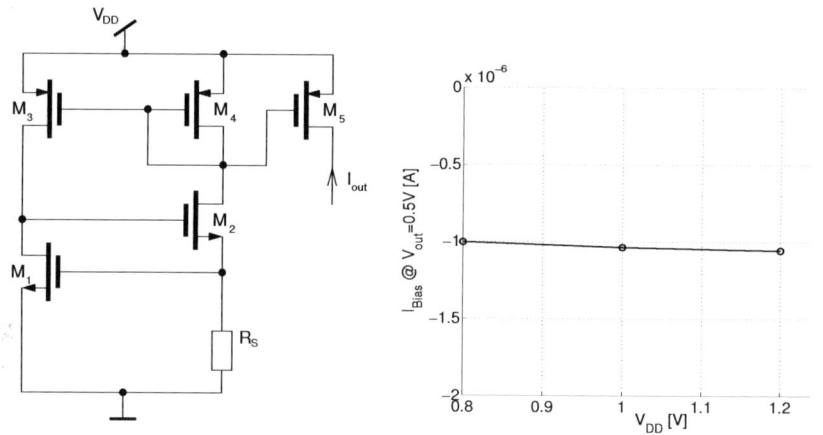

**Fig. 7.28.** Schematic (left) and measured $V_{DD}$ dependence of MuGFET implementation of $V_T$ based current reference (right).

### 7.3.2.2 Bandgap voltage reference

Bandgap based voltage reference circuits are intended to generate a reference voltage $V_{ref}$ independent of $V_{DD}$, temperature and process variations. A clever way for getting rid of any temperature dependence of $V_{ref}$ is to sum up two voltages with positive and negative temperature coefficient respectively. Usually the voltage across pn-diodes or parasitic bipolar transistors ($V_{be}$) is used for the negative temperature coefficient, whereas the difference of two diode voltages with unequal areas serves for the positive temperature coefficient. Fig 7.29 shows a possible implementation of such kind of bandgap reference that is suited for low supply voltages around 1V.[30] $V_{ref}$ is given by:

$$V_{ref} = R_3 \left( \frac{V_{D1a}}{R_{1a}} + \frac{\Delta V_D}{R_2} \right) \qquad (7.8)$$

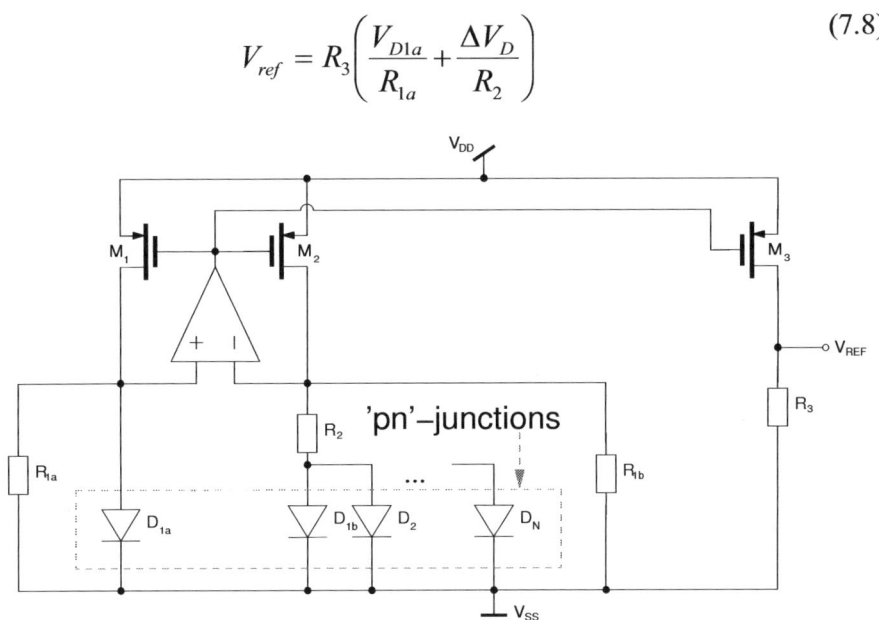

**Fig. 7.29.** Schematic of bandgap reference for low supply voltages.

The need for pn-diodes to realize bandgap references represents a serious problem in fully depleted multi-gate SOI technologies, because intrinsic pn-junctions are not available. A possible alternative is the use of gated diodes [31], sometimes also called lubistors. They consist of standard n- or p-type transistors where one of the source/drain implants is inverted. One can obtain either $p^+$-i-$n^+$ or $p^+$-p/n-$n^+$ diodes, depending on the doping of the fin. Alternative solutions are based on silicide blocking masks to generate lateral $p^+n^+$-junctions.

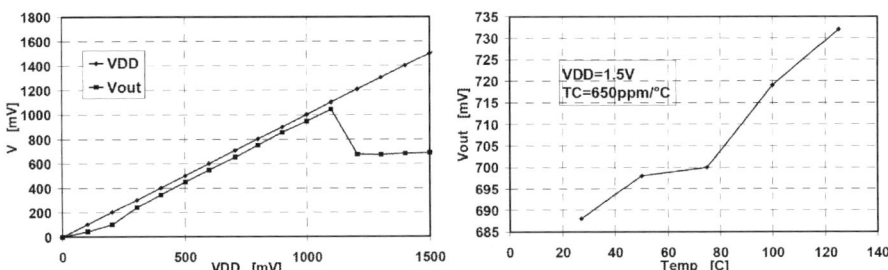

**Fig. 7.30.** Measured $V_{DD}$ and temperature dependence of the reference voltage $V_{out}$. Copyright© 2005 IEEE.

Figure 7.30 shows measured $V_{DD}$ and temperature dependence of a MuGFET bandgap reference using gated diodes.[31] In this case an early multi-gate technology containing polysilicon-gate devices with doped fins is used.[32] Above 1V the regulation loop begins to work and the output voltage is kept constant. The temperature dependence is not fully canceled, however the feasibility is proven.

### 7.3.2.3 Operational amplifier

Operational amplifiers (OPAs) are important building blocks for many applications, *e.g.* for AD or DA converter circuits. As common architecture a two stage Miller OPA serves as example in this chapter, (Fig. 7.31). Important figures of merit are gain-bandwidth product (*GBW*), open loop voltage gain ($A_0$) and power consumption (*P*).

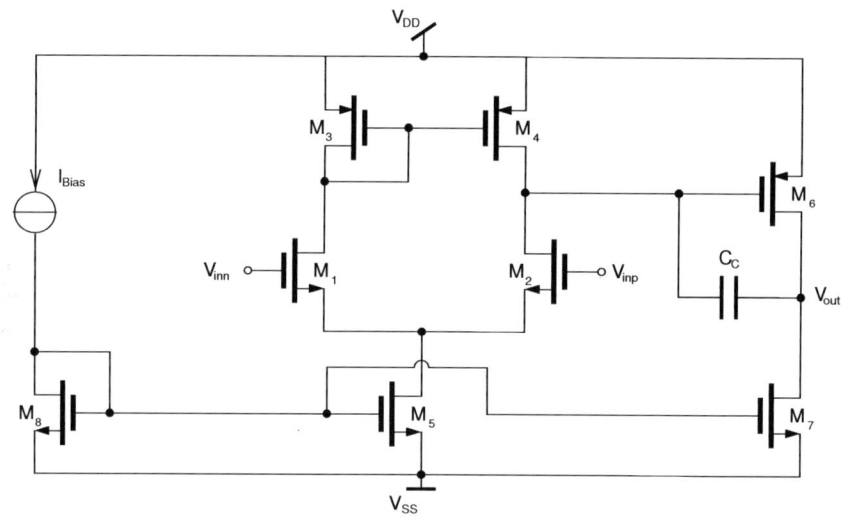

**Fig. 7.31.** Miller compensated two-stage operational amplifier.

Open-loop gain and gain-bandwidth product are determined by the small signal parameters $g_m$ and $g_{ds}$ [29]:

$$GBW = \frac{g_{m1/2}}{2\pi C_c} \quad ; \quad A_0 = \frac{g_{m1/2}}{g_{ds1/2} + g_{ds3/4}} \frac{g_{m6}}{g_{ds6} + g_{ds7}} \qquad (7.9)$$

Obviously the high intrinsic gain of MuGFETs is beneficial for these figures of merit. For a quantitative analysis of the benefits, planar and MuGFET designs are compared.[25] $A_0$, *GBW*, phase margin and load capacitance $C_L$ are specified to be 50dB, 10MHz, 60° and 5pF, respectively. Starting point is a planar design using devices with a channel length of 3 times $L_{min}$ that fulfills the specifications. Converting the design to MuGFET and keeping $L_{gate}$ constant yields a very high open loop gain of 80dB. As this high $A_0$ is not required, $L_{gate}$ can be reduced to obtain higher $g_m$ values. The higher $g_m$ again can be traded against higher GBW (Eqn, 7.9) or less power consumption, (same $g_m$ can be reached with less bias current). Table 7.4 resumes the simulated performance results of planar and MuGFET Miller OPA implementations.

**Table 7.4.** Performance comparison of planar and MuGFET Miller OPAs. Copyright© 2007 IEEE.

|  | Planar | MuGFET | MuGFET | MuGFET |
|---|---|---|---|---|
| $L_{gate}$ | 3 $L_{min}$ | 3 $L_{min}$ | 1.4 $L_{min}$ | 1.4 $L_{min}$ |
| GBW [MHz] | 10.8 | 10.6 | 14.7 | 10.4 |
| $A_0$ [dB] | 48.1 | 81.3 | 47.1 | 47.6 |
| P [uW] | 53.2 | 55.6 | 56.8 | 41.9 |
| FOM [MHz/uW] | 0.203 | 0.191 | 0.259 | 0.248 |
| FOM improvement | ±0 | -6% | +28% | +22% |

### 7.3.2.4 Comparator

Comparators are essential building blocks for AD converters. A simple, robust and widely used comparator circuit is shown in Fig. 7.32. The decision speed, which is simply determined by the bias current through the differential pair and the load capacitance is probably the most important design parameter of the comparator. There is no basic difference between planar and multi-gate comparator design. However, there are some potential issues with multi-gate comparators that designers should be aware of. In the last chapter, different types of transient transistor parameter variations are presented that can degrade comparator operation. Here, the impact of charge trapping on comparator performance is shown as an example. In the simulation the differential input voltage is stepped from a large positive value to a small but still positive residual value as shown in Fig. 7.32. The output voltage of the comparator without charge trapping is shown in Fig. 7.33 (left). As intended, the output voltage follows the input voltage quite fast.

Including the charge-trapping model in the simulation, the situation changes quite dramatically. The equivalent circuit presented in the last section is used with a steady state $V_T$-shift of 3mV and a time constant of 0.3µs. For a certain period of time the comparator delivers a wrong output voltage until it recovers slowly. This behavior is caused by the dynamic mismatch of the threshold voltages of the input devices of the comparator. Before the step of the input voltage $M_1$ is exposed to a big gate voltage and therefore a big $V_T$-shift, while the gate voltage of $M_2$ is very low as well as its $V_T$-shift. The input voltage changes much faster than the $V_T$-shifts of both devices, which slowly drift to their new, approximately equal value. The maximum residual voltage (*i.e.* the achievable resolution) that causes wrong decisions depends on the maximum steady state $V_T$-shift, the time constant of the charge trapping and of course the sampling time, *i.e.* the time when the correct output voltage has to be evaluated.

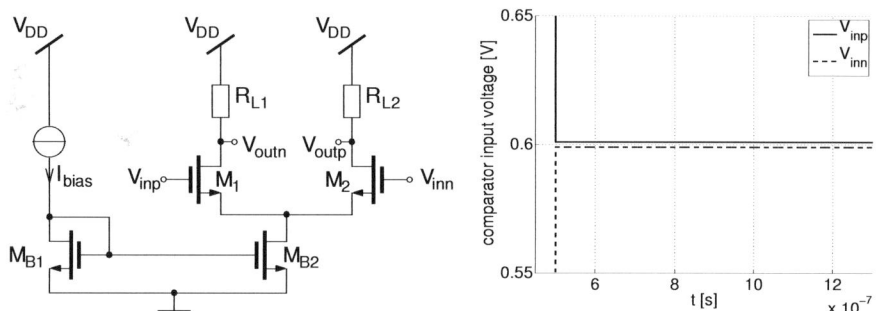

**Fig. 7.32.** Comparator schematic and input voltage for simulation. Copyright© 2007 IEEE.

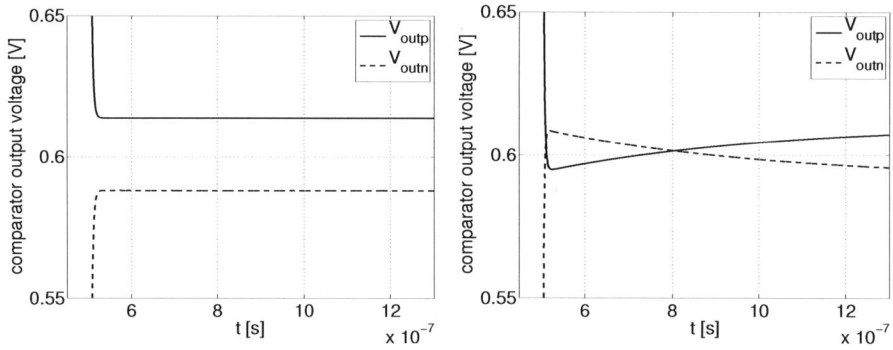

**Fig. 7.33.** Output voltage without (left) and with charge trapping model (right). Copyright© 2007 IEEE.

### 7.3.3 Mixed-signal aspects

In this section some multi-gate specific aspects in the design of mixed-signal circuits are presented.

## 7.3.3.1 Current steering DAC

The idea of current steering digital-to-analog converters is to sum up the currents of binary and/or thermometer coded current sources according to the applied bit pattern (Fig. 7.34). The quality of the current sources in terms of output resistance and matching limits the achievable resolution and yield. The minimum length and area ($WL_{min}$) of the current source devices is given by impedance ($Z_{min}$) and matching requirements [33]:

$$Z_{min} = \frac{NR_L}{4SNR} \quad ; \quad WL_{min} = \frac{1}{2\left(\frac{\sigma(I)}{I}\right)^2}\left(A_{K'}^2 + \frac{4A_{VT}^2}{(V_{GS}-V_T)^2}\right) \quad (7.10)$$

$N$, $R_L$, $\sigma(I)/I$ and $A_{K'}$ represent the number of bits (nominal resolution), the load resistance, the maximum current mismatch (given by resolution and yield specifications) and the mismatch constant of the current factor $\mu C_{OX}$, respectively. It is obvious that the low output conductance as well as the good matching behavior of the MuGFETs can be used to increase the resolution on the one hand or to decrease the power and area consumption of the converter on the other hand.

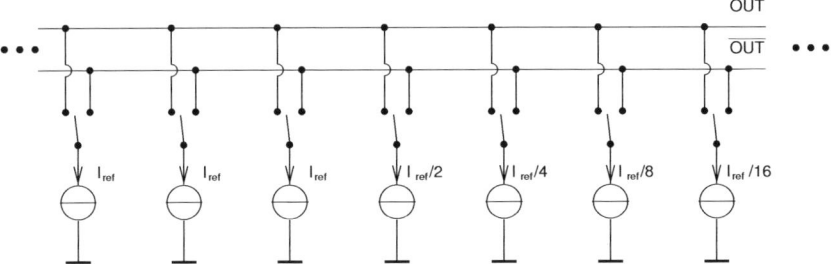

**Fig. 7.34.** Segmented current steering DAC.

## 7.3.3.2 Successive approximation ADC

As an example for analog-to-digital converters a charge distribution based successive approximation (SAR) ADC is shown here, (Fig. 7.35). The design of the main blocks of the ADC, *i.e.* comparator, capacitor network and control logic follows the same rules as in planar bulk technologies. However transient variations due to self-heating or charge trapping can degrade the performance of the SAR ADC.[25]

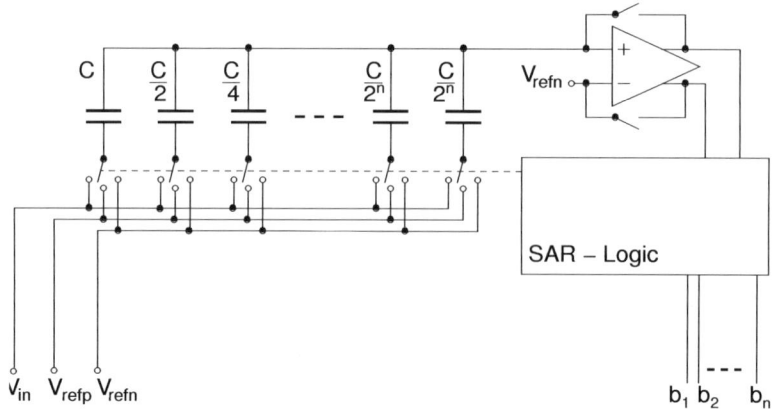

**Fig. 7.35.** DC ramp and SNR degradation due to charge trapping. Copyright© 2007 IEEE.

As shown above, comparator decisions can be influenced by transient variations. The resulting impact on the ADC performance is shown in Figure 7.36. Again charge trapping with a $V_T$-shift of 3mV serves as example. The left side shows the resulting error for a DC sweep of the input voltage. For input voltages lower than $V_{ref}/2$ also almost no errors occur. For input values around $V_{ref}/2$ and $3/4 V_{ref}$ the conversion error is about one LSB. For higher input voltages, the several wrong comparator decisions occur in the conversion cycle revealing a higher conversion error, up to 3 LSB. In addition, the density of faulty codes increases with $V_{in}$.

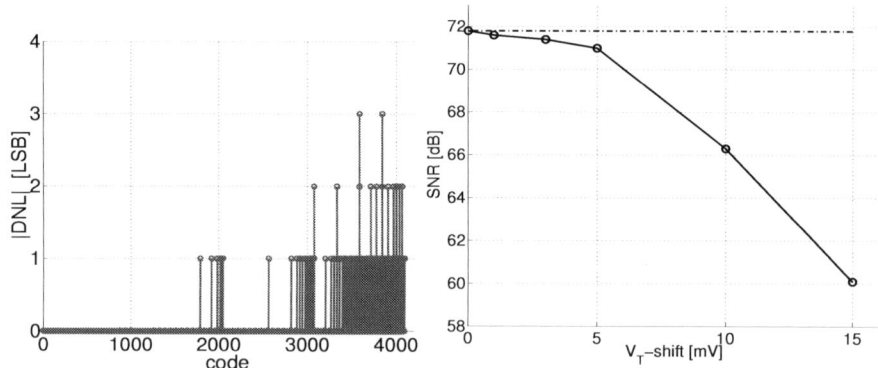

**Fig. 7.36.** DC sweep and SNR degradation due to charge trapping. Copyright© 2007 IEEE.

The right side of Fig. 7.36 shows the signal-to-noise ratio (SNR) for varying steady state $V_T$-shift. For values below 5mV the SNR degradation is quite small, whereas the SNR is reduced by about 6dB and 12dB for a

$V_T$-shift of 10mV and 15mV respectively, corresponding to a degradation of the effective resolution by one to two bits.

### 7.3.4 RF circuit design

For SoC applications the ability to realize also RF circuits is of outmost importance. The most important building blocks for RF systems are LC-VCOs (LC tank Voltage Controlled Oscillators) and LNAs (Low Noise Amplifiers). Fig. 7.37 shows the schematic of an LC-VCO with MuGFET NMOS varactors and MuGFET switching transistors. For the measurement of the LC frequency also output buffers are included. With the current mirror, the power consumption of the VCO core can be changed. Fig. 7.38 shows a chip microphotograph of the VCO including also the pads for the measurement, the details of the varactor array and the switching transistors.

Simulation results with $SiO_2$ gate oxide show a tuning range of > 30% and a phase noise of -132dBc/Hz. The tuning range is already appropriate for GSM applications, but the phase noise performance is not sufficient for such kind of applications. The main reason for this is the limited number of metal layers available for this test chip. The switch to high-k materials, where the flicker noise will most likely increase dramatically will even further degrade the phase noise performance.

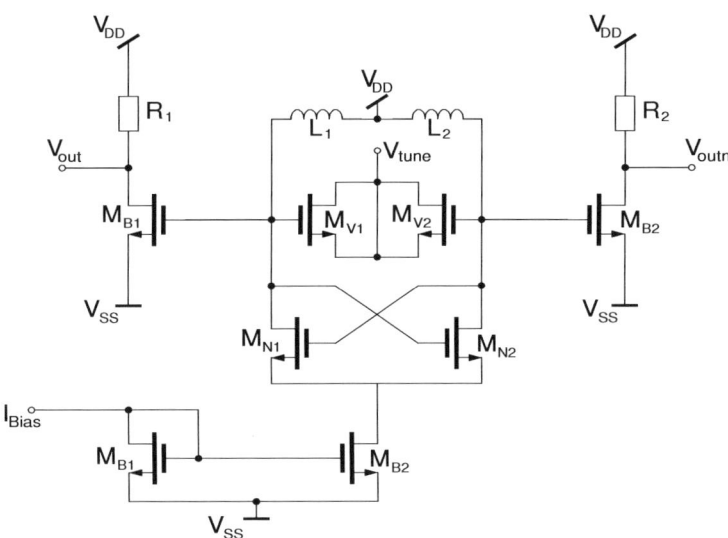

**Fig. 7.37.** Schematic of a LC-VCO including output buffer and current mirror. Copyright© 2007 IEEE.

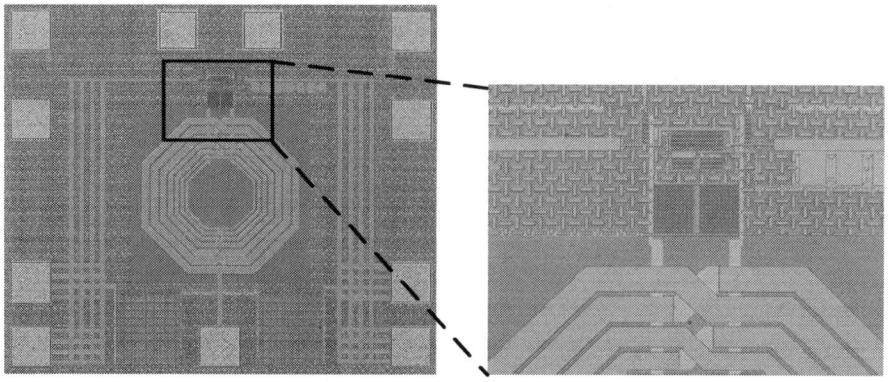

**Fig. 7.38.** Chip microphotograph of the VCO including pad frame and details of the varactor array and switching transistors. Copyright© 2007 IEEE.

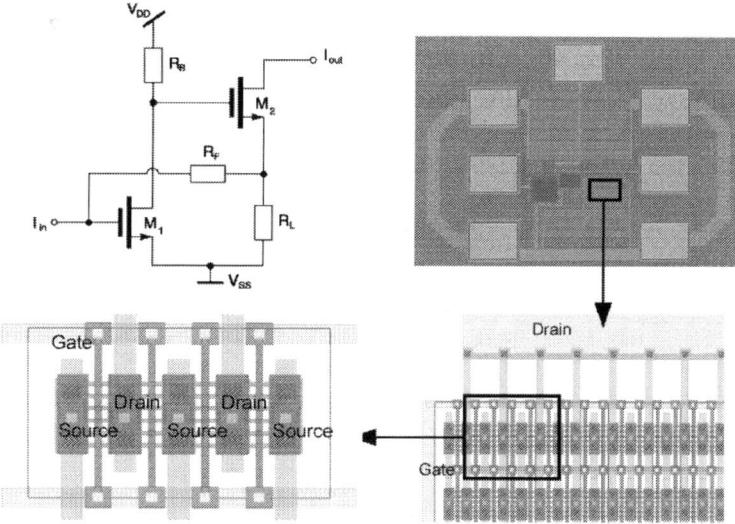

**Fig. 7.39.** Schematic, chip micro photograph of the transimpedance amplifier and details of the optimized MuGFET array layout. Copyright© 2007 IEEE.

Figure 7.39 shows an example for a transimpedance low-noise amplifier realized with MuGFET devices. Because input and output impedance have to be matched to a 50 Ohm environment, the number of fingers in the MuGFET devices is in the range of a few thousands and the current consumption of the circuit is very high (20mA). Therefore the optimization of the MuGFET layout for this kind of applications plays an important role. Also the consideration of self-heating effects in the simulation is very important. Chip microphotograph and an example for an optimized layout of RF optimized transistor unit cells are shown in Fig. 7.39. Large blocking capacitors between the supplies $V_{DD}$ and $V_{SS}$ fill the free area within the pad frame. The simulation based RF figures of merit for this circuit at 2 GHz operation frequency are a noise figure of approximately 2dB for 50 Ohm source impedance and a voltage gain of 15dB, which is comparable to planar designs.[34]

## 7.4 SoC Design and Technology Aspects

The capability of an efficient and heterogeneous integration of SoC applications is a key factor for the success of any advanced technology as MuGFET. A selection of various technology and design related aspects for SoC integration are summarized in Fig. 7.40. Potential benefits for digital, analog and mixed-signal circuit design have already been proven, as presented in the previous chapters. The broad range of different circuit components shows that there are no fundamental roadblocks for MuGFETs to replace planar MOSFETs in a future mainstream CMOS technology.

Furthermore, MuGFETs have shown a good ability of overcoming some potential show-stoppers. For instance, high ESD robustness can be achieved by using gated diodes for the ESD protection circuit and ensuring sufficient heat sinking at IO devices.[35] The reduction of the source/drain resistances *e.g.* by SEG will be a key achievement to resolve digital high performance and RF design challenges. Moreover, full area scaling of digital gates and SRAMs has to be demonstrated at the circuit level and shrinking factors comparable to those attained in planar CMOS devices have been attained. In order to shrink the height of standard cell, small fin pitches, on the order of 100nm are necessary. A reduction of contact-to-gate distances is also required to reduce the cell size in the lateral direction.

Sufficient and cost-efficient process control in terms of systematic and random fin dimension variability is still to be shown for narrowest fins on a large scale. Otherwise, the advantage of combining undoped fins and metal gates to reduce dopant-induced $V_T$-mismatch might be lost to parameter variability due to fin width and height variations.

Tunable or dual-metal gate workfunctions are the preferred solution for achieving low $V_T$'s in high-speed devices, since gate length tuning is only an alternative if cost-efficiency is the main decision criterion. Long-term reliability also needs to be analyzed for a complete assessment of SoC integration capability, especially if the MuGFET gate stack is based on high-k dielectrics. In regard to flicker noise and charge trapping effects, a metal gate/SiON gate stack might be beneficial if analog/mixed signal integration is more important for SoC integration than achieving highest possible digital performance.

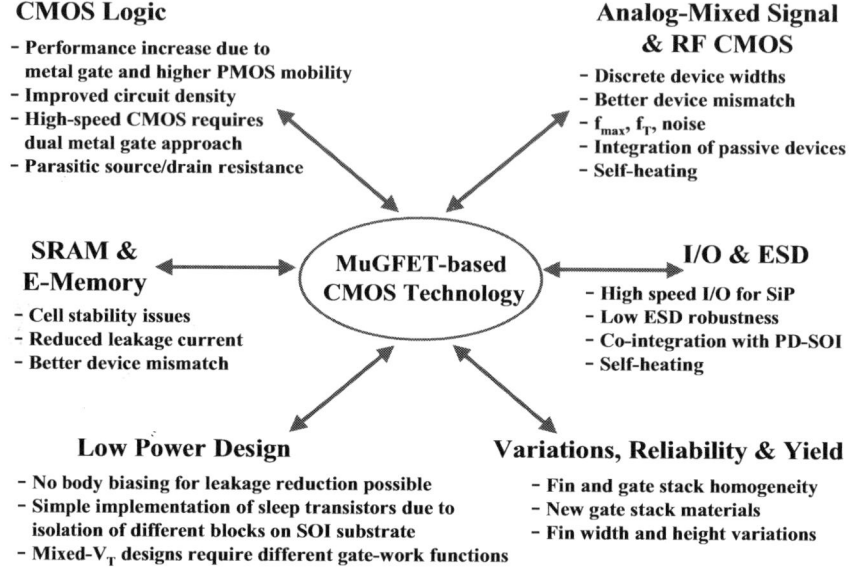

**Fig. 7.40.** Overview of various SoC design and technology aspects. Copyright© 2006 IEEE.

## Acknowledgments

The authors would like to thank the whole joint MuGFET development team of Infineon, Texas Instruments, Advanced Technology Development Foundry (ATDF), and Interuniversity Microelectronics Research Centre (IMEC). Special thanks go to Dr. Klaus von Arnim, Florian Bauer, Dr. Jörg Berthold, and Franz Kuttner for many fruitful discussions on device physics and circuit design. We also thank Dr. Andrew Marshall, Georg Georgakos, Wolfgang Molzer, Dr. Thomas Pompl, Dr. Christian Schluender, Dr. Christian Russ, and Dr. Harald Gossner for supporting us in the fields of circuit design, TCAD, reliability issues, and ESD. The authors are very grateful to Dr. Thomas Schulz, Dr. Weize Xiong, and Dr. Tamer San for technology development and fabricating our circuits, as well as to Dr. Klaus Schruefer and Dr. Rinn Cleavelin for project management and guiding us through exciting and challenging years. We also like to thank Dr. Paul Patruno, SOITEC, for supplying us with SOI wafers.

## References

1. T. Lueftner, J. Berthold, C. Pacha, G. Georgakos, G. Sauzon, O. Hömke, J. Beshenar, P. Mahrla, K. Just, P. Hober, S. Henzler, D. Schmitt-Landsiedel, A. Yakovleff, A. Klein, R. Knight, P. Acharya1, H. Mabrouki, G. Juhoor, Matthias Sauer: 90nm CMOS Low Power GSM/EDGE Multimedia Enhanced Baseband Processor. *ISSCC Dig. Techn. Papers*, 253 (2006)
2. P. Royannez, H. Mair, F. Dahan, M. Wagner, M. Streeter, L. Bouetel, J. Blasquez, H. Clasen, G. Semino, J. Dong, D. Scott, B. Pitts, C. Raibaut, U. Ko: 90nm Low Leakage SoC Design Techniques for Wireless Applications. *Techn. Digest ISSCC*, 138 (2005)
3. H. De Man: Ambient Intelligence: Gigascale Dreams and Nanoscale Realities, *ISSCC Dig. Techn. Papers*, 29 (2005)
4. J. Yuan, S.S. Tan, Y.M. Lee, J. Kim, R. Lindsay, V. Sardesai, T. Hook, V. Sardesai, T. Hook, R. Amos, Z. Luo, W. Lee1, S. Fang, T. Dyer, N. Rovedo, R. Stierstorfer, Z. Yang, J. Li, K. Barton, H. Ng, J. Sudijono, J. Ku, M. Hierlemann, T. Schiml: A 45nm low cost low power platform by using integrated dual-stress-liner technology. *VLSI Tech. Symp.* 100 (2006)
5. V. Chan, K. Rim, M. Ieong, S. Yang, R. Malik, Y.W. Teh, M. Yang, Q. Ouyang: Strain for CMOS performance Improvement. *Proc. IEEE Custom Integrated Circuits Conf. 2005*, 674 (2005)
6. E.P. Gusev, V. Narayanan, M. M. Frank: Advanced high-k dielectric stacks with polySi and metal gates: Recent progress and current challenges. IBM Journal of Research and Development **50-4/5**, 387 (2006)
7. K. von Arnim, E. Augendre, C. Pacha, T. Schulz, K. T. San, F. Bauer, A. Nackaerts, R. Rooyackers, T. Vandeweyer, B. Degroote, N. Collaert, A. Dixit, R. Singanamalla, W. Xiong, A. Marshall, C.R. Cleavelin, K. Schrüfer, M. Jurczak: A Low-Power Multi-Gate FET CMOS Technology with 13.9ps Inverter Delay. *VLSI Tech. Symp.* (2007)
8. C. Pacha, K. von Arnim, F. Bauer, T. Schulz, W. Xiong, K.T. San, A. Marshall, T. Baumann, C.R. Cleavelin, K. Schruefer, J. Berthold: Efficiency of Low-Power Design Techniques in Multi-Gate FET CMOS Circuits. *Proc. of ESSCIRC* (2007)
9. C. Pacha, K. von Arnim, T. Schulz, W. Xiong, M. Gostkowski, G. Knoblinger, A. Marshall, T. Nirschl, J. Berthold, C. Russ, H. Gossner, C. Duvvury, P. Patruno, C.R. Cleavelin, K. Schruefer: Circuit Design Issues in Multi-Gate FET CMOS Technologies, *ISSCC Digest of Techn. Papers*, 420 (2006)
10. K. Chen, C. Hu, P. Fang, M.R. Lin, D.L. Wollesen: Predicting CMOS Speed with Gate Oxide and Voltage Scaling and Interconnect Loading Effects. IEEE Trans. Electron Dev. **44-11**, 1951 (1997)
11. N. Collaert, A. De Keersgieter, K.G. Anil, R. Rooyackers, G. Eneman, M. Goodwin, B. Eyckens, E. Sleeckx, J.-F. de Marneffe, K. De Meyer, P. Absil, M. Jurczak, S. Biesemans: Performance Improvement of Tall Triple Gate Devices With Strained SiN Layers. IEEE Electr. Dev. Lett. **26-11**, 820 (2005)

12 W. Xiong, C.R Cleavelin, P. Kohli, C. Huffman, T. Schulz, K. Schruefer, G. Gebara, K. Mathews, P. Patruno, Yves-Matthieu Le Vaillant, I. Cayrefourcq, M. Kennard, C. Mazure, K. Shin, T.-J. King Liu: Impact of Strained-Silicon-on-Insulator (sSOI) Substrate on FinFET Mobility, IEEE Electr. Dev. Lett. **27-7**, 612 (2006)
13 G. Knoblinger, C. Pacha, F. Kuttner, A. Marshall, C. Russ, P. Haibach, P. Patruno, T. Schulz, K. v. Arnim, J.P. Engelstaedter, L. Bertolissi, W. Xiong, C.R. Cleavelin, K. Schruefer: Multi-Gate MOSFET Design. *Proceedings of ESSCIRC*, 66 (2006)
14 H. Shang, L. Chang, X. Wang, M. Rooks,Y. Zhang, B. To, K. Babich, G. Totir, Y. Sun, E. Kiewra, M. Ieong, W. Haensch: Investigation of FinFET devices for 32nm technologies and beyond. *VLSI Tech. Symp.* 54 (2006)
15 S. Hanson, B. Zhai, K. Bernstein, D. Blaauw, A. Bryant, L. Chang, K. K. Das, W. Haensch, E. J. Nowak, D. M. Sylvester: Ultralow-voltage, minimum-energy CMOS. IBM Journal of Research and Development **50-4/5**, 469 (2006)
16 B. Doyle, B. Boyanov, B, S. Datta, M. Doczy, S. Hareland, B. Jin, J. Kavalieros, T. Linton, R. Rios, R. Chau: Tri-Gate Fully-Depleted CMOS Transistors: Fabrication, Design and Layout. *VLSI Techn. Symp.*, 133 (2006)
17 A. Marshall, W. Xiong; C.R. Cleavelin, K. Matthews, G. Knoblinger. C. Pacha, K. von Armin, T. Schulz, K. Schruefer, P. Patruno: Effects of Temperature on Metal Gate FinFET Circuit Performance. *Proc. IEEE International SOI Conference*, 163 (2006)
18 Z. Guo, S. Balasubramanian, R. Zlatanovici, T.-J. King, B. Nikolić: FinFET-Based SRAM Design. *Proc. Int. Symp. on Low Power Electr. and Design*, 2 (2005)
19 L.Witters, N. Collaert, A. Nackaerts, M. Demand, S. Demuynck, C. Delvaux, A. Lauwers, C. Baerts, S. Beckx, W. Boullart, S. Brus, B. Degroote, J.F. de Marneffe, A. Dixit, K. De Meyer, M. Ercken, M. Goodwin, E. Hendrickx, N. Heylen, P. Jaenen, D. Laidler, P. Leray, S. Locorotondo, M. Maenhoudt, M. Moelants, I. Pollentier, K. Ronse, R. Rooyackers, J. Van Aelst, G. Vandenberghe, T. Vandeweyer, S. Vanhaelemeersch, M. Van Hove, J. Van Olmen, S. Verhaegen, J. Versluijs, C. Vrancken, V. Wiaux, P. Willems, J. Wouters, M. Jurczak and S. Biesemans: Integration of tall triple-gate devices with inserted-$Ta_x/N_y$ gate in a 0.274 µm$^2$ 6T-SRAM cell and advanced CMOS logic circuits. *VLSI Tech. Symp.*, 106 (2005)
20 V. Subramanian, B. Parvais, J. Borremans, A. Mercha, D. Linten, P. Wambacq, J. Loo, M. Dehan, N. Collaert, S. Kubicek, R.J.P. Lander, J.C. Hooker, F.N. Cubaynes, S. Donnay, M. Jurczak, G. Groeseneken, W. Sansen and S. Decoutere: Device and circuit-level analog performance trade-offs: a comparative study of planar bulk FETs versus FinFETs. *Tech. Digest of IEDM*, 851 (2005)
21 P. Kinget and M. Steyaert: Impact of transistor mismatch on the speed-accuracy-power trade-off of analog CMOS circuits. *Proc. IEEE Custom Integrated Circuits Conf.*, 333 (1996)
22 M. Pelgrom, A. Duinmaijer and A. Welbers: Matching properties of MOS transistors. IEEE Journal of Solid-State Circuits **24**, 1433 (1989)

23 C. Gustin A. Mercha, J. Loo, V. Subramanian, B. Parvais, M. Dehan, S. Decoutere: Stochastic Matching Properties of FinFETs. IEEE Electr. Dev. Let. **27**, 846 (2006)
24 T. Schulz, W. Xiong, C.R. Cleavelin, A. Chaudhry, A. Woo and J.P. Colinge: Fin thickness asymmetry effects in multiple-gate SOI FETs (MuGFETs). *Proc. IEEE International SOI Conference*, 154 (2005)
25 M. Fulde, A. Mercha, C. Gustin, B. Parvais, V. Subramanian, K. von Arnim, F. Bauer, K. Schruefer, D. Schmitt-Landsiedel, G. Knoblinger: Analog Design Challenges and Trade-Offs using Emerging Materials and Devices, *Proc. of ESSCIRC* (2007)
26 M. Fulde, D. Schmitt-Landsiedel, G. Knoblinger: Transient Variations in Emerging SOI Technologies: Modeling and Impact on Analog/Mixed-Signal Circuits. *Proc IEEE International Symposium on Circuits and Systems*, 1249 (2007)
27 W. Molzer, Th. Schulz, W. Xiong, R. C. Cleavelin, K. Schruefer, A. Marshall, K. Matthews, J. Sedlmeir, D. Siprak, G. Knoblinger, L. Bertolissi, P. Patruno, J.-P. Colinge: Self Heating Simulation of Multi-Gate FETs. *Proc. of ESSDERC*, 311 (2006)
28 G. Ribes, J. Mitard, M. Denais, S. Bruyere, F. Monsieur, C. Parthasarathy, E. Vincent, G. Ghibaudo: Review on high-k Dielectrics Reliability Issues. IEEE Transactions on Device and Materials Reliability **5**, 5 (2005)
29 P. E. Allan and D. R. Holberg: *CMOS analog circuit design* (Oxford University Press, 2002)
30 H. Banba H. Shiga, A. Umezawa, T. Miyaba, T. Tanzawa, S. Atsumi, K. Sakui: A CMOS bandgap reference circuit with sub-1-V operation. IEEE Journal of Solid-State Circuits **34**, 670 (1999)
31 G. Knoblinger, F. Kuttner, A. Marshall, C. Russ, P. Haibach, P. Patruno, T. Schulz, W. Xiong, M. Gostkowski, K. Schruefer, C. R. Cleavelin: Design and Evaluation of Basic Analog Circuits in an Emerging MuGFET Technology. *Proc. IEEE International SOI Conference*, 39 (2005)
32 W. Xiong, G. Gebara, J. Zaman, M. Gostkowski, B. Nguyen, G. Smith, D. Lewis, C. R. Cleavelin, R. Wise, S. Yu, M. Pas, T.J. King, J.P. Colinge: Improvement of FinFET electrical characteristics by hydrogen annealing. IEEE Electr. Dev. Lett. **25**, 541 (2004)
33 A. van den Bosch, M. Borremans, M. Steyaert and W. Sansen: A 10-bit 1-GSample/s Nyquist current-steering CMOS D/A converter. IEEE Journal of Solid-State Circuits **36**, 315 (2001)
34 G. Knoblinger, P. Klein and M. Tiebout: A New Model for Thermal Channel Noise of Deep-Submicron MOSFETS and its Application in RF-CMOS Design. IEEE Journal of Solid-State Circuits **36**, 831 (2001)
35 H. Gossner, C. Russ, F. Siegelin, J. Schneider, K. Schruefer, T. Schulz, C. Duvvury, C. R. Cleavelin, W. Xiong: Unique ESD Failure Mechanism in a MuGFET Technology. *Tech. Digest of IEDM*, paper 4.5 (2006)

# Index

## 1
1DEG, 28, 184
1-T memory cell, 16

## 2
2DEG, 28

## 4
4T-MuGFET, 82

## A
AD converter, 325
asymmetrical gates, 147, 149, 174, 176, 216

## B
bandgap reference, 322
band-to-band tunneling, 79
bandwidth, 316, 324
bipolar amplification, 276
bipolar gain, 281
bit line, 313
Boltzmann transport equation, 196, 217
BOX, 6, 268, 269
bulk FinFET, 15, 65, 145
bulk inversion, 58
Büttiker probes, 239

## C
carrier confinement, 200
Chemical Mechanical Polishing, 84
CMP, 84
comparator, 325
corner effect, 26, 64, 145, 185
Coulomb scattering, 196, 205
crystal orientation, 24, 218, 221
CV/I, 297
CYNTHIA, 13

## D
degeneracy factor, 157
DELTA, 11, 49
density of states, 35, 158
density-gradient model, 285
DGMOS, 11
DGSOI, 192, 245
DIBL, 3, 175, 297
digital-to-analog converter, 327
DOS, 35, 158, 243
double gate, 4

## E
effective mass, 29, 156, 202, 218, 221, 230
Effective Number of Gates, 26
Electrostatic Integrity, 3, 5, 17
energy ellipsoids, 157, 219
energy subbands, 32, 200, 201, 212, 219, 220, 228
ENG, 22
epitaxial growth, 90, 100
equivalent number of gates, 22, 284
ESD protection, 331

## F
fan-in, 300
fan-out, 300
FDSOI, 9
Fermi-Dirac function, 159
field penetration length, 142
fin pitch, 296
FinFET, 192, 220, 283
flip-flop, 303
$f_{max}$, 319
$f_t$, 319
fully depleted SOI, 5
FUSI gate, 76

## G

$G^4$ FET, 15
GAA, 13, 234, 237, 238, 245, 267, 271
gate stack, 51
gated diode, 323
GIDL, 75, 79
gradual-channel approximation, 117

## H

$H_2$ annealing, 65
halo, 295
Heisenberg uncertainly principle, 180
$HfO_2$, 318
HfSiON, 318
high-current regime, 264
high-k dielectric, 268, 320
high-k dielectrics, 197, 224
hysteresis, 321

## I

impact ionization, 266
interface scattering, 195
inter-subband scattering, 35, 185, 241
intervalley scattering, 197, 227, 230
intra-subband scattering, 241
intrinsic gain, 315
ISSG oxidation, 64
ITFET, 14
ITRS, 1

## K

Kubo-Greenwood formula, 239

## L

lattice mismatch, 95
LET, 274
linear energy transfer, 274
LOCOS, 260
low-power circuits, 293
lubistor, 323

## M

MASTAR, 6
matching, 317
Mathiessen's rule, 207
Maxwell-Boltzmann distribution, 161
MBCFET, 13
MBU, 276
McFET, 15
metal gate, 75
MFXMOS, 12
midgap gate, 74, 295, 299
MIGFET, 82, 174
mismatch, 317, 326
mobility, 31, 61, 198, 214, 223
Monte Carlo simulation, 195, 199, 240
Moore's law, 1, 2
MSD effect, 17
multiple-bit upset, 276
multi-$V_T$, 311

## N

NAND gate, 304
Nano-Beam Stacked Channels, 13
nanowire, 67, 233
natural length, 20, 22, 52, 66
NEGF, 195, 239
noise margin, 314
NOR gate, 304

## O

one-dimensional electron gas, 28
open-loop gain, 316, 324
operational amplifier, 324
output conductance, 144
overlap integral, 241, 244

## P

PDSOI, 9
phonon limited-mobility, 199
phonon scattering, 196, 197, 200, 201, 203, 224
piezoresistance, 92
pitch, 24
PLAD, 86
planarization, 70
Poisson equation, 117, 133, 147, 156, 162, 215, 235
polar phonons, 224
poly depletion, 75
polysilicon gate, 74
primed subbands, 230
punchthrough, 53

## R

radius of curvature, 26
raised source and drain, 87, 100
remote polar phonon scattering, 225
resonant tunneling, 181
RF circuits, 329
RIE, 60
ring oscillator, 301

## S

scattering mechanisms, 196
SCE, 52, 193
SCHRED, 132
Schrödinger equation, 156, 215, 236
SEG, 90
self-assembly, 69
self-heating, 320
SET, 272
SEU, 271
short-channel effects, 2, 137, 143
SiGe, 94, 97, 227, 230
silicides, 78
Silicon-On-Nothing, 12
Silicon-On-Sapphire, 9
single-event latchup, 271
single-event transient, 272
single-event upset, 271

single-transistor latch, 260, 264
SiON, 318
snapback, 260
SoC, 293, 331
SOI, 2
SON, 12
SONOS, 16
SOS, 9
spacer-defined fin formation, 60
SRAM cell, 62, 313
sSOI, 97
STI, 260
straggling, 279
strained silicon, 197, 226
stress, 93
subbands, 32, 35, 133, 134, 169, 180
subthreshold slope, 56, 125, 173, 176, 297, 301
surface roughness, 196, 205, 208, 245
surrounding-gate MOSFET, 13
symmetrical gates, 176, 216

## T

threshold voltage, 71, 215
TMAH, 63
total-dose latch, 260, 264
transconductance, 31, 315
transconductance efficiency, 144
trigate FET, 9, 12, 71
TSNWFET, 13
tunneling, 181, 183
two-dimensional electron gas, 28

## U

UNIBOND$^{TM}$, 261
unprimed subbands, 230
UTB SOI, 242

## V

VCO, 329
virtual gate, 270
Voltage-Doping Transformation, 3
volume inversion, 28, 29, 58, 124, 171, 193, 214, 216, 225, 238

## W

wavefunction, 225
workfunction, 73

## Z

ZRAM, 16

## Π

Π-gate MOSFET, 12, 234

## Ω

Ω-gate MOSFET, 12, 71, 234, 268, 269, 284

*Continued from page ii*

Leakage in Nanometer CMOS Technologies
Siva G. Narendra and Anantha Chandrakasan
ISBN 978-0-387-25737-2, 2005

Statistical Analysis and Optimization for VLSI: Timing and Power
Ashish Srivastava, Dennis Sylvester, and David Blaauw
ISBN 978-0-387-26049-9, 2005

Made in the USA
San Bernardino, CA
03 July 2019